Nuclear and Chemical Dating Techniques

Interpreting the Environmental Record

Lloyd A. Currie, EDITOR

National Bureau of Standards

Based on a symposium jointly
sponsored by the Divisions of
Nuclear Chemistry and Technology,
Geochemistry, and
History of Chemistry
at the 179th Meeting of
the American Chemical Society,
Houston, Texas, March 24–25, 1980.

ACS SYMPOSIUM SERIES **176**

AMERICAN CHEMICAL SOCIETY
WASHINGTON, D. C.   1982

6554-2484
CHEM

Library of Congress CIP Data

Nuclear and chemical dating techniques.
(ACS symposium series, ISSN 0097-6156; 176)

"Based on a symposium jointly sponsored by the
Divisions of Nuclear Chemistry and Technology, Geochemistry, and History of Chemistry at the 179th meeting of the American Chemical Society, Houston, Texas,
March 24–25, 1980."

Includes bibliographies and index.

1. Geological dating—Congresses. 2. Archaeological
dating—Congresses.
I. Currie, Lloyd A., 1930–   . II. American Chemical
Society. Division of Nuclear Chemistry and Technology. III. American Chemical Society. Division of
Geochemistry. IV. American Chemical Society. Division of the History of Chemistry. V. Series.

QE508.N8   551.7'01   81-20649
ISBN 0-8412-0669-4   AACR2
                        ASCMS 8 176 1–516 1982

To the memory of
Willard Frank Libby
The source of much of the inspiration and insight
recorded in this volume

ACS Symposium Series

M. Joan Comstock, *Series Editor*

FOREWORD

The ACS Symposium Series was founded in 1974 to provide a medium for publishing symposia quickly in book form. The format of the Series parallels that of the continuing Advances in Chemistry Series except that in order to save time the papers are not typeset but are reproduced as they are submitted by the authors in camera-ready form. Papers are reviewed under the supervision of the Editors with the assistance of the Series Advisory Board and are selected to maintain the integrity of the symposia; however, verbatim reproductions of previously published papers are not accepted. Both reviews and reports of research are acceptable since symposia may embrace both types of presentation.

CONTENTS

PREFACE

When I was invited to plan a symposium on nuclear dating, it immediately occurred to me to invite my former professor, Bill Libby, to present the keynote lecture. Libby's impact on the field had been enormous, with basic contributions ranging from cosmic ray physics to the history of modern man. His discovery of the first cosmogenic nuclide in nature followed by his development of the method of radiocarbon dating resulted in Libby's being awarded the Nobel Prize in Chemistry in 1960. Since publication of his classic volume on the subject ("Radiocarbon Dating," University of Chicago Press, 1952), bienniel or trienniel International Radiocarbon Conferences have taken place—the first in Andover, Massachusetts in 1954, and the most recent in Bern and Heidelberg in 1979. (These last proceedings are published in *Radiocarbon,* Volume 22, Numbers 2 and 3, 1980.)

Although Libby initially agreed to speak at the symposium, he was unable to attend for reasons of health, and Professor H. Oeschger kindly agreed to present the keynote. Following the meeting I informed Libby of my intention to dedicate this volume to him. He was pleased and graciously submitted the historical perspective that appears at the beginning of the volume. Libby passed on in September of 1980. We shall miss him, but we shall continue to be inspired by his enthusiasm, his insight, and his breadth of interest and knowledge.

The initial title for the symposium has had a twofold expansion, to incorporate *chemical* dating techniques and *interpretation* (or modeling). The relevance of chemical dating is clear when one considers the three kinds of geophysical "clocks"—those depending on (a) the rates of nuclear transformations, (b) the rates of chemical transformations or transport, and (c) natural cycles or accumulation processes (e.g., tree rings, ocean sediment). Also, the chemical properties of the nuclear species themselves are crucial in our approach to and the applicability of nuclear dating schemes, as Libby noted in his remarks with reference to ^{14}C and ^{187}Re.

With respect to interpretation, the existence of alternative dating techniques has made clear the necessity for and the difficulty of this step. That is, nature seldom provides ideal dating systems with fixed injection rates, negligible losses, and constant temperature. As a result, simple dates based upon observed isotopic ratios and nuclear half-lives, for example, frequently require cautious interpretation before they can serve as accurate measures of age; and in the absence of adequate models, alternative dating techniques will give discrepant results. The subject of this volume thus transcends dating. As put by H. Oeschger in his keynote lecture, a simple date (or observed radioisotope concentration) is but one factor to be considered in interpreting the current or past state of the environmental system. An adequate representation (model) of the system is required, as are sufficient isotopic and physicochemical data to yield reliable estimates for the parameters of the model. Thus, there is a dualism in that an accurate age cannot generally be deduced without a suitable environmental model, but simple dates help us to construct such models and to learn more about the state of the system than simply its age.

Isotopic and chemical patterns used in conjunction with absolute or relative dating techniques are also providing extremely interesting insights into the nature of geophysical or archaeological systems at various points in time. Such patterns may reflect physicochemically induced fractionation or composition variations indicative of natural or human activities [c][1]. Some of the examples explored herein include: ^{14}C production and carbon cycle perturbations [2,13], mixing of hydrological reservoirs [2,11], the history of climate [14,15], variations and sources of atmospheric dust [10,15], sources of ancient organic matter [19], an extraterrestrial cause of the Cretaceous extinction [20], and the identification of manufacturing sources in an ancient culture [21]. Geophysical modeling, chemical pattern recognition, and time series analysis make important contributions to such investigations; and one important outcome is chronological refinement.

A principal reason for organizing the symposium at this particular time was the recent occurrence of significant advances in dating techniques. Enormous improvements have taken place in minimizing chemical contamination, and in both the measurement of extremely small differences in isotopic ratios [a], and the separation and measurement of tiny quantities of inorganic [15] and very similar organic species [19,d]. Important progress is taking place in the measurement of very small quantities of long-lived radionuclides by means of direct high-energy (accelerator) ion counting, high-sensitivity microprobe and noble gas mass spectrometry, and ultralow-level counting [Section III]. Among the most important benefits from these advances will be the ability to date samples that are quite rare or difficult to obtain (deep ice cores, precious artifacts, cometary dust, etc.) and an increase in the reliability or information content of the dates through high spatial, temporal, or chemical resolution. (The ability to date less than a milligram of carbon, for example, makes it interesting to

[1] Figures in brackets refer to chapter numbers; letters refer to Symposium papers abstracted in the ACS Book of Abstracts. The Appendix provides further notes and a classification of these references according to technique and application.

determine the radiocarbon age of individual amino acids in bone [23].)
Additional information concerning accelerator mass spectrometry and
selected chemical dating techniques is given in the Appendix.

Progress in the application of a multiplicity of advanced dating techniques to a given problem together with sophisticated modeling [2,11,18]
promises to give us reliable information on the state and age of the system
under consideration, as well as some extra degrees of freedom for model
verification. When applied to natural archives [2; Section IV; c], such
studies can provide vital insight concerning the present and past states of
the environmental (geophysical) system; of critical importance may be
information on the relative influence of man's activities and natural events
on environmental contamination and climate. Finally, questions involving
the history and prehistory of man and the evolution and extinction of life
[Section V] are in many respects the most interesting to examine with
these techniques, and they are certainly among the most challenging.

The efforts of all authors, reviewers, and other symposium participants
are gratefully acknowledged. Special thanks are due Vic Viola, Juan Carlos
Lerman, and Chet Langway for their assistance with the meeting. Credit
for their excellent work in preparing the manuscripts for publication goes
to Joy Shoemaker and Teresa Sperow of the Text Editing Facility of the
Center for Analytical Chemistry, National Bureau of Standards.

LLOYD A. CURRIE
Center for Analytical Chemistry
National Bureau of Standards
Washington, D.C. 20234

August 27, 1981

Nuclear Dating

An Historical Perspective

W. F. LIBBY

University of California—Los Angeles, Institute of Geophysics and Planetary Physics, Environmental Science and Engineering, Los Angeles, CA 90024

I want to thank the symposium members and the Chairman, Dr. Lloyd Currie for the honor you do me.

During my work with nuclear dating in 1930, I built a Geiger Counter, and using air as a gas which is undoubtedly one of the worst possible gases, by brute force got it going using a string electrometer to detect the pulses. In a few months, however, we built a simple electronic circuit which detected and registered them on moving photographic paper using a tiny mirror glued to the needle of a microammeter. Our voltage supply was a motor-generator set and, needless to say, I several times found myself on the floor of the laboratory as a result of having touched the output electrodes. We take a great deal for granted today in electronics and voltage supplies and detection equipment. All this had to be developed, and development was initiated in this period of the early 1930's.

My professor, Wendell Latimer and I, decided that we would use the Geiger Counters to test for natural radioactivity in ordinary elements. We thought it would be a good idea to begin with the Rare Earths since as far as we could tell no one had looked at them. It had been known for many years that potassium was radioactive, as was rubidium. So the idea of radioactivity in the ordinary elements was hardly new. But a systematic search with as sensitive an instrument as the Geiger Counter seemed to be a good idea. We looked at the rare earths through the kindness of Herbert N. McCoy, who had been in the rare earth business in Chicago. Through him, we obtained samples of considerable purity of the various rare earths.

So, in the fall of 1931, spreading into 1932, we finally got going with the samples of the rare earth oxides. We built a counter with a screen instead of a solid cylinder as its wall to permit soft radiation to enter the counting volume.

The first sample we tested was samarium oxide, and we found it to be very radioactive. Now we know today that the radioactive isotope is samarium-147 with a half life of 105,000,000,000 years, and that the radiation is an emission of helium ions (alpha

0097-6156/82/0176-0001$05.00/0

particles). It is a most remarkable radioactivity and now seems to be a useful tool in dating the earth.

We were quite elated, and it appeared that it was a rich field. Now, fifty years later, I must say that it wasn't as rich as we thought. But we have over the years discovered half a dozen natural radioactive elements, and two of these, the samarium-147 with its decay to neodymium-143 and rhenium-187 with its decay to osmium-187, prove to be of use in Nuclear Dating. The importance of rhenium is that it is iron soluble while the other radioactivities are insoluble in metallic iron. In fact, the best half life we have for rhenium-187 was obtained by measuring the osmium-187 to rhenium-187 ratio in iron meteorites which had been dated by other methods. This work was started many years ago by Dr. Herr and others in Germany. The half life is 43,000,000,000 years.

The other natural activities such as indium-115, which has a half life of 10^{15} years, are interesting in their nuclear properties but are too long-lived to be useful in Nuclear Dating. Rhenium-187 in radioactive decay has the least energetic of all known nuclear transformations — its total disintegration energy is not known but it is probably in the range of 2 to 3 kilovolts. If the electronic binding energies were not included the transformation might not occur, hinting strongly that there is a close connection between nuclear transformations and the external electrons in the atom.

In the 1930's we began a search for carbon-14 to fill an obvious blank in the isotope table, for the reason that carbon is so very important in biology — a radioactive carbon isotope longer lived than a few minutes would be very valuable as a tracer. Due to an error in theoretical judgment we failed to find carbon-14. It was later discovered by Samuel Ruben working with Martin Kamen. An interesting point about the failure was that we, Kamen, Ruben, and I, guessed the half life to be about three months. As you all know, it is 5730 years. This meant that we did not make enough of it to detect it. Kamen and Ruben bombarded graphite with a deuteron beam — a sledge hammer approach — and found it.

This development was interrupted by World War II. We resumed work on it in about 1945 at the end of the war when we went to the University of Chicago. Again, we used the screen wall counter together with a new trick to shield the mesons. We surrounded the dating counter with a cylindrical shield consisting of Geiger counters, perhaps a dozen, some two inches in diameter; this arrangement is electronically connected so that if a cosmic ray meson triggers one of the shielding counters, the dating counter is turned off. The carbon radiation is so very weak there is no possibility that it could itself trigger the shielding counters. This whole bundle was put inside a massive iron shield, and in this way we were able to measure the natural radiocarbon and to measure the radiocarbon age.

The generation process for radiocarbon in the atmosphere makes CO_2 which enters the biosphere; because of the long lifetime the mixing is essentially perfect. We assumed the rate of production to be constant which turns out to be somewhat incorrect. Variations of about 10 percent can be seen back in time to early Egyptian periods and before. The earth's magnetic field was apparently weaker then as the cosmic rays delivered to the surface and the atmosphere were more intense.

Now we have many tens of thousands of radiocarbon dates from many laboratories throughout the world and the results continue to proliferate.

One of the most interesting of the geophysics results from radiocarbon dates is the history of the sun. Apparently, it is registered in fluctuations of the cosmic ray intensity. These are fluctuations of rather short duration in terms of the radiocarbon lifetime, perhaps a century or so, and apparently they are caused by variations in the solar wind due to long-term changes in the solar emissions. This idea has been developed in some detail recently by Dr. Lal and his collaborators. It promises to give us a way of watching the history of the sun over tens of thousands of years. This fine structure on the curve of calibration was discovered by Dr. Suess and others.

In archaeology there are many applications. They are very gratifying and successful.

We have seen the development recently of a new method of measuring radioactive isotopes which promises to evaluate smaller samples than we needed before. The present method requires perhaps 10 g of wood, oil, or charcoal or whatever the material is. The newer method of measuring the carbon-14 is by direct counting of the carbon-14 atoms instead of its decays. This should allow us to use only a few milligrams. This is a wonderful development which may allow us to make major advances in many important areas where the available samples were previously too small. One important case is the organic matter in sea cores. (Many investigators think the organic matter is more reliable for dating than shell.) Measurements of variations in the carbon-14 concentration in this organic matter may allow evaluation of the history of the solar fluctuations. Other small samples of special interest are works of art and religious artifacts which are too valuable to date by the conventional method. They may be datable now.

Nuclear dating has been most helpful in establishing the history of the earth and of the moon and of the meteorites. The fact is, there is no other way of measuring their ages. Prior to the discovery of natural radioactivity in the late 19th century, indirect methods were used to estimate the age of the earth, but there were no real answers until the radioactivity of thorium, uranium, and potassium were discovered and we began to understand atomic structure and to realize that nuclear transformation was essentially independent of the chemical form.

4

In addition, other exciting and interesting approaches such as fission track dating and dating by means of chemical reactions occurring under proper conditions are rapidly developing. Of course, there may be as yet undiscovered techniques of great importance.

Once again, I want to thank the members of this symposium and the Chairman, Dr. Lloyd Currie.

RECEIVED July 7, 1981.

The Contribution of Radioactive and Chemical Dating to the Understanding of the Environmental System

H. OESCHGER

University of Bern, Physics Institute, CH-3012 Bern, Switzerland

Radioactive and chemical dating methods are yielding most valuable information on the history of the earth and the planetary system. In this paper mainly methods using cosmic ray produced isotopes are discussed.

During the recent past, fluctuations in radioisotopes produced by cosmic radiation in the earth's atmosphere have been observed, the most convincing example being the fluctuations of the $^{14}C/C$-ratio observed in tree-ring samples. Such fluctuations complicate the interpretation of radioactive ages in terms of absolute ages, and their interpretation asks for the development of models considering not only isotope production variations but also the geochemical behavior of the isotopes of the different elements. For this purpose, it is useful to distinguish between noble gas radioisotopes (e.g., ^{39}Ar, ^{81}Kr), radioisotopes which get incorporated in molecules of gases and vapors (^{14}C, ^{3}H), and radioisotopes of solids (^{10}Be, ^{36}Cl) which get attached to aerosol particles and are deposited with precipitation.

In polar ice sheets air gets continuously trapped, and ice cores obtained by drilling through the ice caps therefore constitute a continuous set of ancient air samples. $^{39}Ar/Ar$ and $^{81}Kr/Kr$ measurements on these samples primarily reflect the production rates of these radioisotopes averaged over a few half-lives.

It is expected that due to the short residence time of Be and Cl in the atmosphere, ^{10}Be and ^{36}Cl measurements on ice cores will directly reveal isotope production variations. Due to dilution in the CO_2 exchanging system the atmospheric $^{14}C/C$-ratio shows a dampened response to ^{14}C production rate variations. In contrast to the noble gas radioisotopes the size of the effective dilution reservoir — atmosphere plus parts of the ocean and biosphere — depends on the characteristic

0097-6156/82/0176-0005$09.50/0

times of the production rate variations. In addition, $^{14}C/C$ variations in atmospheric CO_2 may be caused by variations in the CO_2 exchange dynamics, as indicated by the observation of changes in the atmospheric CO_2 concentration in ice cores.

Finally a strategy for the study of the environmental system and its history is proposed. Dating methods provide the time scale for ancient system states, and fluctuations in the parameters used for dating point to important changes in system processes. Recent developments in field and analytical methods as well as model calculations promise accelerated progress regarding a quantitative understanding of processes determining our environment. This is badly needed in view of possible natural and/or anthropogenic changes with effects on society.

Radioactive and chemical dating methods have not only provided unique information on the history of man and his environment, but also on processes in the solar system and their history. It has been found however that the assumptions on which these dating methods were based are only partly fulfilled. During recent years strong emphasis has been given to studies of some deficiencies of these dating methods and their causes. They have yielded most valuable results on natural processes; an example is the ^{14}C-variations which are attributed to variations in the isotope production rate by cosmic rays on the one hand and to fluctuations in the global CO_2 exchange on the other.

During the last several decades the natural systems have been disturbed by human activities. Natural and anthropogenic disturbances of the environmental system are discussed in terms of models, and answers regarding possible negative consequences of human interactions with natural processes are searched for.

Again the atmospheric $^{14}C/C$ ratio is an excellent example. Man-induced disturbances of the environmental system lead to changes in the $^{14}C/C$ ratio which are of the order of magnitude of the natural fluctuations or even larger: the emission of ^{14}C-free CO_2 from fossil energy consumption leads to a decrease, and the emission of man-made ^{14}C from nuclear weapons testing, to an increase of the atmospheric $^{14}C/C$ ratio.

In this article, we first discuss basic dating principles and then studies based on isotopes produced by cosmic radiation in extraterrestrial matter and in the earth's atmosphere. The discussions are intended to illustrate how analytical physical and chemical studies contribute to the understanding of processes in the environmental system and their history.

Principles of Radioactive and Chemical Dating

Periodic ($\sim e^{i\omega t}$) and aperiodic (e.g., $\sim e^{-\lambda t}$) processes in nature can be used for dating. For the following we mainly concentrate on aperiodic processes, changing the state of a system in nature as well-known functions of time. The present (or end) state of the system is experimentally determined and the initial state of the system is estimated. The time function for system changes then enables us to calculate the age, i.e., the time elapsed between the initial and final states.

System changes which can be used for dating include:

- The decay of long-lived radioisotopes still remaining from the nucleosynthesis. An example is the creation of ^{40}Ar by ^{40}K decay. Based on the measured ratio of the concentration of daughter ^{40}Ar to parent ^{40}K nuclei, the time during which Ar was accumulated in the system (e.g., lunar material after a meteorite impact) can be calculated using the law of radioactive decay with the appropriate decay constant.

- The decay of radioactive isotopes created in the earth's atmosphere by the interaction of cosmic rays with atomic nuclei of atmospheric constituents. After such nuclei (e.g., ^3H as ^3HHO or ^{14}C as $^{14}CO_2$) are removed from the atmosphere, e.g., fed into a groundwater system (^3H) or built into a living organism (^{14}C), their number decreases according to the law of radioactive decay.
  The time elapsed since separation from the atmosphere is calculated from the ratio of the activity at the time of sampling to the estimated activity in the atmosphere at the time of separation.

- Solar energy enables the creation of high order chemical and physical systems in atmosphere and biosphere. Examples are amino acids in living matter and high order crystal arrays of snow flakes. After these systems are withdrawn from the source of high order they start continuous transition into more probable states: amino acids tend to get equally distributed between left and right-orientation and the structure of the snow flake gets less and less complex till finally firn grains of approximately spherical shape are formed.

Processes in nature correspond generally only in a first approximation to what is postulated in the principles of dating methods. An exception is radioactive decay which is almost independent of variations in the environmental conditions, since energy differences are involved which are large compared to differences of thermal excitation in the environment. This is in contrast to chemical and physical processes which do depend on environmental parameters such as temperature. An example showing some of the complications involved in a dating method is ^{14}C dating of groundwater. Firstly the atmospheric $^{14}C/C$ ratio is not constant with time, as shown by tree-ring measurements. For a given atmospheric $^{14}C/C$ ratio chemical processes which probably also were not constant with time determine the $^{14}C/C$ ratio of newly formed groundwater. During groundwater flow the $^{14}C/C$ ratio decreases due to radioactive decay but also due to exchange with the surface material of the aquifer. In addition the water gets dispersed and water masses from different origin may get mixed.

Summarizing, we can distinguish the following idealized dating concepts:

-   A low entropy system gets closed, and, according to a process which can be quantitatively described, goes to states of increasing entropy (radioactive decay, racemization, crystal growth, diffusion).

-   A high entropy system gets exposed to negative entropy (neg-entropy) influx resulting in an entropy decrease according to a process which can be described quantitatively. Typical sources of negentropy are cosmic radiation (isotope production) and solar radiation (creation of high order chemical and physical systems).

Interaction of Cosmic Rays with Meteorites, The Moon and The Earth's Surface

Galactic cosmic rays (mainly protons and α-particles with energies above 1 GeV), and protons emitted from the sun (with energies in general below 10 GeV) interact with nuclei of meteorites, of the surface of the moon and of the earth's atmosphere and produce isotopes. The solar proton contribution to isotope production is detectable in an upper layer of ~1 mm of lunar surface material. For the deeper layers of meteorites and the lunar surface and for the atmosphere, however, the contribution by the galactic cosmic radiation dominates, except for extremely large solar flares. The solar proton flux is related to solar flares and is strongly varying with time. The galactic cosmic rays are modulated due to shielding effects by interplanetary magnetic fields carried outward by the solar wind plasma. The earth's surface is further shielded against charged particles by the geomagnetic field.

---

[1]For highly relativistic particles the momentum P expressed in $\frac{eV}{c}$ is numerically equal to its energy, expressed in eV.

For a given latitude only particles with a momentum higher than a minimum cut-off momentum (P_{min}) can penetrate the geomagnetic shield. This cut-off momentum[1] depends on latitude according to

$$P_{min} = 15 \frac{GeV}{c} \cos^4 \lambda; \quad \begin{array}{l} c: \text{ velocity of light} \\ \lambda: \text{ geomagnetic latitude} \end{array}$$

Figure 1 schematically shows the modulation effects on galactic and solar particles and their interactions.

Whereas isotopes produced in the solid matter of meteorites or the lunar surface material remain at their production sites, isotopes produced mainly in the earth's atmosphere are separated according to the geochemical properties of the different elements.

Studies Based on Isotope Production in Meteorites and on the Lunar Surface

Exposure ages

The penetration depth of cosmic radiation is of the order of 1 m and therefore isotopes are produced by spallation only in the surface layers of meteorites and the moon. After collisions of meteorites with each other or with the moon, newly formed surfaces get exposed to cosmic radiation and production of stable and radioactive isotopes starts. If P_{st} is the production rate of a stable isotope and if P_{rad} is the production rate of a radioactive isotope and if both rates are constant, then the numbers N_{st} and N_{rad} are given by

$$N_{st} = P_{st} t_{exp} \quad \text{and} \tag{1}$$

$$N_{rad} = \frac{P_{rad}}{\lambda} (1 - e^{-\lambda t_{exp}}) \tag{2}$$

(see also figure 2)

with λ = decay constant and t_{exp} = exposure time

If $\lambda t_{exp} \gg 1$, then $\lambda N_{rad} \simeq P_{rad}$ i.e., the decay rate $A_\lambda = \lambda N_{rad}$ equals the production rate P_{rad}. If the ratio P_{rad}/P_{st} is known, the exposure age can then be calculated according to

$$t_{exp} = \lambda^{-1} \frac{N_{st}}{N_{rad}} \frac{P_{rad}}{P_{st}} \tag{3}$$

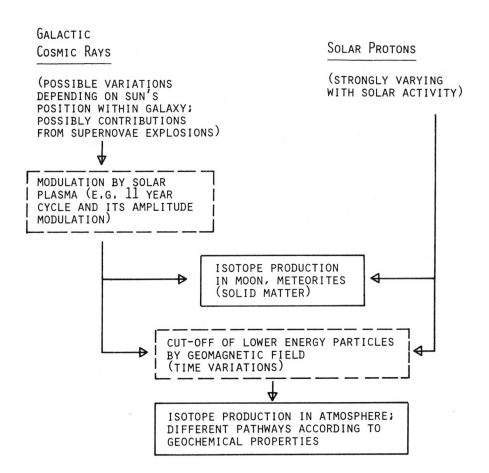

Figure 1.   Cosmic radiation/modulations and interactions.

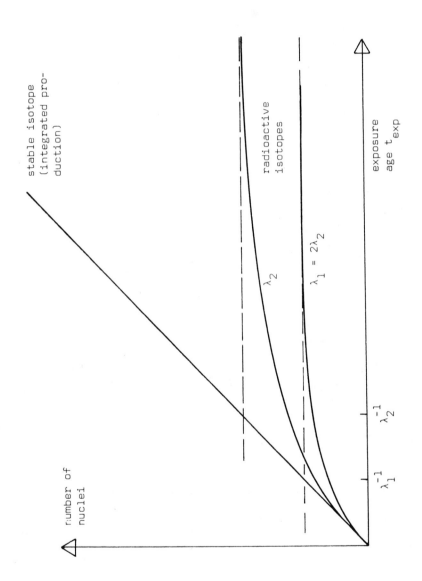

Figure 2. Increase of the number of stable and radioactive nuclei with exposure time.

Pairs of isotopes suited for the determination of exposure ages are e.g., ^{80}Kr (stable) and ^{81}Kr($T_{1/2}$ = 2.13·10^5 y) [1-2][2], ^{3}He (stable) and ^{3}H($T_{1/2}$ = 12.4 y) [3], and ^{41}K (stable) and ^{40}K($T_{1/2}$ = 1.27·10^9 y) [4,5]. The determination of exposure ages is an example of a dating technique which uses the opening of a system to creation of information. Exposure ages of a great number of meteorites of different classes have been measured and the question of the grouping of exposure ages, indicating production of a greater number of meteorites in a single collision event has been discussed. In addition information on lunar surface dynamics has been obtained.

Constancy of cosmic radiation

Comparison of the activity $A_\lambda = \lambda N_{rad}$ of radioisotopes with different decay constants allows one to study the question whether radioisotope production and therefore also cosmic radiation have been constant with time or have shown significant variations. This is a crucial question for all dating techniques based on cosmic ray produced isotopes. It will be taken up again with special respect to dating methods used for terrestrial problems. Measurements of isotopes with half-lives up to 4·10^{6}y (e.g., ^{53}Mn) on meteorite and lunar samples suggest that cosmic radiation did not vary more than a factor of 2 [5-7]. For a varying production rate $P_\lambda(t)$ the number of nuclei of a radioactive isotope i can be calculated according to

$$\frac{A_\lambda(t)}{\lambda} = N_{rad}(t) = \int_0^\infty P_i(t-T)e^{-\lambda T}dT = \mathring{P}_i(\lambda,t) \qquad (4)$$

(T = age, $\mathring{P}_i(\lambda,t)$ = Laplace transform of the production rate $P_i(t)$).

Equation (4) shows that the activities $A_\lambda(t)$ essentially correspond to the production average over the last one to two mean lives. Some information on constancy of cosmic radiation is gained, e.g., a systematic increase or decrease by a factor of three or so should be visible. We will later see, however, that on earth geochemical processes make much better resolved information available.

---

[2]Figures in brackets indicate the literature references at the end of this paper.

The Use of Cosmic Ray Produced Isotopes for Studies of the Environmental System and Its History

Cosmic ray produced isotopes are ideal tracers to study processes in the environmental system (see figure 3). They yield information on mixing and circulation in the atmosphere and the ocean, on air-sea exchange and on global cycles such as the hydrological and the carbon cycles. They also provide time scales to reconstruct the history of environmental processes stored in, for example, glaciers and ice caps, tree-rings, peat bogs, and lake and ocean sediments. But, as mentioned before, there are irregularities in the radioisotope records on which much attention has recently been focussed. These irregularities reflect not only variations in the production rates, but also changes in the dynamics of processes in the environmental system (see also Castagnoli and Lal [8]).

Interaction of cosmic rays with the atmosphere

Radioisotopes are continuously produced in the atmosphere, the stratospheric production being roughly twice the tropospheric one. As shown in figure 4 it is useful to distinguish between isotopes of noble gases, isotopes incorporated in molecules of gases or vapors, and isotopes which get attached to aerosols.

Radioactive noble gas isotopes

Radioactive noble gas isotopes used for the study of processes in atmosphere and ocean are given in Table 1. Over the last 12 years ^{37}Ar has been measured in atmospheric samples (e.g., Loosli, et al., [9]). ^{39}Ar activity has been determined in samples from the atmosphere, ocean, groundwater, and ice. The initial results suggest that ^{39}Ar might become a very interesting dating tool; see below. Until now ^{81}Kr has only been measured in present day atmospheric samples.

Noble gas radioisotopes with $T_{1/2}$ >> atmospheric mixing times

^{39}Ar and ^{81}Kr decay with half-lives (269 y and 2.1×10^5 y), which are long compared to the atmospheric mixing times. There are only negligible amounts of ^{39}Ar in reservoirs other than the atmospheric one. Therefore, we essentially have cosmic ray produced ^{39}Ar and ^{81}Kr in one well mixed atmospheric box (figure 5).

Since the atmospheric composition regarding Ar and Kr contents has probably been constant during the last 10^6 y we get for the activities of ^{39}Ar and ^{81}Kr

$$\frac{A(t)}{\lambda} = N(t) = \int_0^\infty P(t-T)e^{-\lambda T}dT = \bar{P}(t,\lambda)$$

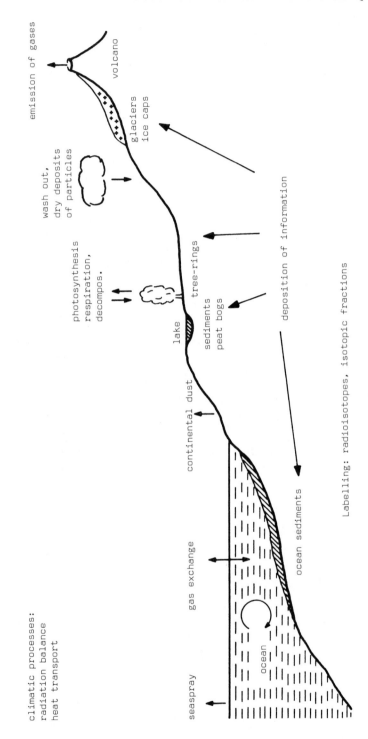

Figure 3. Environmental system.

Table 1.   Cosmic Ray Produced Noble Gas Radioisotopes $T_{1/2} > 1$ month.

Isotope	$T_{1/2}$	Specific Activity	Comments
^{37}Ar	35.1 d	~0.003 dpm/L Ar	Additional production by underground nuclear reaction ($^{40}Ca(n,\alpha)^{37}Ar$) up to 0.2 dpm/L Ar
^{39}Ar	269 y	~0.1 dpm/L Ar	Anthropogenic contribution <5 percent
^{81}Kr	2.1×10^5 y	0.1 dpm/L Kr	
^{85}Kr	10.6 y	Cosmic ray produced ?	Cosmic ray produced component not yet identified
		Anthropogenic ~3×10^4 dpm/L Kr	Fission product

See also Loosli and Oeschger [10,11]; Loosli et al., [9]; Kuzimov and Pomansky [12].

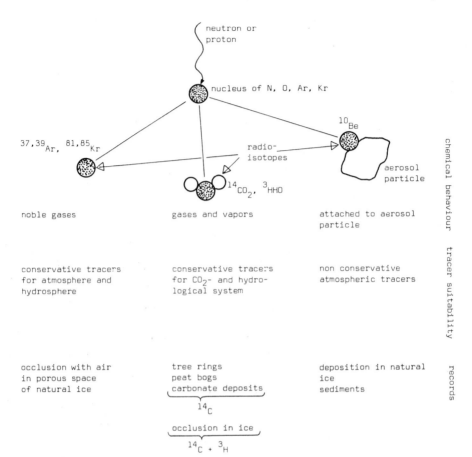

Figure 4.   Pathways and applications of radioisotopes in the environmental system.

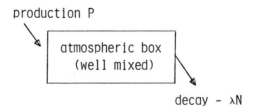

Figure 5.   Radioisotopes with $T_{1/2} >>$ atmospheric mixing time.

i.e., we get the same equation as for radioisotopes produced in extraterrestrial matter. Only the two noble gas isotopes mentioned here provide a direct comparison based on this simple formalism. Regarding the study of constancy of cosmic radiation based on today's spectrum of radioisotopes, we have less information on earth than in extraterrestrial matter. On the other hand, the physical and chemical processes on earth enable us to get information on the history of $A(t)$. An almost ideal process of continuous sampling of atmospheric air takes place in polar ice sheets. At the transition of firn to ice (at a typical depth of 70 m) the air filling the pore space between the firn grains gets pinched off. The bubbles thus formed constitute ideal air samples. They get buried deeper and deeper into the ice, according to the rheological principles. Ice cores retrieved by drilling into the deepest strata of polar ice sheets contain air that may be 10^5 or more years old. In the following text we calculate the ^{39}Ar/Ar variation for a given cyclic ^{39}Ar production variation.

We can express cyclic production rate variations superimposed on a constant term by

$$P(t) = P_0 + P_1 e^{i\omega t} \quad \text{and} \tag{5}$$

expect a corresponding variation in the complex number $N(t)$ of ^{39}Ar nuclei in the atmosphere:

$$N(t) = N_0 + N_1 e^{i\omega t} \tag{6}$$

For a one box system we have the differential equation

$\dot{N} = P - \lambda N$, and with the above "Ansatz"

$$i \omega N_1 e^{i\omega t} = P_0 + P_1 e^{i\omega t} - \lambda N_0 - \lambda N_1 e^{i\omega t} \tag{7}$$

leading to

$$\lambda N_0 = P_0 \quad \text{and} \quad (\lambda + i\omega) N_1 = P_1$$

The relative variation in the ^{39}Ar inventory N_1/N_0 is related to the relative production rate variation P_1/P_0 by

$$\frac{N_1}{N_0} = \frac{\lambda}{\lambda + i\omega} \frac{P_1}{P_0} \tag{8}$$

We can define $D(\omega,\lambda) = \dfrac{\lambda + i\omega}{\lambda}$ as a damping factor and obtain

$$D(\omega,\lambda) = \sqrt{1 + \frac{\omega^2}{\lambda^2}} \; e^{i \; \arctan \omega/\lambda} \tag{9}$$

From observed variations of $^{14}C/C$ in tree-ring samples we have indications of a 200 y cycle in the radioisotope production and P_1/P_0 is suggested to be of the order of 25 percent [13]. For the damping factor for ^{39}Ar ($\lambda = 1/388$ y and $\omega = 2\pi/200$ y) one obtains:

$$|D| = 12.2 \quad \text{and} \quad \arctan \omega/\lambda = 85°$$

Therefore, for the production rate variations mentioned above, we obtain a relative amplitude of the activity $A_1/A_0 = N_1/N_0 \cong 2\%$.

In addition it is calculated that the atmospheric activity variations lag behind the production rate variations by 47 y.

Changes in the ^{39}Ar level during the last 100 years therefore cannot be detected with the present precision of ~5 percent in ^{39}Ar measurements.

The ^{39}Ar dating method

^{39}Ar in atmospheric samples was measured for the first time in 1968 [10]. Its modern net activity has been determined to be 0.112 ± 0.010 dpm/L Ar. ^{39}Ar is produced mainly by (n, 2n) reactions with ^{40}Ar. Because of its low specific activity it is very difficult to measure. Compared to ^{14}C in "modern" samples, the specific activity of ^{39}Ar is smaller by a factor of 65. Therefore the ^{39}Ar dating method starts with specific activities corresponding to that of ca. 35,000 years old radiocarbon samples. At present, the minimum sample size required for a measurement is ~400 mL Ar. For samples of that size a modern net effect of 0.036 cpm and a background of 0.030 cpm (in an underground laboratory) are measured. The dating range at present is 30 to 1200 y, and 270 y (one half-life) old samples can be measured with a statistical error of ±30 y. Of great importance for the method is the observation that the contribution from nuclear weapon tests is less than 5 percent. For application of ^{39}Ar to oceanic circulation studies we therefore can assume a steady state distribution and do not need to distinguish pre-nuclear and nuclear components as in the case of ^{14}C.

A test of the ^{39}Ar method is the measurements on Ar extracted from air occluded in polar ice. On the occasion of several polar projects at Byrd Station, Antarctica, under the auspices of

USARP[3], and in Greenland under the auspices of GISP[4] (collaboration between USA, Denmark, and Switzerland) samples were obtained by in situ melting of about 3 tons of ice in 400 m deep bore holes. Figure 6 shows the ^{39}Ar ages obtained for samples collected in 1974 at Station Crête in Central Greenland [14]. The ages are plotted versus depth and compared with those obtained by annual layer counting based on data on the seasonal δ^{18}O variations. As expected the age of zero years corresponds to the depth of gas occlusion ($\sim$70 m). The good agreement between ^{39}Ar and δ^{18}O ages is a very valuable confirmation of the ^{39}Ar dating method. It shows that the difficult experimental steps: sample collection in the field, separation of Ar from the air (especially from Kr with at present relatively high specific ^{85}Kr activity) and counting of the very low activity are under good control. Application fields of the ^{39}Ar dating method are glaciology, e.g., dating of cold glaciers with complex accumulation and ablation characteristics, and hydrology, e.g., dating of groundwater, especially in comparison with the ^{14}C dating technique. For hydrological dating it is assumed that newly formed groundwater contains Ar with the atmospheric specific ^{39}Ar activity. If the groundwater is no longer in contact with the atmosphere its ^{39}Ar activity decreases according to the law of radioactive decay. ^{39}Ar measurements on groundwater in aquifers with relatively high U and Th contents, however, show that underground production of ^{39}Ar may lead to specific ^{39}Ar-activities which are even higher than that of the atmosphere [15]. On the other hand groundwater was found with ^{39}Ar activity below the detection limit, indicating the existence of aquifers for which the ^{39}Ar dating technique provides useful information. The ^{39}Ar dating method is very promising regarding oceanic mixing and circulation studies, since its half-life compares well with the characteristic ocean mixing times. At the ocean surface, exchange with the atmosphere brings the ^{39}Ar activity of the dissolved Ar close to that of the atmosphere. If a water mass moves from the surface to deeper strata, its ^{39}Ar decays. It is expected that the ^{39}Ar activities of the dissolved Ar cover a range of 100 percent (ocean surface) to 10 percent (deep ocean currents). Of special interest will be the comparison with ^{14}C data, since radioisotopes with different half-lives weigh the age components differently[5], and therefore the ^{39}Ar-^{14}C comparison will give information on the age distribution of water masses.

---

[3]United States Antarctic Research Program.
[4]Greenland Ice Sheet Program.

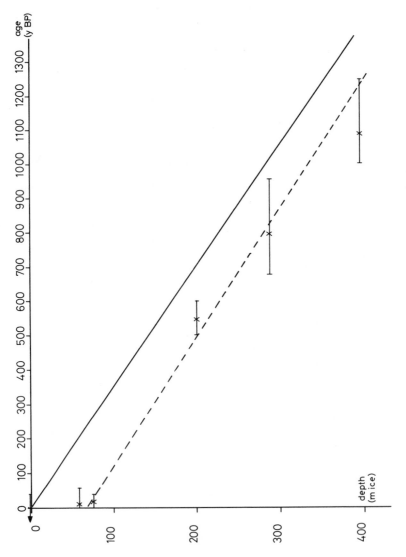

Figure 6. Argon-39 ages vs. depth for a borehole at Station Crête in central Greenland. Key: —×—, ^{39}*Ar-age;* ———, $_{\delta}^{18}O$ *(H. Clausen, Copenhagen); and* – – –, *linear regression.*

At ORNL (Oak Ridge National Laboratory) a group of scientists is developing a method to measure the ^{39}Ar and ^{81}Kr atoms directly with good efficiency, e.g., by mass-spectrometry after multiple electromagnetic enrichment steps. For such measurements a few kilograms of ice would provide enough argon and krypton [16]. If an accuracy of one percent could be obtained, it would be possible to measure variations of the ancient atmospheric activities. Because of the simple geochemical behavior of Ar and Kr these variations could unequivocally be attributed to production rate variations.

^{14}C dating method

In 1947 W. F. Libby and collaborators [17-19] measured for the first time ^{14}C produced by cosmic radiation in the atmosphere. He then proposed the use of this radioisotope for dating of organic material. A unique constellation of factors makes the ^{14}C dating technique a most fascinating and powerful instrument for studies of the last 50,000 y:

- ^{14}C is mainly produced by (n,p) reactions with nitrogen, an element which is ~5000 times more abundant than carbon in the atmosphere. This leads to a relatively high specific activity, thus facilitating measurements.

---

[5]As an example we assume a well mixed reservoir with an average age of water T_r, i.e., the age distribution is

$$g(T) = \frac{1}{T_r} e^{-T/T_r}$$

For a radioisotope with decay constant λ the activity in the reservoir is calculated to be

$$A = A_o e^{-\lambda T}app = A_o/(\lambda T_r + 1)$$

$$\text{and } T_{app} = \frac{1}{\lambda} \ln(\lambda T_r + 1) \quad (T_{app}: \text{ apparent radioactive age})$$

For $T_r = 1000$ years: $T_{app}(^{39}Ar) = 496$ and $T_{app} (^{14}C) = 947$ years. In the case of piston flow the radioactive ages of the two isotopes would be equal.

- After its creation ^{14}C gets oxidized to $^{14}CO_2$ and enters the biosphere; it is present in the environment in organic form (plants and animals) and inorganic form ($^{14}CO_2$ in atmosphere, $H^{14}CO_3^-$ in water, $^{14}CO_3^=$ in sediments, etc.)

- The ^{14}C half-life of 5730 y permits dating of samples not only from the historically documented period but also from the last glacial period.

- The atmospheric $^{14}C/C$ ratio during the last 50,000 y was sufficiently constant to make radiocarbon a remarkably reliable dating tool. Evidence for fluctuations of the $^{14}C/C$ ratio could be found by high precision measurements on samples of known age. These fluctuations can be attributed to variations of processes in the solar system (solar activity) and on earth (fluctuations of CO_2 distribution among the atmospheric, oceanic, and biospheric reservoirs). Both fluctuations of solar activity and of the atmospheric CO_2 content may have contributed to past climatic changes.

- During the industrial era man has influenced the atmospheric $^{14}C/C$ ratio. By 1950 input of CO_2 from combustion of fossil fuel had led to a decrease in this ratio of about 2 percent. By 1963 due to nuclear weapon tests, however, the atmospheric ^{14}C level in the northern hemisphere had increased by about 100 percent. The present excess is still ∿30 percent. There is also an input of ^{14}C from nuclear power and reprocessing plants.

For information on the ^{14}C dating method see also *Radiocarbon* 1980 [20].

Carbon Cycle Models and Disturbances

As partly mentioned before, natural and anthropogenic induced variations of the atmospheric CO_2 concentration and of the $^{14}C/^{12}C$ and $^{13}C/^{12}C$ ratios have been observed. For a quantitative discussion of these variations in relation to possible causes, models for the carbon cycle dynamics have been developed [21-25]. Compared to the noble gas radioisotopes ^{39}Ar and ^{81}Kr, for which we only have to consider a well mixed atmospheric reservoir, we have a much more complicated system for ^{14}C. The CO_2 in the atmosphere exchanges with the carbon in the biosphere and with the CO_2, HCO_3^- and $CO_3^=$ in the ocean. Figure 7 shows the different reservoirs with their relative amounts of carbon. ^{14}C is produced in the atmosphere, and by exchange and mixing it gets distributed in the entire carbon system. The preindustrial $^{14}C/C$ ratio of the carbon in the mixed ocean surface layer is estimated to have been only 95 percent of the atmospheric ratio, and that of the average carbon in the deep ocean, only 84 percent of the atmospheric one. This indicates that in the mean life ^{14}C (∼8000 y) the carbon in the ocean is not well mixed. To discuss disturbances of the system it is therefore necessary to divide it into subsystems. The following discussions are based on the box-diffusion model developed by Oeschger et al., [26], since this model seems to consider the most important characteristics of the carbon cycle without being too complicated. The model responses to a variety of system disturbances can be analytically expressed. As indicated in figure 7 in this model the atmosphere, biosphere, and ocean surface mixed layer are assumed to be well mixed reservoirs. The exchange fluxes between them are assumed to obey first order kinetics. The vertical mixing below the ocean surface is simulated by eddy diffusion with constant eddy diffusivity K. From the preindustrial ^{14}C distribution in deeper ocean strata a value of 4000 m^2 yr^{-1} for K has been derived which will also be used for the discussion of system disturbances. It is further assumed that for the type of disturbances which are of interest here, the exchange with the atmosphere is negligible. The average depth of the ocean H_{oc} is 3800 m, that of the mixed layer h_m, 75 m. The equivalent depth of the atmosphere h_a, defined as thickness of an ocean layer containing the same amount of carbon as the preindustrial atmosphere, is taken as 58 m. (Some numbers, especially h_a, do not correspond to the best possible estimates; they are chosen to keep consistency with references, see also Siegenthaler and Oeschger [27]).

In the following section we will now briefly describe natural and anthropogenic disturbances of the CO_2 system which will later be discussed based on the box-diffusion model.

Observed changes in the atmospheric CO_2 and in its isotopic composition

^{14}C fluctuations

During the last 15 years intensive studies on the history of the atmospheric ^{14}C/C ratio have been performed on tree-rings [20]. The results can be summarized as follows: from 7000 BP (before present) to 2000 BP the average atmospheric ^{14}C/C ratio had decreased by about 10 percent. Superimposed on this general trend are secular variations (Suess-Wiggles) of the order of 1 to 2 percent. In some time intervals a basic period of about 200 years is visible.

The statistical error of the measurements by Suess and collaborators [13,28] is of the order of 0.5 to 0.7 percent, i.e., not much smaller than the observed variations themselves. During the last several years, however, several laboratories confirmed

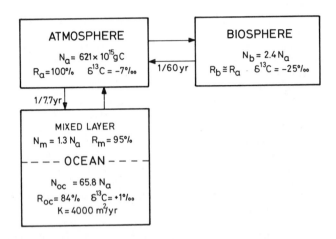

Figure 7.   Main exchanging C reservoirs: atmosphere, biosphere, and ocean; and exchange fluxes.

Carbon content of reservoirs: atmosphere (N_a), biosphere (N_b), mixed layer (N_m), and ocean (N_{oc}). R is the ^{14}C concentration of C in the reservoir; and atmospheric concentration is defined as 100 percent. The ^{14}C concentrations are corrected for isotopic fractionation to a common $\delta^{13}C = -25$ per mil. K is the eddy diffusivity, and $\delta^{13}C$ is the ^{13}C concentration deviation from a standard.

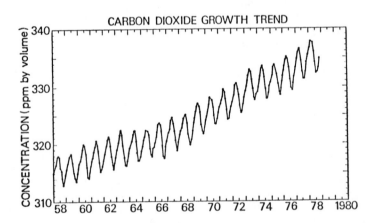

Figure 8.   The CO_2 concentrations observed at Mauna Loa, Hawaii by C. D. Keeling and coworkers.

these characteristic ^{14}C variations by high precision measurements [29-34]. The long-term ^{14}C fluctuations can be approximated by a sine function of time with a period of about 8000 y.

The atmospheric CO_2 increase due to the fossil fuel CO_2 input

The atmospheric CO_2 concentration has been rising during the last hundred years mainly due to the CO_2 input from fossil fuel combustion and partly due to CO_2 release from deforestation. The increase of the atmospheric CO_2 concentration has been continuously monitored by Keeling and co-workers [35-37] at Mauna Loa, Hawaii, and at the South Pole. The Mauna Loa record covering the period 1958 to 1978 is given in figure 8. During that period the yearly average CO_2 concentration increased from 315 ppm to 335 ppm. During the same period an amount of CO_2 corresponding to 36 ppm has been released into the atmosphere due to fossil fuel combustion and cement manufacture [38]. An apparent airborne fraction for the fossil CO_2 can be defined as

$$F_a = \frac{\text{atmospheric } CO_2 \text{ increase}}{\text{fossil } CO_2 \text{ input}}$$

For the period 1958-1978, the apparent airborne fraction is

$$F_a = \frac{20 \text{ ppm}}{36 \text{ ppm}} = 0.56$$

The expression "apparent" is used, since the fossil CO_2 input does not represent the total anthropogenic CO_2 input into the atmosphere. It provides, however, the greatest part of it. $1/F_a$ is the system dilution factor $D_{CO_2,S}$; it expresses the CO_2 uptake capacity of the system in units of the atmospheric uptake capacity. This quantity will later be estimated using the box-diffusion model.

The ^{14}C-dilution due to the fossil CO_2 input (Suess-effect)

The fossil CO_2 brought into the atmosphere does not contain ^{14}C and leads to a ^{14}C dilution. Without exchanges between the atmosphere and the other reservoirs, a fossil CO_2 input of 10 percent until 1950 would have led to a decrease of the $^{14}C/C$ ratio by 10 percent. The actually observed reduction of the $^{14}C/C$ ratio, however, is of the order of 2 percent, i.e., much smaller. This is to be explained by the exchange with the biospheric and oceanic reservoirs. Again a dilution factor $D_{^{14}C,S}$ of the system can be calculated.

Calculation of dilution factors with the box-diffusion model

For the following we assume that the atmospheric variations in CO_2 and in its carbon isotopic composition are entirely due to atmospheric system disturbances, such as the input of ^{14}C-free CO_2 from fossil CO_2 production, and deviations from the average rate of ^{14}C production by cosmic radiation. The system dynamics, i.e., the exchange coefficients and the eddy diffusivity are kept constant. We approximate the fossil CO_2 input p(t) by

$$p_{CO_2}(t) = p_{1,CO_2}e^{\mu t}, \text{ with } \mu = 1/28 \text{ y } [39].\qquad(10)$$

We attribute the long-term and short-term ^{14}C variations to cyclic variations in the ^{14}C production rate $p_{14C}(t)$ which are approximated by $p_{14C}(t) = p_0 + p_{1,14C}e^{i\omega t}$, with the characteristic frequencies $2\pi/10{,}000$ y and $2\pi/200$ y. It thus happens that the CO_2 exchange system has been and is subjected to exponential (fossil CO_2 input) and cyclic (short-time ^{14}C variations) disturbances with characteristic times of the order of 30 y. In addition the system has been exposed to a quasi cyclic disturbance of the ^{14}C production rate which is at present in general attributed to changes in the Earth's magnetic dipole moment. Paleomagnetic data indicate that 7000 y ago the value of the dipole moment was only about half of that of 2000 y ago, and estimates using a model for the geomagnetic modulation of radioisotope production and a carbon cycle model show that this increase in the magnetic field could well have caused most of the observed decrease in the atmospheric ^{14}C/C ratio[6].

In the following we shall discuss the short-term system disturbances. Figure 9 shows how we can understand the penetration of atmospheric disturbances into other reservoirs. If the disturbances have characteristic times which are long compared to the exchange and mixing, the entire system is responding and we have a dampening effect corresponding to that of a one box system, as in the case of [39]Ar. For the kind of disturbances discussed here (characteristic times $\sim$30 y) only fractions of the biosphere and the ocean are responding. In figure 9 this is schematically explained and definitions which are used for the following are given.

As in the case of the [39]Ar production variations a dampening factor $D = (P_1/P_0)/(N_1/N_0)$ can be introduced. It is the product of the dampening $D_{atm}(\frac{\mu}{i\omega})$ we calculate if only the atmosphere were

---

[6](e.g., Sternberg and Damon, reference [40]).

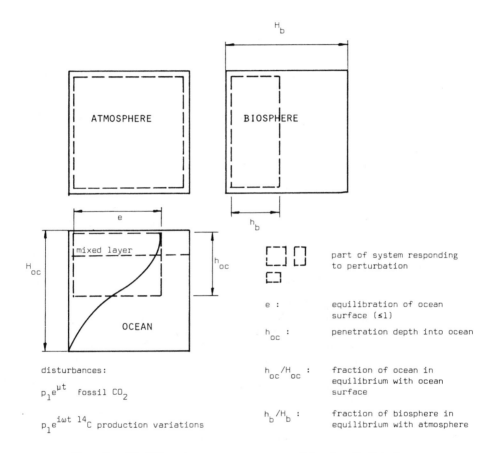

Figure 9.   The CO_2 system response to exponential and cyclic disturbances.

responding and the dilution factor $D_s(^\mu_{iw})$ for the system's response. The total dampening becomes

$$D_{tot}(^\mu_{iw}) = D_{atm}(^\mu_{iw}) \cdot D_s(^\mu_{iw})^7 \qquad (11)$$

The factor for the atmosphere $D_{atm}(\mu)$ for the fossil CO_2 input $P_1 e^{\mu t}$ is considered by expressing the CO_2 production and CO_2 increase in ppm of air. $D_{atm}(iw)$ is $\frac{iw+\lambda}{58 \lambda}$ for cyclic variations in the ^{14}C production $p_1 e^{iwt}$. This expression corresponds to that obtained for the one box model, considering that only 1/58 of the total ^{14}C in the CO_2 exchange system is in the atmosphere. Using the formalisms explained in Oeschger et al., [26], and Oeschger et al., [39], the dampening factor for the system $D_s(^\mu_{iw})$ is

$$D_s(^\mu_{iw}) = 1 + e(^\mu_{iw}) \frac{h_{oc}(^\mu_{iw})}{H_{oc}} \frac{N_{oc}/\xi}{N_a} + \varepsilon \frac{h_b(^\mu_{iw})}{H_b} \frac{N_b}{N_a} \qquad (12)$$

      (atmos-   (ocean uptake in   (biosphere uptake in
      phere)   atmospheric units)   atmospheric units)

For the following we assume that the perturbations are relatively small and characteristic times $<< \frac{H_{oc}^2}{K}$, i.e., the disturbance does not effectively penetrate to the ocean bottom. The parameters ξ and ε are to be set equal to one if perturbations in the $^{14}C/C$ ratio are considered. They differ from one if the perturbation is a change in the CO_2 content of the atmospheric reservoir, since the fluxes to the other reservoirs do not change in proportion to the ratio of the new atmospheric CO_2 concentration to the steady state concentration. An increase in atmospheric CO_2 and a dissolved CO_2 gas in the oceans brings about a shift in the chemical equilibria between dissolved CO_2, HCO_3^- and $CO_3^=$, resulting in an increase of the total CO_2 concentration $(CO_2+HCO_3^-+CO_3^=)$ which is smaller than that of the dissolved CO_2 gas alone. This is taken into account by introducing a buffer factor ξ: if the dissolved CO_2 increases by p percent, the total CO_2 concentration

---

[7]In the expression $D(^\mu_{iw})$ the variation is μ for an exponential disturbance and iw for a cyclic disturbance.

of seawater increases quasi stationarily only by p/ξ percent. For average surface seawater and a CO_2 level of ~ 300 ppm, $\xi \cong 10$; ξ increases for higher CO_2 pressure [41].

Since the biospheric growth rate depends, among other factors, on the CO_2 supply, it is probable that the CO_2 increase induces, at least for part of the biosphere, an increased growth rate ("CO_2 fertilization"). A simple concept to take this into account is the introduction of a biota growth factor ε: if the atmospheric CO_2 pressure increases by p percent, the CO_2 flux to the biosphere increases by εp percent. Typically, values for ε between 0 and 0.5 have been used in carbon cycles models [26,41].

In Table 2 the formulas for $e(\frac{\mu}{iw})$, $h_{oc}(\frac{\mu}{iw})/H_{oc}$ and $h_b(\frac{\mu}{iw})/H_b$ for the different perturbations are given.

First we apply the formalism to <u>the short-term ^{14}C varia-</u><u>tions</u>. If we assume that they are caused by quasi cyclic ^{14}C production rate variations with a period of 200 y [42] we get a dilution factor

$$D_{^{14}C,tot}(iw) = \underbrace{\frac{iw}{\lambda \cdot 58}}_{D_{^{14}C,atm}} \underbrace{\left\{ 1 + e(iw) \cdot \frac{h_{oc}(iw)}{H_{oc}} \cdot \frac{N_{oc}}{N_a} + \frac{h_b(iw)}{H_b} \cdot \frac{N_b}{N_a} \right\}}_{D_{^{14}C,S}} \quad (13)$$

$D_{^{14}C,tot}(iw)$ is a complex number expressing both the ratio of the relative amplitudes and the phase shift between ^{14}C production and observed ^{14}C concentration

$$D_{^{14}C,tot}(i/28\ y) = |D_{^{14}C,tot}|\ e^{i\phi} \quad (14)$$

$|D_{^{14}C,tot}| \cong 20$; phase angle $\phi = 36°$ corresponding to a lag of ~ 20 y.

For many years, this calculated strong model attenuation was the reason why the existence of the secular ^{14}C fluctuations was doubted: variations in ^{14}C of two percent would correspond to ^{14}C production rate variations of ~ 40 percent, compared to variations of the order of 10 to 20 percent as predicted from ^{14}C production models. Stuiver and Quay [31], used the box diffusion model to calculate the ^{14}C production rate variations observed in tree-rings. Figure 10 shows these production rate variations plotted together with inverse sun spot numbers. A good correlation is obtained. Stuiver and Quay compared the magnitude of these geochemically derived production changes with ^{14}C production changes derived from atmospheric neutron flux measurements. A

Table 2.   Calculations Regarding System Response to Exponential and Cyclic Disturbances.

Perturbation	Parameters	$e(\frac{i\omega}{\mu})$	$\dfrac{h_{oc}(\frac{i\omega}{\mu})}{H_{oc}}$	$\dfrac{h_b(\frac{i\omega}{\mu})}{H_b}$
^{14}C production variation	$\xi = 1$	$\dfrac{h_a k_{am}}{h_a k_{am} + h_{oc} i\omega}$	$\dfrac{h_m + \sqrt{K/i\omega}}{H_{oc}}$	$\dfrac{k_{ba}}{k_{ba} + i\omega}$
$p_1 e^{i\omega t}$	$\varepsilon = 1$			
$\omega = \dfrac{2\pi}{200\ y}$				
CO_2 input				
$p_1 e^{\mu t}$	$\xi \cong 10$	$\dfrac{\xi h_a k_{am}}{\xi h_a k_{am} + \mu h_{oc}}$	$\dfrac{h_m + \sqrt{K/\mu}}{H_{oc}}$	$\dfrac{k_{ba}}{k_{ba} + \mu}$
	$\varepsilon = 0.2$ (assumption)			
$\mu = 1/28\ y$				
^{14}C dilution (Suess Effect) due to fossil CO_2 input	$\xi = 1$	$\dfrac{h_a k_{am}}{h_a k_{am} + \mu h_{oc}}$	$\dfrac{h_m + \sqrt{K/\mu}}{H_{oc}}$	$\dfrac{k_{ba}}{k_{ba} + \mu}$
	$\varepsilon = 1$			
$p_1 e^{\mu t}$				
$\mu = 1/28\ y$				

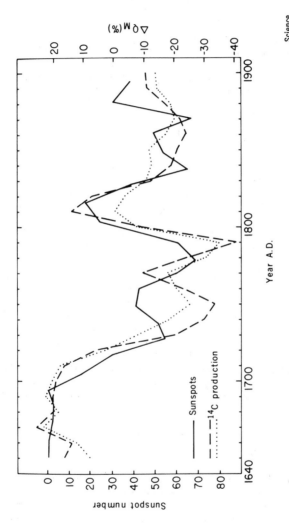

Figure 10. *Carbon-14 production as calculated from tree-ring ^{14}C concentrations by means of the box-diffusion model (31).*

The dashed and the dotted curves correspond to two different assumptions for the biosphere. The good correlation with sunspot numbers (———) suggests that indeed solar modulation of the cosmic radiation causes the ^{14}C variations.

good agreement is obtained for the 20th and 19th century data, but for the Maunder Minimum period, A.D. 1654 to 1741, a greater dependence on sun spot numbers is suggested, which Stuiver attributes to an additional ^{14}C production increase during periods when sun spots are absent. A generally satisfactory agreement is obtained, suggesting that for cyclic perturbations with a characteristic time of 30 years the model seems to give reasonable answers.

In the second place we apply the formalism to the CO_2 increase due to the fossil CO_2 input. As mentioned before, for the period 1958 to 1978 an apparent airborne fraction of 0.56 corresponding to a dilution factor of 1.79 has been observed.

For $\mu = 1/28$ y we obtain with $\xi = 10$, $h_{oc} = 410$ m, $h_{oc}/H_{oc} = 0.108$, e = 0.837, $h_b/H_b = 0.318$, finally a dilution factor

$$D_{CO_2,S} (\mu = 1/28 \text{ y}) = 1 + 0.60 + \begin{cases} 0 & (\varepsilon = 0) \\ 0.15 & (\varepsilon = 0.2) \end{cases} \qquad (15)$$

The model predicted dilution factors are within the error limits of the observed value, D = 1.79 ± 10 percent, for both values of ε. For optimum agreement, the biota growth factor ε should be 0.25.

There would be, however, a considerable discrepancy between the model-calculated dilution factor and the dilution factor required if the biospheric CO_2 input were of comparable size as the fossil CO_2 input as stated by biologists [43]. For a discussion of this question see also Oeschger et al., [39]. Thirdly, we calculate the ^{14}C dilution corresponding to the CO_2 increase. In 1950, before the nuclear weapon tests, the integrated CO_2 production amounted to about 10 percent of the preindustrial atmospheric CO_2 content. If there had been no exchange with other reservoirs, a decrease of the $^{14}C/C$ ratio by 10 percent would have resulted. Tree-ring ^{14}C measurements indicated, however, a decrease by only about 2 percent. Again we calculate the system dilution. In a first approximation ξ and ε are set equal to one and we obtain

$$D_{^{14}C,S} (1/28 \text{ y}) = 1 + 2.4 + 0.8 = 4.2 \qquad (16)$$

Thus the ^{14}C dilution including system dilution in 1950 is estimated to be -10 percent/4.2 = -2.4 percent.

Using the actual CO_2 production history, Oeschger et al., [26] calculated for the Suess effect in 1950 a value of -2.0 percent. This is essentially in agreement with the measurements [44], though ^{14}C fluctuations due to ^{14}C production rate variation make a precise determination of the Suess effect difficult.

Changes in the CO_2 system at the end of last glaciation

Compression of the ^{14}C time scale around 10,000 BP

Dendrochronologically dated tree-rings for the observation of ^{14}C variations are available for about the last 8000 y. A comparison of the ^{14}C variation record with the climatic history record suggests the existence of relations between mechanisms producing ^{14}C variations and those responsible for climatic change. If such a relation indeed existed, especially pronounced variations would be expected to have occurred during the transition period from Glacial to Postglacial, i.e., from about 14,000 to 9000 BP. When dating peat bog samples covering the end of the Younger Dryas cold phase, as determined by pollen analysis, we observed irregularities in the ^{14}C time scale.

Detailed ^{14}C analyses covering this period were then performed on samples from a peat bog near Wachseldorn (Switzerland) which from pollen analyses is known to have grown continuously during the whole Late Glacial and Postglacial. The samples from the second half of the Younger Dryas cold period to the beginning of the Preboreal show rather constant ^{14}C concentrations over a period for which, based on the assumption of constant peat growth, one would expect a change by about 7 percent, corresponding to a difference of age of about half a millenium [45]. These results must be confirmed by additional studies. At present we are measuring the $^{14}C/C$ ratio on lake chalk samples covering the period of interest. Already, now we consider it as very probable that strong ^{14}C variations occurred during this period of major climatic change. The question is what has caused them. Variations of the atmospheric $^{14}C/C$ ratio can be caused either by changes in the ^{14}C production rate or by changes in the terrestrial carbon system or both. Changes in the terrestrial carbon system leading to $^{14}C/C$ ratio changes might have been induced either by changes in the partitioning of the CO_2 among the atmospheric, biospheric, and oceanic reservoirs or by changes in the dynamics of ocean mixing. Assuming a constant galactic cosmic radiation production rate, we would expect variations to be caused mainly by changes in the earth's magnetic field with its shielding properties and by modulation of the galactic cosmic radiation by solar plasma magnetic fields.

During the late pleistocene the geomagnetic field strength seems to have been generally lower and the atmospheric $^{14}C/C$ ratio therefore higher than in the holocene [46]. Barbetti therefore notices that there should be a compression in the ^{14}C time scale from 12,000 to 10,000 BP. But at the end of the last glacial period there also might have been a change in solar activity. Periods of cold climate coincide with periods of high ^{14}C production (Maunder Minimum and Little Ice Age). The ^{14}C plateau in samples covering the Younger Dryas-Preboreal transition therefore might reflect a switching from low to high solar activity.

However, changes in the carbon cycle during the period of interest cannot be excluded, as is discussed in the next paragraph.

The last 30,000 years history of the atmospheric CO_2 content

For an assessment of the CO_2 problem a knowledge of the history of the atmospheric CO_2 content would be of great value. What was the pre-industrial atmospheric CO_2 content? Did it fluctuate or was it rather constant? The answer to such questions would help on the one hand to improve our knowledge on the carbon system and its response to disturbances, and on the other hand provide information regarding the sensitivity of the climate system to atmospheric CO_2 changes.

Probably the only possibility to reconstruct directly the history of the atmospheric CO_2 content (and $^{13}C/^{12}C$ and $^{14}C/C$ ratios) are measurements on natural ancient ice samples. Ice formed by sintering of dry cold snow contains air with atmospheric composition in its bubbles. Until a few years ago attempts to determine ancient atmospheric CO_2 contents by measuring CO_2 contents of the air occluded as air bubbles in natural ice seemed to provide unreliable results [47-49]. CO_2/N_2 ratios much higher than the atmospheric values were found indicating the presence of additional CO_2 of undefined origin. In the last few years, however, the extraction technique has been further developed and considerably improved by two laboratories [50,51]. The group in Bern obtained CO_2 concentrations for air occluded in young ice samples of 270 to 370 ppm, with an average of 310 ppm [52]. This value is close to the assumed pre-industrial atmospheric CO_2 level and supports the theory that on ice samples from very cold accumulation areas, CO_2/N_2 ratios can be measured which do indicate the atmospheric composition at the time of ice formation. The extraction procedure is as follows:

Samples of 300 g of ice are melted in vacuum and the gases produced by the explosion of air bubbles on the melting ice surface are extracted. This so called first extraction fraction is considered to be representative for the composition of the gases in the bubbles. Extraction of the gases is then continued (second extraction fraction) for several hours until no more CO_2 is collected. Based on the analyses of the two extraction fractions an estimate can then be made of how much CO_2 is contained in the air bubbles and how much in the ice lattice. Analysis of the gas composition (N_2, O_2, Ar, CO_2) is made by gas chromatography.

Our experience to-date suggests that current measurement and analysis techniques allow the reliable detection of variations in the atmospheric CO_2 content of 30 percent or more. Our first measurements have been made on samples covering the last 30,000 years, a period of major climatic change which might possibly have led to a change in the atmospheric CO_2 content. Ice cores suited for such a study were available from the successful drilling by U.S. scientists in 1966 to the bedrock of the Greenland ice cap in northwest Greenland (Station Camp Century) and in 1967/1968 to the bedrock of the West Antarctic ice shield at Byrd Station.

In figure 11, CO_2 contents measured in the first extraction fraction are plotted together with the $\delta^{18}O$ record for the Camp Century and the Byrd Station core [53,54]. The dating is based on model calculations given in references [54,55]. Both records show similar trends: low values during the last glaciation, and then, parallel to the $\delta^{18}O$ transition, an increase to higher holocene CO_2 values. For both cores similar minimum values (200 to 230 ppm) are found for the last glaciation. First fraction measurements on young ice (shown here only for the Byrd core) yield values as in the holocene; for the Camp Century core higher values are obtained. The question has to be answered whether the general CO_2 trend in the ice core is mainly due to a change in atmospheric CO_2 content or a climatic effect on CO_2 enclosure process. The most probable explanation for the general trend — low values during Glacial and higher values during Postglacial — is a corresponding change in the atmospheric CO_2 content. The difference between the two profiles in the holocene may partly be due to a contribution of CO_2 trapped in melt layers during the climatic optimum or due to another climatic effect on CO_2 enclosure at Camp Century.

For the climatic optimum we consider the values measured on the Byrd Station core as the more reliable ones and do not exclude the possibility that during that period the atmospheric CO_2 content was significantly higher than at present[8]. A possible explanation for a change in atmospheric CO_2 content is that it strongly depends on the total CO_2 content of the ocean surface. Due to the buffer effect, a relative change in total CO_2 in the ocean surface leads, for assumed constant alkalinity, to a tenfold relative change in the atmospheric CO_2 content. The total CO_2 is determined partly by marine biospheric activity which leads to a depletion at the ocean surface compared to the ocean average. A relatively slight change in biospheric activity could, therefore, lead to a significant decrease in atmospheric CO_2. For an interesting discussion of CO_2 variations due to $PO_4^=$ variations in the ocean, see Broecker [56].

---

[8]Recent measurements indicate that the relatively high CO_2 concentrations determined for part of the holocene ice probably are due to contamination of the ice core which for this age range shows many small cracks.

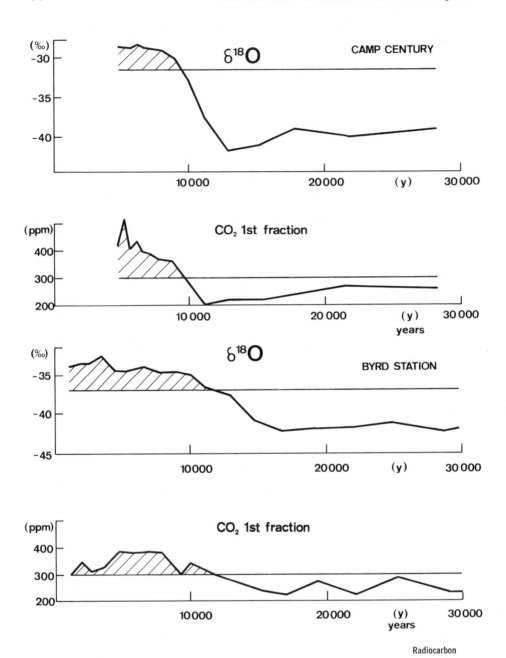

Radiocarbon

Figure 11.   Camp Century, Greenland and Byrd Station, Antarctica ice cores: CO₂ contents of the first gas extraction fraction and δ¹⁸O profiles (50). The δ¹⁸O profiles are from Dansgaard and coworkers, and the ages are calculated according to Ref. 54.

Measurements of radioisotopes of solids, directly deposited on polar ice caps

One is compelled to pose the question if experimentally it will become possible to decide whether the ^{14}C variations observed on tree-ring samples, peat bogs, sediments, etc., are primarily caused by an external forcing of the system (production rate variations) or by an internal one. Recent progress in detection of small numbers of nuclei of an isotope by mass spectrometry based on the use of a particle accelerator [57,58] make it possible to measure the cosmic ray produced [10]Be or [36]Cl deposited in only 1 kg of ice. These isotopes get attached to aerosol particles and deposited with them.

Their residence time in the atmosphere is relatively short (months to a few years). Changes in their production rates are therefore relatively unattenuated and reflected in precipitation with good time resolution.

Several laboratories therefore intend to measure profiles of these isotopes on ice cores drilled in polar ice caps. The new technique also makes possible measurements of the ^{14}C in the CO_2 occluded in about 30 kg of ice. From the ^{14}C measurements again twofold information is expected:

- dating of ice of cores drilled into the polar ice caps but also of surface samples collected in their ablation area.

- studies of ^{14}C variations in ice core samples from areas with regular stratigraphy enabling independent dating.

It is possible to obtain such samples covering the full range of the ^{14}C time scale, i.e., more than 50,000 years back in time. Comparison of ^{14}C and [10]Be variations will enable us to disentangle the external and internal causes of the ^{14}C variations: the [10]Be and [36]Cl variations will serve as a measure for the ^{14}C production rate variations. To a first approximation we may assume that they are proportional to the ^{14}C production variations. Based on the [10]Be and [36]Cl measurements we therefore will approximately know the ^{14}C production rate variation $p(t)$. Based on tree-ring measurements the atmospheric ^{14}C variations of the last 8000 years are known and from measurement on CO_2 extracted from ice cores it hopefully will be possible to get information on atmospheric ^{14}C/C variations over a large time range. $P(t)$ and [14]C(t)/C are then related to each other via the response of the CO_2 system to a [14]C δ-input being named $R(\tau,t)$; we get the following relation

$$^{14}C(t)/C = {_0}\int^{\infty} P_{14_C}(t-\tau)R(\tau,t)d\tau \tag{17}$$

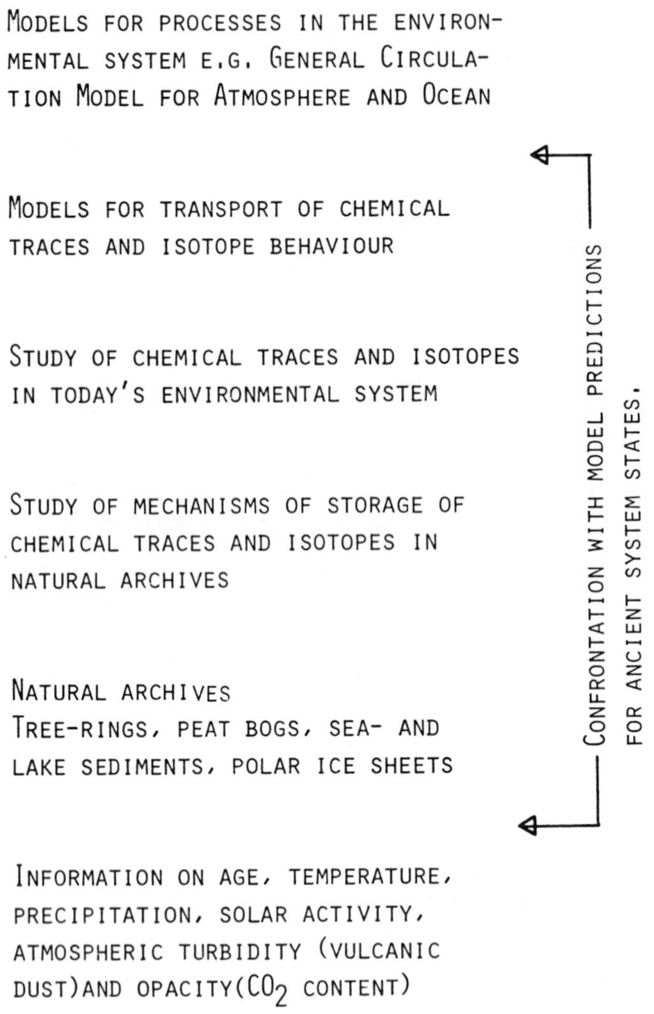

MODELS FOR PROCESSES IN THE ENVIRON-
MENTAL SYSTEM E.G. GENERAL CIRCULA-
TION MODEL FOR ATMOSPHERE AND OCEAN

MODELS FOR TRANSPORT OF CHEMICAL
TRACES AND ISOTOPE BEHAVIOUR

STUDY OF CHEMICAL TRACES AND ISOTOPES
IN TODAY'S ENVIRONMENTAL SYSTEM

STUDY OF MECHANISMS OF STORAGE OF
CHEMICAL TRACES AND ISOTOPES IN
NATURAL ARCHIVES

NATURAL ARCHIVES
TREE-RINGS, PEAT BOGS, SEA- AND
LAKE SEDIMENTS, POLAR ICE SHEETS

CONFRONTATION WITH MODEL PREDICTIONS
FOR ANCIENT SYSTEM STATES.

INFORMATION ON AGE, TEMPERATURE,
PRECIPITATION, SOLAR ACTIVITY,
ATMOSPHERIC TURBIDITY (VULCANIC
DUST)AND OPACITY(CO_2 CONTENT)

Figure 12.   Strategy for environmental system studies.

Measurements of ^{14}C and ^{10}Be variations therefore will reveal:

- history of the isotope production rate by cosmic radiation in the atmosphere (^{10}Be variations)

- information on the steady state $R(\tau,t_o)$, i.e., the response of the CO_2 system to atmospheric ^{14}C input variations alone

- information on variations of $R(\tau,t)$ with time in case ^{14}C variations would be observed without indications of triggering ^{14}C production rate variations based on ^{10}Be data.

Conclusions

   In view of decisions regarding man's future it is an urgent task for the natural sciences to improve the quantitative knowledge on environmental processes. Of special interest are the studies of past environmental system states. On one side it can be checked how well models are able to describe system states which deviate from the present one. On the other side past system states may to some degree be analogues for future ones. Physical and chemical analytical studies provide the time scale for the information on the environmental system stored in natural archives. They also provide direct information on externally and internally forced system changes. Examples are modulations of solar activity and changes in the atmospheric CO_2 content. Figure 12 shows a strategy for environmental system studies emerging from the discussion in this paper.
   Essential for the rapid development of this field is the progress in low level counting (^{39}Ar, ^{81}Kr), the development of accelerator based mass spectrometry (^{10}Be, ^{36}Cl) and hopefully in the near future the application of single atom detection using laser resonance ionization. Also achievements in field techniques such as ice core drilling and study are important contributions, besides the increasing efforts in computer data handling and simulation of environmental processes with numerical models.
   The research program displayed here is especially rewarding due to its interdisciplinary character and the attempt to make use of a very broad spectrum of methods in the natural sciences.

   This paper is intended to be a synthesis of ideas developed and work performed during many years not only together with my colleagues and collaborators in Berne but also together with a large group of U.S. scientists. Special thanks are due to Dr. L. A. Currie, who inspired me to write this paper and to Dr. C. C. Langway, Jr., who opened up for us the field of isotope applications to glaciology. I am also much obliged to the Swiss and U.S. National Science Foundations for the continuous support of work along the lines discussed here.

References

[1] Eugster, O., Eberhardt, P., Geiss, J., Earth Planet. Sci. Lett., 1967, 2, 77.
[2] Marti, K., Phys. Rev. Lett., 1967, 18, 264.
[3] Begemann, F., Geiss, J., Hess, D. C., Phys Rev., 1957, 107, 540.
[4] Voshage, H., Hintenberger, H., Z. Naturforsch., 1961, 16a, 1042.
[5] Voshage, H., Z. Naturforsch., 1962, 17a, 422.
[6] Arnold, J. R., Honda, M., Lal, D., J. Geophys. Res., 1961, 66, 3519.
[7] Geiss, J., Oeschger, H., Schwarz, U., Space Sci. Rev., 1962, 1, 197.
[8] Castagnoli, G., Lal, D., Radiocarbon, 1980, 22, 133.
[9] Loosli, H. H., Oeschger, H., Studer, R., Wahlen, M., Wiest, W., 1973, Proc. Noble Gases Symposium, Las Vegas, Sept. 24-28, R. E. Stanley, A. A. Moghissi, eds., 24.
[10] Loosli, H. H., Oeschger, H., Earth Planet. Sci. Lett., 1968, 5, 191.
[11] Loosli, H. H., Oeschger, H., Earth Planet. Sci. Lett., 1969, 7, 67.
[12] Kuzimov, V. V., Pomansky, A. A., Radiocarbon, 1980, 22, 311.
[13] Suess, H. E., Radiocarbon, 1980, 22, 200.
[14] Loosli, H. H., Eine Altersbestimmungsmethode mit Ar-39, Habilitationsschrift, Universität Bern, Switzerland, 1979.
[15] Loosli, H. H., Oeschger, H., Radiocarbon, 1980, 22, 863.
[16] Hurst, G. S., Payne, M. G., Kramer, S. D., Chen, C. M., Physics Today, Sept. 1980.
[17] Libby, W. F., Phys. Rev., 1946, 69, 671.
[18] Anderson, E. C., Libby, W. F., Weinhouse, S., Reid, A. F., Kirschenbaum, A. D., Grosse, A. V., Phys. Rev., 1947, 72, 931.
[19] Libby, W. F., Anderson, E. C., Arnold, J. R., Science, 1949, 109, 227.
[20] Radiocarbon 1980, 22, Proceedings of the Tenth International Radiocarbon Conference, Bern and Heidelberg, M. Stuiver, R. Kra, eds.
[21] Craig, H., Tellus, 1957, 9, 1.
[22] Revelle, R., Suess, H. E., Tellus, 1957. 9, 18.
[23] Keeling, C. D., Chemistry of the Lower Atmosphere, 1973, S. I. Rasool, ed., 251.
[24] Björkström, A., The Global Carbon Cycle, 1979, B. Bolin, et al., eds., John Wiley & Sons, 403.
[25] Broecker, W. S., Peng, T. H., Engh, R., Radiocarbon, 1980, 22, 56.
[26] Oeschger, H., Siegenthaler, U., Schotterer, U., Gugelmann, A., Tellus, 1975, 27, 613.
[27] Siegenthaler, U., Oeschger, H., Science, 1978, 199, 388.

[28] Suess, H. E., Radiocarbon Variations and Absolute Chronology, 12th Nobel Symposium, Uppsala, 1971, Proc: Stockholm, Almquist and Wiksell-Gebers Forlag AB, 1969, 303.

[29] de Jong, A. F. M., Mook, W. G., Becker, B., *Nature*, 1979, 280, 48.

[30] de Jong, A. F. M., Mook, W. G., Radiocarbon, 1980, 22, 267.

[31] Stuiver, M., Quay, P. D., Science, 1980, 207, 11.

[32] Stuiver, M., Quay, P. D., Radiocarbon, 1980, 22, 166.

[33] Bruns, M., Münnich, K. O., Broecker, B., Radiocarbon, 1980, 22, 273.

[34] Pearson, G. W., Radiocarbon, 1980, 22, 337.

[35] Keeling, C. D., Bacastow, R. B., Bainbridge, A. E., Ekdahl, C. A., Guenther, P. R., Waterman, L. S., Chin, J. F. S., Tellus, 1976, 28, 538.

[36] Keeling, C. D., Adams, J. A., Ekdahl, C. A., Guenther, P. R., Tellus, 1976, 28, 552.

[37] Keeling, C. D., The Influence of Mauna Loa Observatory on the Development of Atmospheric CO_2 Research, 1978, J. M. Miller, ed., NOAA Special Report, Air Resources Laboratory, Silver Spring.

[38] Rotty, R. M., Experientia, 1980, 36, 781.

[39] Oeschger, H., Siegenthaler, U., Heimann, M., Proceedings of the International Workshop on Energy-Climatic Interactions, Münster, Germany, 3-7 March 1980, in press.

[40] Sternberg, R. S., Damon, P. E., 1979, Radiocarbon Dating, Proceedings of the Ninth International Conference, Los Angeles and La Jolla, 1976, R. Berger, H. E. Suess, eds., 691.

[41] Bacastow, R. B., Keeling, C. D., Carbon and the Biosphere, 1973, G. M. Woodwell, E. V. Pecan, eds., USAEC, Springfield, Virginia, 86.

[42] Neftel, A., Suess, H. E., Oeschger, H., 1980, in preparation.

[43] Woodwell, G. M., Whittaker, R. H., Reiners, W. A., Likens, G. E., Delwiche, C. C., Botkin, D. B., Science, 1978, 199, 141.

[44] Tans, P. P., de Jong, A. F. M., Mook, W. G., *Nature*, 1979, 280, 826.

[45] Oeschger, H., Welten, M., Eicher, U., Moell, M., Riesen, T., Siegenthaler, U., Wegmüller, S., Radiocarbon, 1980, 22, 299.

[46] Barbetti, M., Radiocarbon, 1980, 22, 192.

[47] Scholander, P. F., Hemmingsen, E. A., Coachman, L, K., Nutt, D. C., J. Glaciology, 1961, 3, 813.

[48] Matsuo, S., Miyake, Y., J. Geophys. Res., 1966, 71, 5235.

[49] Raynaud, D., Delmas, R., 1977, Proceedings of the Grenoble Symposium 1975, Isotopes and Impurities in Snow and Ice, IAHS-AISH publication, 118, 377.

[50] Berner, W., Oeschger, H., Stauffer, B., Radiocarbon, 1980, 22, 227.

[51] Delmas, R. J., Ascencio, J. -M., Legrand, M., *Nature*, 1980, 284, 155.

[52] Berner, W. , Stauffer, B. , Oeschger, H. , <u>Nature</u>, 1978, <u>275</u>, 53.
[53] Dansgaard, W. , Johnsen, S. J. , Clausen, H. B. , Langway, Jr. , C. C. , 1971, The Late Cenozonic Glacial Ages, K. K. Turekian, ed. , New Haven and London, Yale University Press, 37.
[54] Johnsen, S. J. , Dansgaard, W. , Clausen, H. B. , Langway, Jr. , C. C. , <u>Nature</u>, 1972, <u>235</u>, 429.
[55] Hammer, C. U. , Clausen, H. B. , Dansgaard, W. , Gundestrup, N. , Johnsen, S. J. , Reeh, N. , <u>J. Glaciology</u>, 1978, <u>20</u>, 3.
[56] Broecker, W. S. , 1980, Glacial to Interglacial Changes in Ocean Chemistry, to be published in the CIMAS symposium volume.
[57] Nishiizumi, K. , Arnold, J. R. , Elmore, D. , Ferraro, R. D. , Gove, H. E. , Finkel, R. C. , Beukens, R. P. , Chang, K. H. , Kilius, L. R. , <u>Earth Planet. Sci. Lett.</u>, 1980, in press.
[58] Raisbeck, G. M. , Yion, F. , Fruneau, M. , Lieuvin, M. , Loiseaux, J. M. , <u>Nature</u>, 1978, <u>275</u>, 731.

RECEIVED September 9, 1981.

III. ADVANCED DATING TECHNIQUES: COSMOCHEMICAL AND GEOCHEMICAL APPLICATIONS

The Application of Electrostatic Tandems to Ultrasensitive Mass Spectrometry and Nuclear Dating

K. H. PURSER, C. J. RUSSO, and R. B. LIEBERT—General Ionex Corporation, 19 Graf Road, Newburyport, MA 01950

H. E. GOVE, D. ELMORE, and R. FERRARO—University of Rochester, Nuclear Structure Research Laboratory, Rochester, NY 14627

A. E. LITHERLAND, R. P. BEUKENS, K. H. CHANG, L. R. KILIUS, and H. W. LEE—University of Toronto, Department of Physics, Toronto, Ontario, Canada

A review is presented of a new ultra-sensitive particle identification technique which uses as one component an electrostatic tandem accelerator. This technique has applications to nuclear dating in both direct detection of rare atoms and the measurement of precision isotopic ratios. Experimental procedures and fractionation problems associated with the detection of ^{14}C, ^{10}Be, ^{26}Al, ^{36}Cl and ^{129}I at concentrations close to $1:10^{14}$ will be described. For ^{14}C, accuracies approaching 1 percent have been achieved for milligram samples, with dates being achieved for samples as small as 200 micrograms. Some applications of the technique are discussed when it is applied to microprobe mass spectrometry and dating by isotopic ratio measurements.

Widespread interest has developed during the last few years in a new analytical technique which uses a tandem electrostatic acclerator as one element of a double mass spectrometer [1-19][1]. With this technique, individual atoms of many nuclear species from most areas of the periodic table can be identified with good efficiency and in the presence of an almost zero background of unwanted atoms and molecular species of the same mass.

Since May 1977 a consortium of scientists from the University of Rochester, the General Ionex Corporation and the University of Toronto (RIT) have explored some of the features of this new spectrometry. The scope of this program includes:
1. Detection of rare atoms, both stable and radioactive
2. The measurement of precision isotopic ratios
3. An exploration of controlled molecular fragmentation for organic and inorganic analysis.

---

[1]Figures in brackets indicate the literature references at the end of this paper.

0097-6156/82/0176-0045$07.50/0

Up to the present, most of the work of the consortium has focussed attention on the detection of the long-lived radionuclides [10]Be, ^{14}C, [26]Al, [36]Cl, and [129]I. Most of these have been detected at levels of 1 part in 10^{14} or lower. One reason for this attention to radioactive nuclei results from the ease of preparing low concentration (1 in 10^{13}) samples. In much of the work relating to ^{14}C detection that is reported here, Dr. M. Rubin, head of the radiocarbon laboratory of the United States Geological Society in Reston, Virginia, was a collaborator. The U. S. G. S. has been very helpful in preparing and providing samples that have already been dated.

The present paper describes the status of this work on the direct detection of radionuclei. It also describes, in a general manner, apparatus that is being built for dedicated detection facilities that are being installed at the Universities of Arizona, Oxford, Toronto and Nagoya.

The Direct Detection of Radioactive Atoms

It has been hoped [20,21] that a method could be developed which would directly detect the radioatoms that are present in nature by an efficient ultra-sensitive mass spectrometer technique which would not itself depend upon the fact that the atoms being investigated are radioactive. The advantage of an efficient mass spectrometer system for long-lived radioisotopes can be seen from the equation for calculating the number of atoms present in a sample from its measured radioactive decay rate:

$$N_p = (dN/dt) \times (t_{(1/2)}/0.693)$$

Here N_p is the number of atoms present in the sample;

dN/dt is the decay rate in disintegrations/sec

$t_{(1/2)}$ is the half-life in seconds

The natural cosmic ray background and the environmental radioactivity set a lower limit on measurable counting rates and thus the minimum number of radioactive atoms in the sample. This minimum number increases linearly with the half-life. For ^{14}C the number of atoms present in a sample is given by:

$$N_{C-14} = (\text{Disintegrations/hour}) \times (7.2 \times 10^7)$$

Clearly, in a practical ^{14}C experiment only about 1 atom in 10^7 contributes to the measurement.

In contrast, the direct detection technique for ^{14}C can count approximately one percent of the ^{14}C atoms that are present in a 200 μg sample with virtually zero background in times that are of the order of a few hours. Thus, for milligram quantities of carbon, the improvement in sensitivity of direct counting over radioactivity is of the order 7.2×10^5.

The first problem which has previously prevented the use of such mass spectrometer methods is that for long-lived beta-active isotopes there is almost always a stable daughter isotope whose mass very nearly equals that of the radioactive atoms itself. For example, ^{14}C has as its stable daughter the isobar ^{14}N, which differs in mass by only 1 part in 84,000. If these are to be mass separated (and there is inevitably an enormous background of ^{14}N atoms), a mass resolution, $M/\Delta M$, greater than 500,000 is essential. Such a mass spectrometer would involve the use of extremely large magnetic elements with very small acceptance angles and narrow defining slits; these all lead to low particle transmission and low sensitivity.

The second serious problem of a high transmission instrument relates to the presence of molecular fragments. These molecules, which originate in the source, have nearly the same mass as the wanted ions. For ^{14}C measurements, the mass difference between this atom and the molecule $^{12}CH_2$ is approximately 0.1 percent. The resolution needed to separate these components is about 6,000, because there may be 10^9 $^{12}CH_2$ molecules for each ^{14}C atom.

An example is shown in figure 1 of the molecular interferences which must be dealt with around mass 87 if one wishes to use a mass spectrometer for rubidium/strontium measurements in a geological sample [22]. The major elements in this lunar sample all have mass numbers less than 48. Thus, the mass 87 region should be completely free of atomic peaks except for the minor components such as rubidium and strontium. This is clearly not the case and at most mass numbers in the rubidium region there are major interferences from molecules.

Until recently the only satisfactory way to separate these molecular interferences has been on the basis of nuclear mass defects, i.e., the mass of molecules having the same mass number differs from that of the atoms of the same mass number. Figure 2 shows the resolution that is needed to resolve the molecular impurities present in the previous example. Clearly, an unambiguous identification can be made, and all molecular fragments can only be eliminated for an instrument with resolution $M/\Delta M$ approximately 20,000. Once again, the need for high resolution will cause the transmission efficiency to be low.

Features of the New Spectroscopy

The basic features of the new spectroscopy which eliminate these problems are:
1. Negative ions are used at the ion source as one step in the discrimination against backgrounds; i.e., in ^{14}C measurements the ion N^- is not stable, so ^{14}C can be analyzed completely free of its daughter, ^{14}N.
2. Molecular interferences can be completely eliminated by exploiting the fact that multiply charged molecules fragment with 100 percent probability, because of the internal coulomb forces, when several electrons are removed [1].

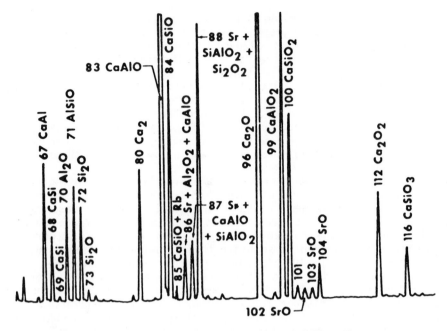

Figure 1.   Partial mass spectrum of Plagioclase 15415.

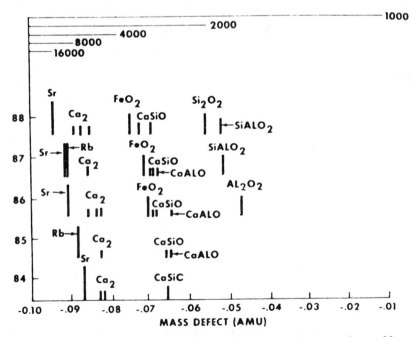

Figure 2.   Resolution required to resolve molecular impurities around mass 88 on the basis of mass defect.

3. Nuclear detectors are used to provide detailed information about each individual ion regarding its energy, rate of energy loss, mass and atomic number [17]. These are information dimensions which are not available in classical mass spectrometry and allow each ion which does arrive at this detector to be thoroughly identified.

Molecular Elimination

The evidence available at this moment indicates that any process which removes more than two electrons from a neutral molecule can be used to eliminate the molecule from detection by a double mass spectrometer [1,23]. The principle of the process is shown in figure 3. Here, a particle from the ion source and having mass M_1 has been selected. This particle is then stripped of several outer electrons so that it has a positive charge state of 3+. Molecular dissociation occurs. If the resultant products are now reanalyzed, non-fragmented atoms will have the same mass M_1 while the separated molecular fragments have a lower mass, and so are completely rejected by the second mass analyzer. It should be emphasized in such a system that the resolution needed at both mass selections is only about one mass unit, making possible an instrument with large acceptance angles, wide slits and consequently having high transmission efficiency.

The most efficient way to remove two or more electrons from an atom or molecule is by stripping the particles after they have been accelerated to a velocity such that on average three electrons are removed after passage through a thin foil or an appropriate thickness of gas [23]. For light elements, this requires ion energies of about 3 MeV; for heavy elements about 6 MeV is needed to reach the optimum efficiency. Figure 4 shows the efficiency with energy; it can be seen that 3 MeV represents a useful compromise between efficiency and cost for elements across the whole periodic table.

Equipment Description

Only a short description of the equipment will be made here because several recent review papers [17,36] describe in detail the use of electrostatic devices as ultra-sensitive particle identifiers. The principles of the dedicated systems being constructed for the Universities of Arizona, Toronto and Nagoya are shown schematically in figure 5. It can be seen that for ^{14}C detection the signal to background counting rate at various parts of the machine changes by 17 orders of magnitude; from 1 to 10^{13} at the ion source to 1 in 10^{-4} at the detector. This tremendous discrimination is made possible by the use of negative ions at the ion source; the destruction of interfering molecular species by the fragmentation of molecules; and finally the use of nuclear detectors to identify individually the accelerated ions.

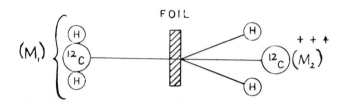

Figure 3.   Schematic of molecular dissociation process. When three or more electrons are removed, molecule breaks up with 100% certainty.

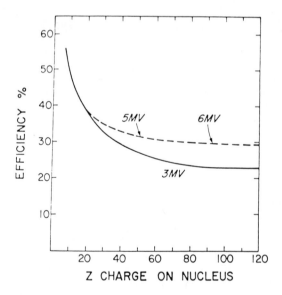

Figure 4.   The efficiency of charge exchange to 3+ state as a function of atomic number and tandem terminal voltage. Key: – – –, maximum efficiency.

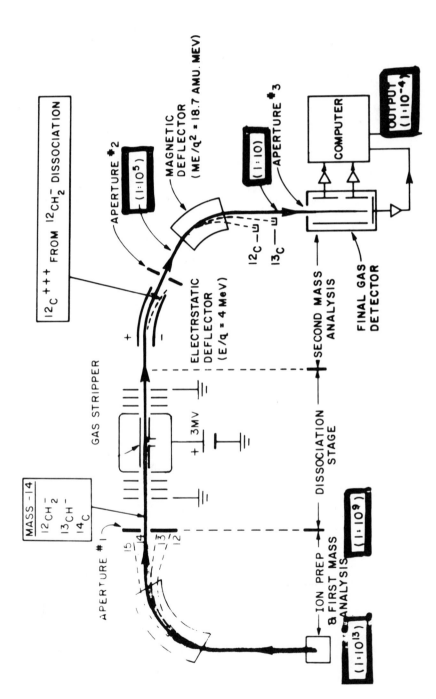

Figure 5. Schematic of tandem ultrasensitive mass spectrometer showing the reduction in background at various points through the system.

The sample is inserted as a solid at the anode of a cesium sputter ion source [24,25]. In such a source, a 30 keV focussed cesium beam impinges on the surface to produce negative ions, both molecular and atomic. These ions are extracted and mass analyzed by the first ($M/\Delta M = 110$) mass analysis stage. After this stage, the ratio of ^{14}C to background particles has been increased by approximately four orders of magnitude. The selected mass-14 particles are then accelerated by a high current, highly stable 3 MV tandem, and stripped of outer electrons in a differentially pumped gas cell where, as a consequence, the molecules also dissociate. The multiply charged positive ions are then accelerated again back to ground potential, where ions having charge state 3+ and the correct energy are selected for final analysis. After this selection, the ratio ^{14}C/background has been improved to 1 in 10^5, with the background consisting mainly of ^{12}C and ^{13}C ions that have been formed from the dissociation of mass-14 molecules. The final element of the second mass spectrometer stage directs the unwanted ^{12}C and ^{13}C ions into separate cups and reduces the ^{14}C/background ratio to less than about 1 to 10. The final detector is a critical element in the chain, as it permits an increase in ^{14}C/background to better than 1 in 10^{-4} at the computer output.

The final nuclear detector makes possible a separation of isobars based upon the principle that the range and rate of energy loss for particles of a given energy is atomic number dependent. Ions such as ^{14}C and ^{14}N have ranges in solids or gases that differ by over 20 percent at energies of about 14 MeV. The basis for this separation is the Bethe-Bloch equation [26,27], which can be simplified to read:

$$\frac{-dE}{dx} = k \frac{Z^2}{V^2}$$

Because the velocity is implicitly defined [17] when electric and magnetic fields are used in the geometry shown in figure 5, the rate of energy loss is independent of the mass of the particle and is just proportional to the atomic number squared:

$$\frac{-dE}{dx} = k_2 Z^2$$

A schematic diagram of the detector and its electronics is shown in figure 6. The active length of the counter is 500 mm. The window size is limited to a diameter of 5 mm and is covered by a 1 μm Paralene-C foil.

The anode is subdivided into two parts to give independent dE/dx signals from each of two plates. The particles are stopped in a final section where the collecting electric field is rotated by 90° to have its direction along the axis of the counter. This geometry permits a determination of Z by measuring the range for each particle, independently of the two dE/dx measurements.

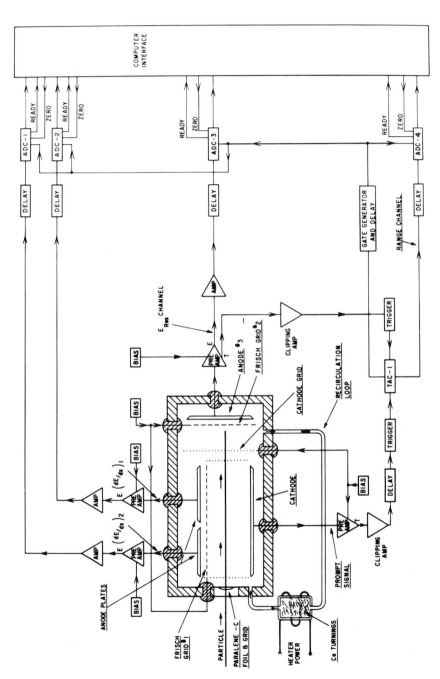

Figure 6.   The final (dE/dx, E) detector.

A spectrum that was derived during ^{14}C measurement from the Rochester multiparameter heavy ion detector is shown in figure 7. The ungated spectrum in the upper half of the figure demonstrates the total dE/dx spectrum of particles that enter the detector. The peak labeled ^7Li arises from molecular $^7Li_2^-$ which is dissociated at the terminal of the machine into two independent $^7Li^{2+}$ ions. Because of the dynamics of electric and magnetic deflection fields, these $^7Li^{2+}$ particles have half the mass of ^{14}C, half the charge and half the energy at all points throughout the system, and have identical trajectories to the ^{14}C ions [17]. It is interesting to notice that the peak labeled $2(^7Li)$ arises from two ions which travel independently but as a pair from the dissociation process at the terminal, and enter the detector simultaneously. Clearly, the difference in dE/dx for these particles is substantial, and the detection apparatus can easily separate out these unwanted groups. The ^{14}C peak is completely clean and shows virtually zero background for particles that do not come from the ion source.

The Need for Fast Cycling

A critical problem for precision measurements is ensuring the constancy of ion source output and the constancy of the transmission coefficient of the instrument itself. Because for age determinations the ratios $^{14}C/^{12}C$ and $^{13}C/^{12}C$ are the only quantities needed, the effects from changes in source efficiencies and system transmission efficiencies can be eliminated by measuring these ratios frequently by accelerating ^{12}C and ^{13}C. For the systems being installed at Arizona, Toronto and Nagoya, it will be possible to switch rapidly between beams of ^{12}C, ^{13}C and ^{14}C through the complete system. With the post-acceleration arrangement chosen, high-speed switching is only needed at the ion source because the DC post-acceleration optics deflects the different beams to either precision faraday cups or the gas counter [28].

Data

The results of some 1977 measurements on ^{14}C made by the RIT group are shown in figure 8. The geological samples had been dated previously and the graph shows that at that stage of development there was good (5-10%) agreement between the two methods.

Because of the ease of making such measurements, many descriptions of the technique have tended to emphasize its impact as a new ^{14}C dating method. We believe that this emphasis on radiocarbon dating has been somewhat premature, as there is an enormous practical difference between the detection of ^{14}C atoms in samples at natural concentrations using existing equipment and the measurement of $^{12}C/^{13}C/^{14}C$ isotopic ratios with the necessary precision for dating to better than one percent accuracy. In this

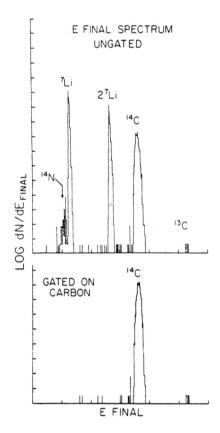

Figure 7. Spectrum of pulses from final detector.

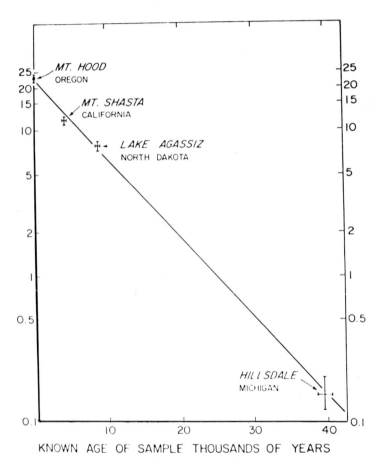

Figure 8.   *Data from preliminary dating measurements.*

regard, the assay of other radioactive isotopes such as [10]Be is much easier because it is not necessary to compete with an accurate, well-developed existing method.

The RIT group has tried to address this problem and understand some of the differential fractionation effects peculiar to accelerator dating. Table 1 shows some recent measurements which are beginning to show excellent agreement with known ages. One interesting example is that of the bull mummy shroud. Initially the isotopic concentration of ^{14}C appeared to be in error by about 2.5 percent from the predictions based upon the known age and from comparisons with known tree ring ^{14}C concentrations. Much of this discrepancy was eliminated when it was recognized that the isotopic fractionation between ^{14}C and ^{12}C is different between the formation of wood standards and the linen flax used by the weavers of the bull mummy shroud. The accuracy to which the Mt. Shasta samples can now be measured reproducibly approaches one percent. This is a substantial improvement over the five percent values previously reported and shown in figure 8.

In June 1977 in intact, completely frozen, baby wooly mammoth (named Dima) was uncovered by a bulldozer operation in the Magadan region of northeastern Siberia. A sample of 1.3 grams of muscle was made available to M. Goodman of the Anatomy Department of Wayne State University School of Medicine. At his request, a sample of about .9 milligrams, obtained from 4 milligrams of muscle, was dated to have an age of 27,000 years.

Finally, in Table 1 are shown the Rochester results for an Australian sucrose sample which is being considered as a new contemporary standard and which is being measured by all the major radiocarbon dating laboratories worldwide. The ^{14}C/^{12}C ratio compared to that for the ratio in A.D. 1950 is shown along with an indication of the range of values obtained at other values. The sample sizes employed range from about 1 to 10 milligrams.

Sample Preparation

The ion source presently used operates with a reflected cesium beam geometry and accommodates very small samples. Samples as small as 200 micrograms have been dated successfully. An ideal sample preparation method has not yet been devised. In principle, the ideal separation preparation should optimize the conversion efficiency of the source material to graphite, and the conversion should be free from laboratory contamination. Up to the present, the method which has best met the above requirements is the dissociation of acetylene (C_2H_2) [39]. Acetylene can be made by converting carbonaceous material into lithium carbide (Li_2C_2) which is subsequently reacted with water to produce C_2H_2 and lithium oxide. This part of the synthesis process has a conversion efficiency of better than 95 percent and has been used extensively in the synthesis of benzene for radiocarbon dating.

Table 1. Recent Tests Made by RIT Group.

Run Date	Sample[a]	Weight	Run Time[b]	$^{14}C/^{12}C$ Ratio Measured	$^{14}C/^{12}C$ Ratio Expected	Age BP[c] Measured	Age BP[c] Expected
June 1978	Mt. Shasta	9 mg	249.4	6.74×10^{-13} ($\pm4\%$)	6.78×10^{-13} ($\pm3\%$)	4640 ± 400	4590 ± 250 (USGS)
June 1978	Mt. Shasta	9 mg	102.5	6.79×10^{-13} ($\pm1.1\%$)	6.78×10^{-13} ($\pm3\%$)	4580 ± 90	4590 ± 250 (USGS)
Aug. 1978	Mt. Shasta	9 mg	155.3	6.41×10^{-13} ($\pm1.1\%$)	6.78×10^{-13} ($\pm3\%$)	5000 ± 140	4590 ± 250 (USGS)
Sept. 1978	Egyptian Bull[e] Mummy Wrap	0.9 mg	117.4	9.11×10^{-13} ($\pm1.9\%$)		2200 ± 150	2050 ± 200 (USGS)
March 1979	ANU Sucrose		47	1.77×10^{-12} ($\pm4.5\%$)			
March 1979	Russian Baby Wooly Mammoth	0.9 mg	189	6.81×10^{-14} ($\pm6\%$)		>23000	
Sept. 1979	Russian Baby Wooly Mammoth	0.9 mg		3.7×10^{-14} ($\pm12\%$)		27000 ± 1000	

The most satisfactory mechanism used to date to convert the acetylene gas into graphite-like carbon has been electrical dissociation. The apparatus is shown in figure 9. The efficiency of the mechanism of dissociation is strongly dependent upon gas pressure, and after experimentation a pressure of about 10 torr has been selected as optimum. At this pressure, a hard black carbon deposit is observed on the substrate with a thickness of more than 0.1 mm. In the current source, samples produce several microamperes after about 1 hour of running, and this current remains very steady for several hours of operation.

Recently a proposal for preparing carbon for radiocarbon dating was made by Ringwood [29], who has conducted experiments which demonstrate a technique by which graphite, suitable for ion source use, can be readily produced. In this technique the cleaned material is loaded into a small platinum cylinder which is sealed and subjected to pressures of several thousand bars at a temperature of 1500 °C. At these temperatures, platinum becomes largely transparent to oxygen, hydrogen, and nitrogen, and the carbon present condenses into a graphite-like material. This conversion takes a time of about 15-20 minutes. The technique has several advantages, one that the prepared carbon may be kept sealed until the actual $^{14}C/^{12}C$ measurements are to be made; secondly, the conversion cycle is rapid, thus breaking the carbon preparation bottleneck which appeared to be the limiting factor in high throughput for accelerator-based carbon dating mechanisms; thirdly, fractionation should be minimal.

Fractionation Effects

Measurements of the $^{12}C/^{13}C/^{14}C$ ratios are necessary to estimate a date. The $^{12}C/^{13}C$ ratio is needed to correct for isotopic fractionation of the sample. Machine fractionation due to the ion source and other parts of the accelerator is as much a problem in ultra-sensitive mass spectrometry as it is in conventional mass spectrometry. In addition, there are other machine fractionation contributions in this new technology which are not present in conventional mass spectrometry. For example, the electron stripping cross sections at the terminal of the accelerator are velocity dependent. This, because the ions are stripped of electrons at constant energy rather than at constant velocity, the probability of generating a given charge state is not the same for all isotopes. This effect can be seen in figure 10. It can be seen that at a negative ion energy of about 2.7 MeV the correction will be zero for $^{14}C/^{12}C$, with a small but finite correction for the ratio $^{13}C/^{12}C$. To correct for this machine fractionation every sample is measured with respect to an appropriate standard.

A potentially more serious problem may be encountered due to isotopic fractionation at the sputter ion source itself. Some

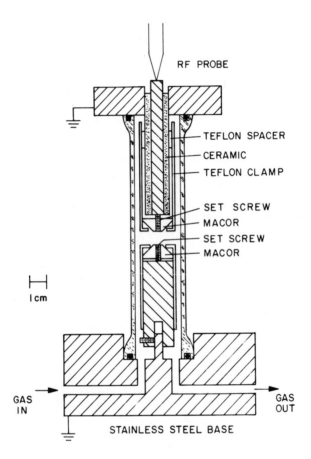

Figure 9. Sample preparation apparatus.

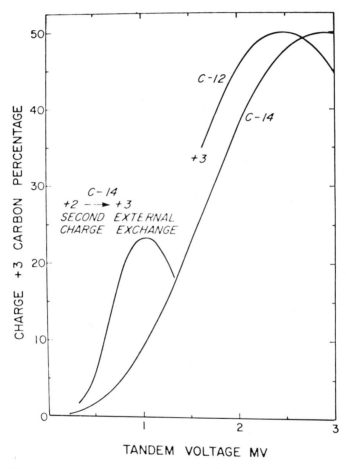

Figure 10. Percentage of 3+ carbon ions vs. terminal voltage. The curve on the left shows the efficiency of forming 3+ ions in a two-stage process where one charge exchange is in the terminal and the second is at ground.

measurements made by Tombrello [30,31] and his colleagues have
shown that the isotopic ratios vary with time as well as the angle
of emission from a surface being sputtered.

Other Applications of the New Technique

The ultra-sensitive mass spectrometry technique discussed
above can be employed to detect many other nuclear isotopes, both
stable and unstable.

Tritium Detection

Tritium is presently being measured in many laboratories
using counter techniques. It is nevertheless a good candidate
for direct detection and was the first natural radionuclide
detected by Muller [3]. Negative hydrogen sources capable of
producing 40-60 microamperes of H^- with short memory effects are
available. At this hydrogen current, 1.5 negative tritons are
produced per hour from hydrogen having a concentration of 1 tritium unit.[2] In such measurements the only backgrounds come from
tritium contamination or from natural helium-3. This helium
background is completely separable by solid state counters at
quite low energies (2-3 MeV). It is expected that there will be
virtually zero backgrounds in such measurements and that counting
rates of events per hour will represent a useful measurement.

^{10}Be Measurements

Raisbeck et al. have reported on the application of the
Grenoble cyclotron for the measurement of ^{10}Be in artifically
enriched samples [9]. Later experiments have measured ^{10}Be in
melted arctic glacier ice cores [10], marine sediments [32] and
ocean surface layers [33]. The Yale group, Turekian et al., [11]
have measured the ^{10}Be content in magnesium nodules and demonstrated that these nodules accrete at the rate of approximately
4.5 mm/10^6 years.
    A $^{10}Be/^9Be$ ratio as low as 2×10^{-17} can be expected at
equilibrium due to neutrons from uranium and thorium within
igneous rocks. Therefore, whenever a chronology based on ^{10}Be is
to be extended to an age of 20 million years or whenever the ^9Be
abundance is high, detection efficiencies capable of handling
trace quantities of Be below 10^{-14} are needed. Successful demon-
strations have been made by the RIT group of the tandem detection of Be at these very low levels [12]. In these measurements,
four spectroscopically pure ^9Be samples were irradiated in the
thermal neutron flux from the University of Toronto Slowpoke
research reactor. A comparison between the ^{10}Be concentration
compared to the neutron flux to which the sample was exposed is

---

[2]One tritium unit is a triton concentration in hydrogen of 10^{-18}.

shown in figure 11. It can be seen that the response ranges over at least three orders of magnitude, with a linearity of better than one percent. It is only fair to comment that these results were taken during a period of good apparatus stability; however, they do demonstrate what can be achieved with an existing tandem accelerator and peripheral equipment, even when this equipment is not precisely tailored to the task.

^{26}Al

^{26}Al has been measured by the RIT group in an Al/Mg alloy which had been exposed some years ago to high energy electrons. The ^{26}Al/Al ratio was determined to be less than 2×10^{11} by gamma counting. Figure 12 shows the two-dimensional spectrum obtained during these measurements. No certain evidence was obtained during the experiment for the formation of Mg ions of sufficient stability to pass through the acceleration system to the final detector, and the present belief is that Mg is not stable. Some ^{26}Mg events can be seen in figure 12, but it is believed that these particles probably originate from within the accelerator due to the injection of molecules such as ^{26}MgH$^-$. The injection of such molecules followed by fractionation in the terminal and charge exchange from the residual gas in the high energy acceleration tube can produce ^{26}Mg having the correct M/q and E/q.

It has been suggested by Lal [34] that ^{10}Be and ^{26}Al taken together form an ideal pair for absolute dating. Their residence time and geochemical behavior in the earth's biosphere are sufficiently similar that one may expect a constant ratio of these radioisotopes in all sediments at the time of deposition. Each of the radioisotopes decays exponentially, so that the age of the sediments can be calculated from:

$$^{26}\text{Al}(t) = N_{Al} \ e^{-t/T_{Al}}$$

$$^{10}\text{Be}(t) = N_{Be} \ e^{-t/T_{Be}}$$

$$^{26}\text{Al}/^{10}\text{Be} = (N_{Al}/N_{Be}) \ e^{-(1/T_{Al} - 1/T_{Be})t}$$

Because the half-lives of these two isotopes vary by more than a factor of two, the method is sensitive. Since the ratio ^{26}Al/^{10}Be decreases by a factor of two every 1.3 million years, it is thought that the method should provide absolute dates up to several million years.

An estimate of the total efficiency of the Rochester system for the detection of Al is close to 10^{-4}. We believe that it should be possible to improve this efficiency substantially by creating the ^{26}Al$^-$ ions by sputtering the Al as a positive ions, followed by acceleration to 20 keV and charge exchange in a

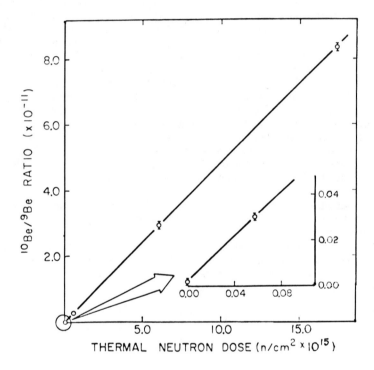

Figure 11.   Comparison between the measured concentration of ^{10}Be and the thermal neutron flux used to produce the samples.

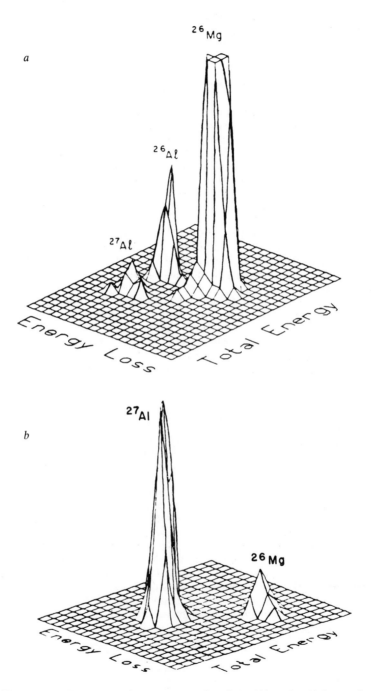

Figure 12. Comparison between (a) irradiated Al (1% Mg) (which contains ^{26}Al*) and (b) pure Al (99.999%).*

potassium or sodium vapor charge exchange canal. Figure 13 shows
the efficiency of production of Al^- in such a cell. One percent
efficiency appears possible and makes feasible the $^{10}Be/^{26}Al$ pair
method.

The Detection of ^{36}Cl from Natural Water

^{36}Cl is a cosmic ray produced radioisotope having a half-life
of 308,000 years. In direct detection methods, ^{36}Cl must be
distinguished from the isobars ^{36}Ar and ^{36}S, as well as from any
other molecules such as $^{12}C_3$; $^{18}O_2$, etc. Also, because the mass
number of ^{36}Cl is divisible by a large number of integers (2, 3,
4, 6, 12, and 18) it is necessary to analyze using positive ion
charge states 5+ or 7+. Otherwise many fragments having identical E/q, M/q can be transmitted through the complete system and
overload the detector.

During the search for ^{36}Cl at Rochester, 5-13 µA of Cl^- ions
were produced from silver chloride in the cesium sputter source.
The detection techniques were basically similar to those discussed for ^{14}C. The samples and the results obtained are given in
Table 2.

Figure 14 shows the Cl data [13] where the events are plotted
on a two-parameter display of range vs. energy. When improved
chemical separation preparation techniques were used in later
experiments on antarctic meteorites [14], the lower limit of Cl
detection that was reached was 2×10^{-16} for $^{36}Cl/Cl$, with a
quoted accuracy between ± 5 percent and ± 10 percent.

^{129}I

^{129}I, with a half-life of 1.6×10^7 years, is one of the
longest lived of the cosmogenic radionuclides. For the detection
of ^{129}I at Rochester, the apparatus used was the same as that in
the previous measurements, except that the final ionization
detector was replaced by a 2.2 meter time-of-flight detector.
^{129}I can be identified on the basis of mass alone, since the only
stable isobar of ^{129}I is ^{129}Xe, a noble gas that is not expected
to form negative ions. Figure 15 shows the time-of-flight spectrum for ^{129}I ions compared to the spectrum from a blank. The
sensitivity achieved was about two orders of magnitude greater
than the best previous measurement [35]. With adequate resolution
of the first mass identifier stage, it is expected that the direct
counting of ^{129}I should have a sensitivity comparable to that of
$^{36}Cl/Cl$. In this regard it has been suggested by Middleton [37]
that LaB_6 surface ionization sources may be very useful for high
sensitivity measurements of the strongly electronegative halogens,
such as ^{36}Cl and ^{129}I.

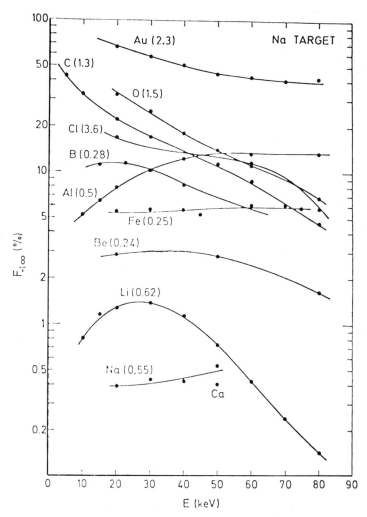

*Figure 13.   The production of negative ions by charge exchange in sodium vapor
for a variety of ion species and exchange energies.*

Table 2. Results for ^{36}Cl Measurements.

Sample	Total Run Time (min.)	Total ^{36}Cl Counts	No. of Runs	Concentration[b] $^{36}Cl/Cl$ $(\times 10^{15})$	Activity[c] (d/m-g Cl)	Cl/H_2O[d] (mg/L)
Groundwater, Rillito River (ephemeral), AZ	90	7681	7	1950 ± 150	0.145	48
Surface water, San Pedro River Tombstone, AZ	204	8916	13	650 ± 110	0.048	4-12
Groundwater, flowing well in St. David, AZ	68	2403	7	400 ± 40	0.030	8
Groundwater, deep well near Tucson, AZ	101	990	4	320 ± 50	0.024	10
Subsurface water, Lake Ontario, NY (5/19/78)	41	4555	4	210 ± 20	0.15	27
Subsurface water, Lake Ontario, NY (6/30/78)	172	2062	9	120 ± 20	0.009	27
AgCl reagent, zone refined	57	32	3	3 ± 1	0.0002	–
NaCl enriched to $^{36}Cl/Cl$ $\approx$ 33000x10^{-15}	13	21799	2	18500 ± 900	1.38	–
NaCl enriched diluted to $^{36}Cl/Cl$ $\approx$ 330x10^{-15}	58	975	4	250 ± 30	0.018	–

[a]Although a few hundred mg of AgCl was obtained from the natural water samples, only 16-66 mg was used in the ion source, and most of this remained after the experiment.

[b]Errors quoted are one standard deviation for several measurements.

[c]Specific activity is that expected based on the ^{36}Cl concentration measured.

[d]The approximate chlorine concentration in the original water sample.

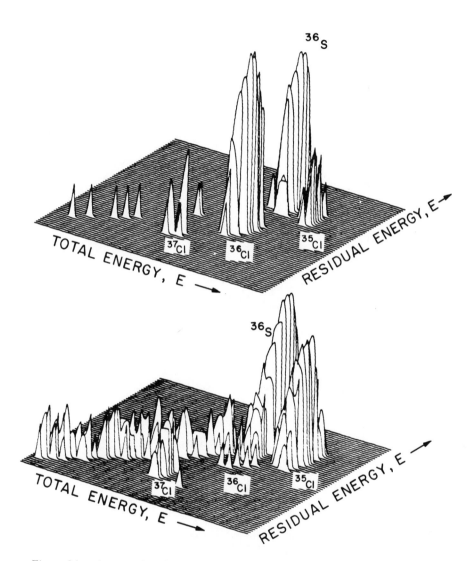

Figure 14. A comparison between a sample known to contain ^{36}Cl at concentrations of $1:10^{13}$ and a sample of salt known to be essentially free of ^{36}Cl.

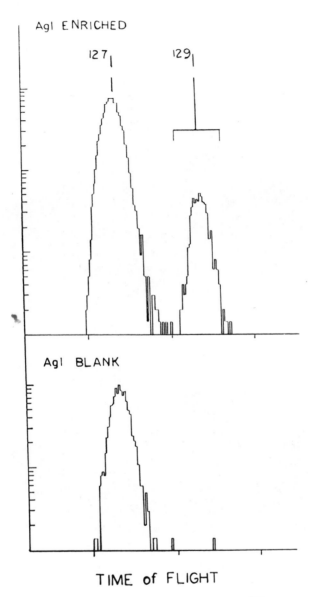

Figure 15.   Time of flight spectrum for ^{129}I.

Applications of the Technique to Age Determinations by Mass Spectrometry

The measurements that have been made at Rochester and the experience that has been gathered over the years on the operation of sputter ion sources [38] indicate that an analytical tool of unprecedented sensitivity and accuracy for isotopic ratio determinations can be constructed by coupling SIMS technology with the new accelerator technique. The only difference in principle between the experiments that have been conducted to date and the technique as it would be applied in secondary ion mass spectrometry is that the primary beam of cesium would be focussed to a fine probe of μm dimensions rather than the spot diameters of approximately 1 mm that have been used to date.

Attempts to use radiogenic isotopes to measure the ages of individual mineral grains by ion microprobe have as yet been unsatisfactory. The main reason for this is the ambiguity inherent in the measurement of secondary ion beams where the possibility of molecular ions prevents confident identification of the desired isotope species. With the molecular disintegrator described in this paper it should be possible to obtain reliable isotope ratios for Pb. The use of the $^{40}K/^{40}Ar$ and $^{87}Rb/^{87}Sr$ geochronometers presents formidable problems even with the accelerator technique. The need to measure the relative intensities of isobars which cannot readily be separated, coupled with the need to measure them simultaneously to achieve the necessary precision, makes the use of these geochronometers very difficult at present.

The $^{207}Pb/^{206}Pb$ clock, which has not been reliably applied with existing ion microprobes because of the existence of ^{206}PbH, can now be used. $^{207}Pb/^{206}Pb$ isotope ratios have the potential to yield accurate dates, and these are best measured in the center of zircons where the Pb is effectively immobile. Diffusion at the edges of the zircons means that whole crystal data may yield erroneous results. The ability to select Pb isotope data from the undiffused plateau region of zircons will be a great boon to those involved in such work.

An example of an exciting application of the new technique in the field of geochronology is in the measurement of the isotopes of Sm and Nd in geological materials. The rare earth elements (REE) have been recognized as being most valuable for tracing crustal evolution and the history of a wide variety of rock types. REE are quite ubiquitous at low (1 ppm) levels in many common materials, and show remarkable stability through geological upheavals. This is in contrast to other marker elements, such as Sr and Rb, which may be disturbed during metamorphism, etc. Measurements of the isotopic ratios of $^{147}Sm/^{144}Nd$ and $^{143}Nd/^{144}Nd$ have been used to date Pre-Cambrian rocks. This method requires a precision in the ratio determinations of better than 1 part in 10^4. The same information should be derivable from precise measurements of ^{147}Sm, ^{146}Nd and ^{143}Nd. As these isotopes

are all unique in their atomic weight, their measurement in an
accelerator mass spectrometer would be totally unambiguous. By
simultaneous measurement of these three isotopes sputtered from
a solid sample, precise dating in the range 1-5 10^9 years should
be feasible. Calculations suggest that for material such as
felspar, between 5 and 250 g (i.e., material with a volume of a
cube with an edge between 0.1 and 0.4 mm) should be sufficient
material for this measurement. Minerals such as felspar typically
carry at least 1 ppm of each REE and the time for analysis might
be expected to be a few hours. The successful accomplishment of
this experiment probably represents the ultimate challenge in the
geological applications of the ion microprobe analyzer.

References

[1] Purser, K. H., U.S. Patent 4,037,100, Filed March 1, 1976,
    Issued July 19, 1977.
[2] Purser, K. H., Liebert, R. B., Litherland, A. E., Beukens,
    R. P., Gove, H. E., Bennett, C. L., Clover, M. R., and
    Sondheim, W. E., Rev. Phys. Appl. 12, 1487 (1977).
[3] Muller, R. A., Science 196, 489 (1977).
[4] Bennett, C. L., Beukens, R. P., Clover, M. R., Gove, H. E.,
    Liebert, R. B., Litherland, A. E., Purser, K. H., and
    Sondheim, W. E., Science 198, 508 (1977).
[5] Bennett, C. L., Beukens, R. P., Clover, M. R., Elmore, D.,
    Gove, H. E., Kilius, L. R., Litherland, A. E., and Purser,
    K. H., Science 201, 345 (1978).
[6] Gove, H. E., Fulton, B. R., Elmore, D., Litherland, A. E.,
    Beukens, R. P., Purser, K. H., and Naylor, H., IEEE Trans-
    actions on Nuclear Science NS-26, 1414 (1979).
[7] Proc. First Conference on Radiocarbon Dating with Accelera-
    tors, ed. H. E. Gove, Rochester, 1978.
[8] Nelson, D. E., Korteling, R. G., and Stott, W. R., Science
    198, 508 (1977).
[9] Raisbeck, G. M., Yioux, F., Furneau, M., and Loiseaux, J. M.,
    Science 202, 215 (1978); see also, Proc. First Conference on
    Radiocarbon Dating with Accelerators, ed. H. E. Gove,
    Rochester, 1978, p. 38.
[10] Raisbeck, G. M., Yioux, F., Furneau, M., Lieuvin, M., and
    Loiseaux, J. M., Nature 275, 731 (1978).
[11] Turekian, K. K., Cochran, J. K., Krishnaswami, S., Lanford,
    W. A., Parker, P. D., and Bauer, K. A., Geophysical Res.
    Lett. 6, 417 (1979).
[12] Kilius, L. R., Beukens, R. P., Chang, K. H., Lee, H. W.,
    Litherland, A. E., Elmore, D., Ferraro, R., Gove, H. E., and
    Purser, K. H., Nucl. Inst. and Meth. 171, 355 (1980).
[13] Elmore, D., Fulton, B. R., Clover, M. R., Marsden, J. R.,
    Gove, H. E., Naylor, H., Purser, K. H., Kilius, L. R.,
    Beukens, R. P., and Litherland, A. E., Nature 277, 22 (1979).

[14] Nishiizumi, K., Arnold, J. R., Elmore, D., Ferraro, R. D., Gove, H. E., Beukens, R. P., Chang, K. H., and Kilius, L. R., Earth and Planet. Sci. Lett. 45, 285 (1979).

[15] Raisbeck, G. M., Yioux, F., and Stephan, C., Journal de Phys. Lett. 40, L241 (1979).

[16] Kilius, L. R., et al., Nature 282, 488 (1979).

[17] Purser, K. H., Litherland, A. E., and Gove, H. E., Nucl. Inst. and Meth. 162, 637 (1979).

[18] Muller, R. A., Stephenson, E. J., and Mast, T. S., Science 201, 347 (1978).

[19] Stephenson, E. J., Mast, T. S., and Muller, R. A., Nucl. Inst. and Meth. 158, 571 (1979).

[20] Oeschger, H., Houtermans, J., Loosli, H., and Wahlen, M., Proc. of the 12th Nobel Symposium on Radiocarbon Variations and Absolute Chronology, ed. I. U. Olsen, Uppsala, 1970, p. 487.

[21] Schnitzer, R., Aberth, W. A., Brown, H. L., and Anbar, M., Proc. of the 22nd Annual Conference on Mass Spectrometry (ASMA), Philadelphia, 1974, p. 64.

[22] Meyer, C., Jr., and Anderson, D. H., Proc. of Fifth Lunar Conference, Geochim. Cosmochim. Acta 1, Suppl. 5, 685 (1974).

[23] Betz, H. D., Rev. Mod. Phys. 44, 465 (1972).

[24] Model 834 HICONEX, manufactured by General Ionex Corporation, Newburyport, Massachusetts 01950.

[25] Middleton, R., Nucl. Inst. and Meth. 144, 373 (1977).

[26] Livingston, M. S., and Bethe, H. A., Rev. Mod. Phys. 9, 261 (1937).

[27] Northcliffe, L. C., and Schilling, R. G., Nucl. Data Tables A7, 233 (1970).

[28] Purser, K. H., and Hanley, P. R., Proc. First Conference on Radiocarbon Dating with Accelerators, ed. H. E. Gove, Rochester, 1978, p. 165.

[29] Ringwood, A., Australian National University (personal communication).

[30] Russell, W. A., Papanastassion, D. A., and Tombrello, T. A., Geochem. Cosmochem. Acta 42, 1075 (1978).

[31] Tombrello, T. A., Proc. Lunar and Planetary Sciences Conference X, Part 3, 1233 (1979).

[32] Raisbeck, G. M., Yioux, F., Fruneau, M., Loiseaux, J. M., Lieuvin, M., Ravel, J. C., and Hays, J. D., Geophys. Res. Lett. 6, 717 (1979).

[33] Raisbeck, G. M., Yioux, F., Fruneau, M., Loiseaux, J. M., and Lieuvin, M., Earth Planet. Sci. Lett. 43, 237 (1979).

[34] Lal, D., Journal Oceanograph. Soc. of Japan, 20th anniversary issue, 600 (1962).

[35] Rook, A. L., Suddueth, J. R., and Becker, D. A., Anal. Chem. 47, 1557 (1975).

[36] Purser, K. H., Litherland, A. E., and Rucklidge, J. C., Surface and Interface Analysis 1, 12 (1979).

[37] Middleton, R., University of Pennsylvania, personal communi-
     cation, 1980.
[38] Middleton, R., Rev. Phys. Appl. 12, 1437 (1977).
[39] Beukens, R. P., Lee, H. W., and Litherland, A. E., Atomic
     Energy of Canada Report TR-39 (1980).

RECEIVED May 23, 1981.

4

Techniques for the Direct Measurement of Natural Beryllium-10 and Carbon-14 with a Tandem Accelerator

J. R. SOUTHON[a], D. E. NELSON, and R. KORTELING
Simon Fraser University, Burnaby, B.C., Canada V5A 1S6

I. NOWIKOW, E. HAMMAREN, J. McKAY, and D. BURKE
McMaster University, Hamilton, Ontario, Canada L8S 4K1

This paper is a review and progress report on our project to adapt a Tandem Van de Graaff accelerator for radiocarbon and radioberyllium dating. This project began in 1977 immediately following Muller's proposal [1][1] to use a cyclotron as a mass-spectrometer of such high sensitivity that many naturally occurring radioisotopes could be detected directly. Measurement by direct detection, if practical, would have many advantages over the conventional specific activity measurement techniques.

Since the type of measurements required for radioisotope dating are identical to those for stable isotope studies, with the additional requirements of much higher sensitivity, we realized that a Tandem Van de Graaff accelerator offered many practical advantages over the cyclotron for such purposes. In principle, it could be operated as a direct analogue of conventional isotope-ratio mass spectrometers. A preliminary experiment [2] showed that a Tandem could provide the necessary sensitivity, and we then decided to modify this accelerator for radiocarbon and radioberyllium dating. This modification was to proceed in stages because of funding and manpower restraints, and because we would then have an opportunity to test each idea in turn and not become locked into an unworkable scheme.

At the same time, another group of researchers [3] (from the Universities of Rochester and Toronto, and the General Ionex Corporation) working quite independently and unknown to us, also showed that a Tandem accelerator offered many advantages for direct detection radioisotope dating. Since those initial experiments, many groups have undertaken similar work, and the first commercial systems should appear within the next year.

The measurement approach we have proposed [2,4] differs in detail from that proposed by the Rockester-Toronto-General Ionex group. Subsequently, most other researchers have adopted this latter technique. In the following sections, we discuss the

---

[1]Figures in brackets indicate the literature references at the end of this paper.

[a] Current address: McMaster University, Hamilton, Ontario, Canada L8I 4K1.

advantages and the disadvantages of each of these techniques, the results we have obtained on our own system, and our plans for further development.

The Measurement Technique

In general, the Tandem systems attain high measurement sensitivity by permitting the identification and counting of rare isotopes in a sea of other particles. This is done by ionizing a sample, accelerating a beam of the rare isotope of interest to an energy in the region 10-40 MeV, sorting the ions with an analyzing magnet and then using specialized detectors to discriminate between any remaining contaminant ions and the ion of interest. The primary requirement is that the numbers of any contaminating ions do not exceed the count-rate capabilities of the detector. In a Tandem system, molecular ions are automatically dissociated in the accelerating process, and isobaric contaminants can be eliminated or reduced in intensity by their instability as negative ions (e.g., ^{14}N), by the use of absorbers in front of the detectors, or by other techniques. The rare isotopes can then be detected with very high sensitivity.

Our system differs from others in the method by which we propose to measure the relative concentrations of the isotopes. In our system, we ionize, accelerate and detect beams of the stable and the rare isotopes simultaneously. In other systems, the isotopes are sequentially accelerated and detected. The rationale underlying our choice is as follows:

In radiocarbon dating, the quantity to be measured is the ratio of the abundances of the rare isotope (^{14}C) to that of the stable isotopes (^{12}C, ^{13}C). These abundance ratios are not measured on an absolute basis, but are compared to that of an internationally-accepted standard. (It is likely that a similar standard will be adopted for ^{10}Be dating.) These measurement requirements have several consequences:

1) it is not necessary that the absolute efficiency of the measurement system be known, since all measurements of unknowns are referenced to a measurement for the standard taken under identical measurement conditions

2) the absolute efficiency need not remain stable in time

3) the efficiency for the rare isotope need not be identical to that for the stable isotope

4) the one requirement is that the measurement efficiency for the rare isotope relative to that for the stable isotope remain stable over times long compared to the time required for measurements of both sample and standard.

Systems with these capabilities will satisfy the requirements for radioisotope dating. If absolute isotopic abundances are required for other work it will be necessary to either use a standard of accurately known concentrations or to determine the efficiencies of the system.

The requirements outlined above are identical to those required for measurements of stable isotopes, such as $^{13}C/^{12}C$ or $^{18}O/^{16}O$. In the conventional mass spectrometers developed for these measurements, very high measurement accuracy ($<$ .1 ^{o}/oo) is achieved by ionizing, accelerating, analyzing, and detecting the two isotopes of interest simultaneously. These devices are designed such that any changes in efficiency (such as beam-intensity from the ion-source) will affect both isotopes equally and thus leave the isotope ratio unchanged. The sample and the standard are alternately measured at short intervals.

Our proposed Tandem system will also produce, accelerate, and analyze the beams of the different isotopes simultaneously. This has the advantage that the instabilities that are common to Tandem accelerators will affect the beams equally and leave the ratios constant. Any changes that do occur will be made readily apparent by monitoring the beam current of the stable isotope. (These changes would be more difficult to detect by monitoring the count-rate for the rare isotope, as they would be perceived as a small change in an already-small rate.) Since some changes in these accelerators take place in times of the orders of seconds or less, this will be an important advantage. Further, for radiocarbon dating, a measure of the constancy of the relative isotopic efficiencies for ^{14}C and ^{12}C can be made by continuously monitoring the $^{13}C/^{12}C$ ratio as the $^{14}C/^{12}C$ measurement progresses. If sufficient measurement accuracy can be obtained, this measurement will also allow the date to be corrected for isotopic fractionation processes. A technical advantage is that the continuous stable isotope beam can be used to control the existing accelerator voltage stabilization system. This operates by sensing the position of the beam on a pair of slits downstream of the analyzing magnet, and feeding back an error signal if the voltage changes and the position of this beam shifts.

The major difficulty to be expected for this mode of operation is that ions from the stable beams could, by a variety of circumstances, be scattered into the beam of the rare isotope. If these cannot be eliminated, the sensitivity of the technique may be lost.

The alternative measurement technique [5] is to select the ions of interest from the ion source and sequentially inject these into the accelerator. As only one beam is in the system at any given time, there is no possibility of inter-beam interferences. With such a system, it will be necessary to ensure that the cycle time between beams is short compared to the time-span over which any efficiency changes could occur. As well, the normal method for stabilizing the accelerator will not be available, and other methods must be developed.

Whatever system is used, it may prove valuable in some circumstances to be able to reduce the beam intensities of the most abundant isotope to reduce accelerator loading. This can in principle be easily accomplished with sequential injection by defining the length and repetition rate of the abundant isotope to give the average beam current desired. However, the effect of introducing short, sharp pulses of beam into the accelerator may in practice cause stabilization difficulty. For multi-beam detection, proposed methods for reducing the stable isotope beam intensities are discussed in a following section.

Our project to test a Tandem accelerator for multi-beam detection was then to modify the accelerator analyzing magnet such that the various beams could be dispersed and detected simultaneously, to design an injection system to appropriately process the different isotopic beams from the ion source for injection into the accelerator, and to construct an ion-source to allow us to routinely handle small samples. Our interests at present lie in radiocarbon and radioberyllium studies, and so we have designed this system to accommodate these isotopes.

Present System

An overall view of our present system is shown in figure 1. Negative ions are sputtered from a small sample in a Cs sputter ion source, extracted by a potential of 25-30 kV, injected into the accelerator via a 20 degree inflection magnet, and accelerated to the high voltage terminal. The ions are stripped of some of their electrons (and molecules are broken up) in a gas and/or foil stripper at the terminal; the resulting positive ions are accelerated further. A 90 degree analyzing magnet selects ions of the appropriate momentum-to-charge ratio and passes the stable isotopes to Faraday cups and the rare isotope to a set of electrostatic deflectors. These deflectors carry out further filtering of the 'rare' beam according to the energy-to-charge ratio of the ions. The ions passing this filter are counted and their masses and atomic numbers are identified by a ΔE-E semi-conductor detector telescope. These detectors can handle counting rates of $\sim$30,000 particles/second. The Faraday cups can be moved to intercept any beam of interest exiting the magnet.

Our present technique for injecting several isotopes simultaneously consists of simply passing them all through the 20 degree inflection magnet at the low-energy end of the accelerator. Although the different isotopes are bent differently, all the emerging beams of interest overlap the acceptance phase space of the accelerator to some extent. This method is far from ideal, as the beams are treated very differently. However, it was a simple and easy technique to use for initial tests. A better injection system, discussed below, is planned but not yet constructed.

Results to Date

Although we originally began our test studies on ^{14}C, our work is now primarily centered around radioberyllium measurements, for several reasons. First, the concentrations of ^{10}Be in natural

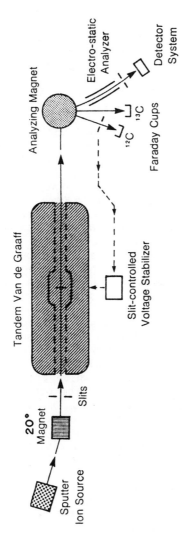

Figure 1. An overall schematic of our present Tandem system.

samples are several orders of magnitude higher than for ^{14}C, which makes the experiments easier to perform. Second, little is known about ^{10}Be in the environment, and even low-accuracy (<10 percent) measurements on natural samples will be of interest. Third, it is possible to easily tune and align the system for ^{10}Be/^{9}Be measurement by using a sample containing both ^{9}Be and ^{10}B. The ^{10}B beam is detected with a Faraday cup in place of the energy telescope, such that both the mass 9 and mass 10 channels can be simultaneously adjusted with easily-measured beams. Radioberyllium measurements thus provide an ideal problem on which to test our concepts.

For these tests, we have irradiated several samples of pure BeO powder to yield ^{10}Be/^{9}Be ratios in the range approximately $10^{-9} - 10^{-13}$. Approximately 30 mg of each sample was mixed with an equal weight of pure silver binder and pressed into a small ring-shaped sample. Beams of BeO ions of intensities of ∿.5 μA were obtained from these samples using the standard Cs sputter source. These ions were accelerated and the ^{10}Be^{4+} ions at 37.3 MeV were selected. The highest charge state was chosen to avoid any background from 'charge-exchange' ^{9}Be. This effect occurs when ions of the lower-mass stable isotope emerge from the accelerator stripper with a high charge state and charge-exchange down to the selected charge state in collisions with residual gas molecules in the high-energy accelerator tubes. Some of these arrive at the analyzing magnet with sufficient extra energy to give them the same momentum as the radioisotope ions, and a few of these pass through the subsequent filtering system and give a background at the detector.

The experimental procedure was to use a BeO-B sample to adjust the total system such that the ^{10}B^{4+} beam was maximized and the accelerator voltage was well-stabilized on the ^{9}Be^{4+} beam. The mass -10 Faraday cup was then replaced by an energy-telescope preceded by a 16 mg/cm^2 absorber, which prevented any ^{10}B and other, higher-mass adventitious ions from over-loading the detectors. (Since the mass-10 cup is rotated into position in front of the detectors, the switch can be done easily, and it is possible to easily check the system periodically.) The irradiated BeO sample with ^{10}Be/^{9}Be ratio of ∿10^{-9} was arbitrarily selected as a standard and measurements of all other samples were made with respect to this standard.

Figure 2 shows a spectrum obtained with the telescope for this standard sample. In this spectrum, the flat background is due to alpha particles created in nuclear reactions in the absorber. These can easily be removed electronically by setting appropriate gates. The ^{9}Be peak we believe arises from BeH$^-$ ions from the source which are scattered into the accelerator from residual gas in the inflection magnet. In figure 3 we give a similar spectrum for a sample with a ^{10}Be/^{9}Be ratio of 10^{-11}. A

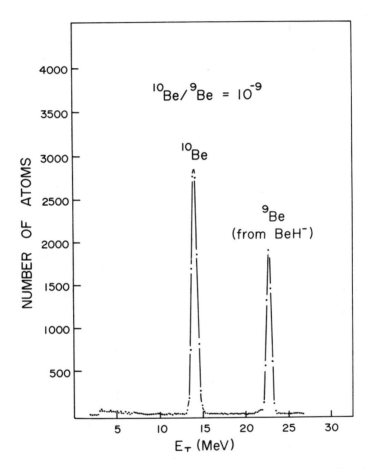

Figure 2. A total energy spectrum of ions from a BeO sample with $^{10}Be/^9Be \sim 10^{-9}$. The ^9Be peak is caused by the gas scattering of BeH$^-$ ions in the inflection magnet (see text).

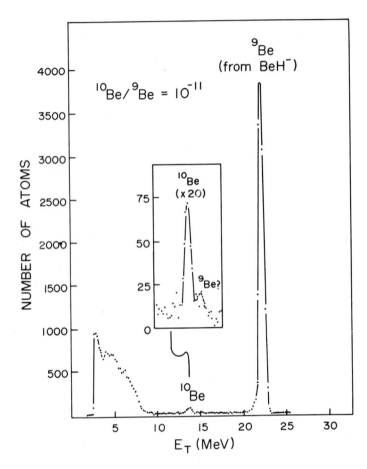

Figure 3. A spectrum from a sample with $^{10}Be/^{9}Be \sim 10^{-11}$. A small peak due to gas scattering of ^{9}Be in the analyzing magnet may be present to the right of the ^{10}Be peak.

very small peak due to ^9Be ions results from a few ions from the primary Be beam being scattered into the detectors from residual gas in the analyzing magnet.

Figure 4 gives a plot of measured vs."true" $^{10}Be/^9Be$ ratios for the irradiated samples. (There is an uncertainty of 10 percent in the absolute magnitudes of the "true" $^{10}Be/^9Be$ ratios of these samples. However, the relative ratios, which depend only on irradiation times, are known to $\sim$2 percent.) Data collection times ranged from a few minutes for the 10^{-9} sample to $\sim$2 hours for the 10^{-12} sample. Measurements of the standard sample were made throughout the experiment and were found to be reproducible to within counting statistics. The limit on sensitivity at $\sim$3 x 10^{-12} is likely due to cross-sample contamination in the ion source (a direct line-of-sight existed between different samples) and to contamination of parts of the source itself by material sputtered from high-concentration samples. The sensitivity was $<10^{-13}$ before this contamination built up, and we expect to equal or better this given better shielding between the samples. These results show that the multi-beam concept is indeed valid, and that even with this present system we can make ^{10}Be measurements in the $10^{-9} - 10^{-11}$ range.

Work on radiocarbon has been limited to a few experiments to test the multi-beam concept. These experiments are considerably more difficult, as the relative concentrations are much lower, our present injection system is very poor for simultaneous injection of ^{14}C, ^{13}C, and ^{12}C and there is no easily-available mass-14 substitute beam on which to tune up the system ($^{14}N^-$ is unstable).

The experiments show that, with our present analyzing system the problem of inter-beam interferences is too large to allow practical ^{14}C measurements. Specifically, when the system is set up to count, say, one $^{14}C^{4+}$ ion/second, there is a beam of $\sim10^{10}$ $^{13}C^{4+}$ ions/second a few cm away in the analyzing magnet vacuum box. A few of the $^{13}C^{4+}$ ions scatter off residual gas molecules and are deflected through small angles towards the deflector plates. These ions have the correct energy-to-charge ratio to reach the detectors and produce peaks in the spectra which are sufficiently large that the tail on the high energy side can obliterate the small ^{14}C peak from a low-concentration sample.

Our preliminary experiments indicate that these gas-scattered ^{13}C ions occur at about the same intensity as ^{14}C ions from modern carbon. Two solutions are possible. The first is to separate the ΔE and E detectors and measure the flight time of the ions. Preliminary tests have indicated that a flight path of 0.5 m will allow us to satisfactorily separate the two isotopes. The other solution is to introduce an additional filtering stage between the analyzing magnet and the detectors. This could be either an additional magnet before the deflector plates, or a magnet or velocity filter after the plates. Our limited resources have not yet allowed us to pursue these alternatives.

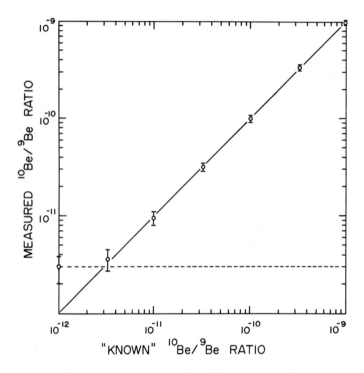

Figure 4.   Results from measurements of $^{10}Be/^9Be$ ratios in BeO samples irradiated for different periods.

The ratio data were normalized by assuming that the highest ratio measured was $^{10}Be/^9Be = 10^{-9}$. (In fact, the $^9Be(n,\gamma)^{10}Be$ cross section is only known to $\pm 10\%$.) The diagonal line represents the response of a perfectly linear system, and the dashed horizontal line gives the present limit of sensitivity.

Current Equipment Development

We have recently commissioned a new reflected beam Cs sputter source [6] to be used in these experiments. This source is better suited to this work than the old non-reflecting source, as the Cs beam is focussed to sputter ions off a very small pill or plug of material. This gives an ion beam with good optical properties (since the source spot size is small) and allows the use of very small samples. We are currently working with samples of 1-5 mg of BeO or graphite.

Typically, beams of BeO^- and C^- are obtained from these samples at intensities of 0.5 µA and 20 µA respectively. The improved shielding between the samples in this new source will hopefully reduce the inter-sample contamination.

Should we wish to improve the precision of our measurements significantly, it seems likely that we will have to find a more satisfactory method of injection. A new injection system which we have designed and are currently building is shown in figure 5. (Details of this system are to be published elsewhere.)

The heart of the system consists of two dipole magnets (B1 and B2 in figure 5), two electrostatic quadrupoles (Q1 and Q2), and an aperture plate (AA). The first magnet separates the ions into beams of different masses, and the following quadrupole lines up the central rays of all beams parallel with each other and perpendicular to the aperture plate. The plate then filters out unwanted beams (except for those few which fortuitously have the same mass as a beam of interest). The second quadrupole magnet-pair is just a mirror image of the first, so that all the beams are recombined and emerge on the axis of the low-energy beam line of the accelerator. Other lenses (not shown in figure 5) are used to provide proper control of the beams entering and leaving the system. Four sets of electrostatic plates (D1 – D4) can be used to displace the beam vertically in the region between the quadrupoles. Thus, by providing different sets of apertures in the "up", "down", or "zero" displaced positions, we can transmit different combinations of beams. By controlling the deflector plates appropriately, the system can be operated to provide full-time simultaneous injection of all isotopes of interest, full-time injection of some plus pulsed injection of others or complete sequential injection. By using restrictive apertures and/or pulsed injection the intensity of one of the beams can be reduced. We expect that this device will allow us to fully investigate these alternative measurement methods such that we can choose that method best suited to any given situation.

Future Work

The results of these studies have shown that this multi-beam concept does provide a practical measurement system. We intend to use our present system in the next few months to measure $^{10}Be/^9Be$

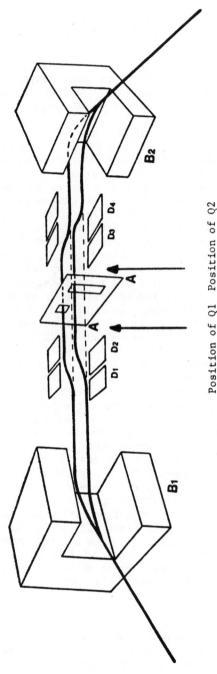

Position of Q1   Position of Q2

Figure 5. An isometric view of the new injection system, showing the dipole magnets B1 and B2, the electrostatic deflectors D1–D4, and the aperture plate AA. The positions of the electrostatic quadrupoles Q1 and Q2 are indicated also. The principal rays for beams of two different isotopes in the "deflected up" position are indicated by the solid lines, with the undeflected rays shown as dashed lines.

ratios in samples derived from deep-sea sediments. Our tests indicate that we should be able to measure these ratios (in the range $10^{-9} - 10^{-11}$ after the addition of ^{9}Be carrier during sample preparation) at a precision of 5 − 10 percent. Higher accuracy measurements will be obtained following the construction, testing, and installation of the new injection system. Further work to choose the best solution to the gas-scattering problem is required such that the system will be useful for radiocarbon measurements as well. This work will likely also result in a considerable increase in the sensitivity for ^{10}Be detection.

We hope that, given this work, we will have constructed a relatively inexpensive radiocarbon and radioberyllium measurement device that is capable of making measurements at high accuracy on very small samples.

We gratefully acknowledge the invaluable assistance of Dr. D. E. Lobb, (University of Victoria) who carried out the ion optics calculations for the new injection system.

We also thank Dr. J. A. Kuehner and his operations staff at the McMaster Tandem for their support and assistance for this project. We are grateful to the staff at the McMaster Reactor and the Toronto Slowpoke reactor for irradiating our Be test samples.

Funds were obtained from grants from the NSERC (Canada), from Simon Fraser University, and in collaboration with T. L. Ku (University of Southern California), from the NSF (U.S.A.) through the MANOP program (NSF grant CE-7901092).

References

[1] R. A. Muller, Science, 196 (1977) 489.
[2] D. E. Nelson, R. G. Korteling, and W. R. Stott, Science, 198 (1977) 507.
[3] C. L. Bennett et al., Science, 198 (1977) 508.
[4] D. E. Nelson et al., Proceedings of the First Conference on Radiocarbon Dating with Accelerators, (H. E. Gove, ed.), University of Rochester (1978) 47.
[5] K. H. Purser and P. R. Hanley, ibid., p. 165.
[6] K. Brand, Nuclear Instruments and Methods, 141 (1977) 519.

RECEIVED August 4, 1981.

Sample Preparation for Electrostatic Accelerator Dating of Radiocarbon

MEYER RUBIN

U.S. Geological Survey, Reston, VA 22092

Graphite produces the large beam currents desirable in electrostatic accelerator dating of radiocarbon. However, samples to be dated can be converted to other forms of carbon with less effort, and still provide satisfactory results.

The extremely rapid advance of techniques in radiocarbon dating with accelerators has been paralleled by the different approaches in sample preparation to achieve a high ^{12}C beam current, essential to measuring samples of low ^{14}C content. The many laboratories involved in accelerator dating have experimented with various forms of sample preparation and at present have not devised one best method. Possibly, different methods will continue to be used in each laboratory and even different methods within one laboratory depending on the initial form of the carbon. This paper discusses some of the attempts at ^{14}C dating by laboratories using electrostatic accelerators, but makes no pretence of being inclusive or final, as the technique is changing weekly.

Requirements of a Good Sample Preparation Method

A good sample preparation would be one that produces from the original sample a form of carbon that 1) yields a large beam current strength 2) with a minimum of handling and processing 3) with small or no isotopic fractionation 4) from a very small (mg) size sample 5) in minimum time.

The method should be able to handle the various standards and background samples that are used, such as the modern standard (oxalic acid) and "dead-carbon" (bituminous coal). Organic gases are sometimes submitted for dating, as well as peat, wood, soil, and organic tissue. Carbonates of differing forms make up about half of the samples run. The method should be applicable to these forms of carbon with minimum conversion. The less the sample is handled, the less chance there is of contamination, a major problem in samples of milligram size.

All carbon samples could easily be oxidized to CO_2, a form of the sample greatly preferred by most users, for many reasons. Neither the on-line or dedicated electrostatic accelerators under construction have succeeded in overcoming the problem of too much memory of CO_2 gas. Consequently, the sample presented to the cesium sputter beam source will probably have to be a solid. The cesium sputter source is the most likely to be used because it can produce negative ions of carbon in microampere beams.

Physical Configuration of Sample Holders

Various means have been adapted for holding the sample in the cesium sputter beam. A circular spindle, containing a number of discs of either stainless steel or pure aluminum, is commonly used. The disc, about 1 cm in diameter, has a 2 mm dimple drilled into its surface to contain the sample carbon and a 2 mm hole through the disc if reflected beam sputtering is used. Originally the dimple was a large cone-shaped indentation with the sample pressed to its walls and open at the end for the cesium to pass through. Newer sample holders may incorporate the discs in a stacked array, with a vacuum lock for rapid changing and with sensitive x and y adjustments to position the sample for maximum exposure to the sputter beam, tuneable for maximum output beam current strength of ^{12}C. Wires and rods have also been used as sample holders.

Forms of Carbon Tested

Three general forms of carbon have been used for sample preparation: charcoal, amorphous carbon, and graphite. It was known from early experiments that graphite gives the strongest beam as measured by a Faraday cup. It has been suggested that the graphite beam be considered the unit for comparison, with other materials rated as fractions of the graphite beam intensity. However, the high temperatures and pressures required for graphite production made other preparations more attractive.

Charcoal

Charcoal gave satisfactory ion beams of a few microamperes, so it was considered a good possible form for sample preparation. Organic matter was charred in an inert gas atmosphere and the resulting charcoal squeezed into a cone. Loss of the volatiles meant some loss of the sample, but yielded improved production of negative carbon ions because less interfering gas was produced by the high-temperature sputtering process. Not only does the cesium beam raise the temperature of the sample a few hundred degrees, but it physically bombards its way through a considerable thickness of the sample, if left on one focus point. For this reason, oxalic acid crystals, or loosely packed sugars or powders get boiled and blasted away. Therefore, for mechanical reasons, as well as for better conductivity, the charcoal was intimately ground with a binder of either powdered copper, silver, NaCl, or KBr in a Wig-L-Bug (a dental mixer) and pressed into the dimple. Sufficient currents in the high nanoampere to low microampere range were obtained with these samples and runs on unknowns performed with precision of 1 - 3 percent.

Amorphous Carbon

In order to handle samples not amenable to charring, amorphous carbon was tested. Carbon can be produced in various ways. One method used by the University of Toronto – University of Rochester-General Ionex group converts the sample by the lithium pot method (proportionately scaled down) to acetylene gas. The acetylene is then cracked or reduced to carbon onto the dimples of the discs by high-voltage sparking by means of a Tesla coil. This produces satisfactory samples, particularly when the disc is then heat-treated in a vacuum to further reduce the carbon. At the U.S. Geological Survey in Reston, Virginia various gases were used for the cracking and it was found that carbon monoxide, methane, and acetylene could be cracked. The efficiency of the cracking was in reverse order. Gold sputtering the end product allowed for non-contaminating handling. The group at the University of Washington, Seattle, has pioneered the carbon monoxide cracking method, using hot zinc to reduce carbon dioxide to carbon monoxide and circulating the gas by convection past a glow discharge where the carbon is deposited on small discs. When AC high voltage is used the deposit appears on adjacent discs, when DC is used, it appears on one disc. The oxygen released by the cracking is picked up by the hot zinc. At Chalk River Laboratory, Canada, carbon soot is produced by reduction of carbon dioxide by magnesium.

The cracking method is possible, as the amorphous carbon does give high nonoampere ion currents. However, the problem of isotopic fractionation, when cracking of the gas is incomplete, has yet to be studied completely. The method is somewhat wasteful, as not all of the gas sample is cracked, and so the full potential of the accelerator-dating method is not realized.

Graphite

The reduction of the sample to graphite has been achieved through pyrolysis of acetylene by the group at Oxford University, England and by using a high-pressure high-temperature press in Canberra, Australia by the Australian National University (ANU) group. The pyrolized graphite is deposited onto a tantalum wire resistively heated to 2000 °C in a flask containing acetylene at 10 – 20 torr. The acetylene is made by a scaled-down lithium pot method. Eighty microamperes of ion current have been measured from this pyrolized graphite routinely and a current greater than 100 microamperes has been achieved, greater than that from natural graphite. A serious drawback of the method is that the carbon deposits in layers with different isotopic fractionation by this thermal reaction. Possibly the rapid switching from ^{13}C to ^{14}C throughout the analysis would allow corrections to apply for any spatial isotopic variations of the sample.

The ANU group has converted samples of wood, cellulose, and sucrose to graphite by encapsulating 10 - 100 mg of the organic matter in a platinum capsule 1/8 - 3/16 inch (outer diameter). After capping by arc welding, the capsule is pressurized to 10,000 - 30,000 atmospheres and heated to 1300 °C for approximately 30 minutes. Some gaseous products diffuse through the platinum, while the carbon forms microcrystalline graphite. Attempts will be made by the ANU group to reduce calcium carbonates in a similar manner.

One such capsule of graphitized sucrose was measured at Rochester and only 20 nanoamperes of ion current were detected. When analyzed by x-ray diffraction, this particular capsule was found not to contain graphite, but some unknown organic crystal. However, the method is still considered attractive and viable as the procedure is simple and inexpensive; two samples per hour may be processed at an overall cost of $20 each. The method is also believed not to fractionate the carbon isotopes.

One unlikely prospect was tried at the University of Rochester at the suggestion of George Holdren of the Department of Geological Sciences, who asked whether lithium carbide would be a possible candidate. The carbide is a mid-way product of the lithium pot acetylene generator, but was believed to be too reactive with moisture. Surprisingly, after a short exposure in the atmosphere and a quick pump-down in the target area, commercial grade carbide gave large nanoampere ion currents for sufficient time for analysis. Lithium carbide was made at the U.S. Geological Survey laboratory in Reston in a lithium pot apparatus by inserting a grain of elemental lithium into the dimple of a disc, heating to 600 °C, and adding a small quantity of CO_2. Gold sputtering sealed the sample with a protective layer for shipping and handling. This method may be practical. Calcium or strontium carbides are possibilities also.

In conclusion, there is no fully acceptable way to prepare samples for radiocarbon accelerator runs, but several ways do the job somewhat satisfactorily. Possibly there will be a choice of several methods depending on the initial material. Solid organic material containing a large amount of carbon may be pressure-temperature treated to graphite; soils, sediments, gases, and carbonates may be burned to carbon dioxide and converted either to amorphous carbon through acetylene or carbon monoxide cracking or to pyrolitic graphite by means of the acetylene-tantalum wire method.   The fractionation problem will probably be solved by corrections applied after measurements of the different isotopes by means of the extremely rapid switching between isotopes designed into dedicated electrostatic accelerators.

RECEIVED October 20, 1981.

Ion Probe Magnesium Isotopic Measurements of Allende Inclusions

IAN D. HUTCHEON

Enrico Fermi Institute, University of Chicago, Chicago, IL 60637

The Mg isotopic compositions of eight Ca-Al-rich inclusions from the Allende meteorite have been measured with an ion microprobe. The microscopic spatial resolution of the ion probe and its ability to analyze isotopically femtogram quantities of Mg permitted exceptionally detailed studies of the distribution of ^{26}Mg on a 10 μm scale and revealed large differences in Mg isotopic composition among petrographically distinct Allende inclusions. Large ^{26}Mg excesses of up to 15 percent found in anorthites from B1 inclusions correlate strictly with $^{27}Al/^{24}Mg$ ratios in anorthite and spinel and provide additional evidence for the in situ decay of ^{26}Al. The observed $(^{26}Al/^{27}Al)_o = 4.6 \times 10^{-5}$ is consistent with other measurements. Ion probe measurements establish the first single crystal Al-Mg isochron and show that all anorthites had initially homogeneous Al isotopic ratios despite zoning of Mg and extensive mineralogical alteration. Anorthites from B2 inclusions also contain large ^{26}Mg excesses but exhibit a heterogeneous isotopic pattern with no unique correlation between $\delta^{26}Mg$ and Al/Mg ratios. $(^{26}Al/^{27}Al)_o$ ratios in B2 inclusions range from 4.6 x 10^{-5} to <5 x 10^{-6} suggesting either late formation of some anorthites, isotopic variation due to secondary alteration, or nonuniformity in the distribution of ^{26}Al. One Type A inclusion is also isotopically heterogeneous. Data from melilite define a good isochron with slope $(^{26}Al/^{27}Al)_o = 2.3 \times 10^{-5}$ but hibonite has a proportionately larger ^{26}Mg excess, suggesting either early formation of hibonite or heterogeneity in $(^{26}Al/^{27}Al)_o$ An igneous-textured inclusion shows only a very small ^{26}Mg excess,

0097-6156/82/0176-0095$08.50/0

corresponding to $(^{26}Al/^{27}Al)_o \sim 6 \times 10^{-6}$, which together with the bulk composition, suggests formation ~ 2 million years after Allende B1 inclusions. The widespread evidence of Al isotopic heterogeneities reinforces the picture of an early solar system which was chemically and isotopically inhomogeneous and seriously inhibits the use of ^{26}Al as a chronometer for early solar system events.

Ca-Al-rich inclusions from Type 3 carbonaceous chondrite meteorites are probably representative of the earliest material condensed from the solar nebula ([1][1] and references therein) and have proved to be an extraordinarily rich source of new discoveries leading to dramatic changes in our perception of the physical and chemical state of the primitive solar nebula. Discoveries by Clayton and coworkers [2] of large oxygen isotopic anomalies and by Gray and Compston [3] and Lee and Papanastassiou [4] of Mg isotopic variations in refractory inclusions from the Allende meteorite destroyed the long-standing belief in an initally homogeneous solar nebula and suggested the addition of extrasolar material preserved in an isotopically heterogeneous solar nebula. Additional studies [5,6] found clear evidence for ^{26}Mg excesses linearly correlated with $^{27}Al/^{24}Mg$ ratios, strongly suggesting that the enrichments in $^{26}Mg/^{24}Mg$ were produced by the in situ decay of now extinct ^{26}Al:

$$^{26}Al \rightarrow {}^{26}Mg + \beta^{+}.$$

The short half-life of ^{26}Al ($T_{1/2} \sim 7.2 \times 10^{5}$ y) implies that ^{26}Al must have been produced within a few million years of the formation of Allende inclusions and makes ^{26}Al especially interesting for planetary evolution. ^{26}Al can serve not only as a chronometer for ancient events but also as a potential heat source for planetary differentiation [7].

If ^{26}Al was present in the solar system at the time refractory inclusions formed, it must have been incorporated into the major minerals together with normal ^{27}Al, initially with an abundance ratio $(^{26}Al/^{27}Al)_o$. Subsequent decay of ^{26}Al to ^{26}Mg produced enhanced $^{26}Mg/^{24}Mg$ ratios, the magnitude of the enhancement being proportional to the Al/Mg ratio in a particular mineral. Thus, for an assemblage of minerals with different Al/Mg ratios which formed simultaneously from a reservoir of uniform $^{26}Al/^{27}Al$ and were not subsequently disturbed, the Mg isotopic composition now is given by:

$$^{26}Mg/^{24}Mg = (^{26}Mg/^{24}Mg)_o + (^{27}Al/^{24}Mg)(^{26}Al/^{27}Al)_o ,$$

[1]Figures in brackets indicate the literature references at the end of this paper.

where $(^{26}Mg/^{24}Mg)_o$ is the initial Mg isotopic composition of the
reservoir. This equation defines a straight line on an Al-Mg
evolution diagram (e.g., figure 8) with slope $(^{26}Al/^{27}Al)_o$ and
intercept $(^{26}Mg/^{24}Mg)_o$. As discussed in detail in [6], a set of
measurements which define a positive linear correlation provides
strong evidence for the in situ decay of ^{26}Al and argues against
the addition of fossil $^{26}\overline{Mg}$ [8].

The presence of ^{26}Mg excesses correlated with Al/Mg ratios in
fifteen Ca-Al-rich inclusions from the Allende and Leoville
carbonaceous chrondrites has provided additional strong evidence
for the in situ decay of ^{26}Al (see [9] for a recent review of
isotopic anomalies). There are also, however, several examples of
minerals whose isotopic compositions depart substantially from a
unique Al-Mg isochron, even within a single inclusion [10,11].
Since deviations from the isochron may reflect either differences
in the formation age of individual minerals or intrinsic heterogeneities in the initial $^{26}Al/^{27}Al$ ratio, the value of the Al-Mg
system as a chronometer for early solar system events remains
unclear.

One of the difficulties in interpreting the isotopic data
from refractory inclusions has been the lack of correlation·with
distinctive mineralogic and textural features. This report
presents new ion probe Mg isotopic measurements in refractory
inclusions from Allende which represent part of our study of
relationships between textures and compositions and the isotopic
structure of primitive solar system material. One important
advantage of the ion probe for this type of study is the ability
to analyze isotopically samples prepared as standard polished thin
sections. This capability enables us to characterize fully the
samples with optical, scanning electron, and luminescence micro-
scopes and the electron probe prior to istopic analysis. The
luminescence microscope and SEM, for example, have proved
extremely useful in revealing textural relationships between
primary and secondary phases not easily observed by common petrographic techniques (see, e.g., [12]). The microscopic spatial
resolution of the ion probe enables us to measure the isotopic
composition of precisely those areas identified as interesting.
We believe that this multifaceted approach is essential to a more
complete understanding of the complex histories of these relics of
the early solar system.

Analytical Procedure

The Mg isotopic measurements were performed with a modified
AEI IM-20 ion microprobe [13,14]. Secondary ions were generated
by bombarding the sample with a focussed ion beam to excavate a
small volume of the sample. A fraction of the sputtered material
is ionized during the sputtering process and is drawn off into
the mass spectrometer. A duoplasmatron ion source produces a

negatively charged primary beam which is mass analyzed to yield a $^{16}O^-$ beam free of impurities (e.g., OH^- and NO^-). The intensity and diameter of the primary ion beam at the sample are controlled by two electrostatic lenses capable of focussing the beam to a diameter of 2-3 μm. Under standard conditions this system produces a $^{16}O^-$ primary beam of 20 keV impact energy and ∿ 10 mA/cm^2 current density. The diameter of the area analyzed ranged from 2-3 μm with ∿ 1 nA beam current for Mg-rich minerals to 10 μm with 10 nA beam current for Mg-poor minerals. Figure 1 shows a typical "burn spot" after several hours of analysis.

Positively charged secondary ions sputtered from the sample are focussed by an electrostatic immersion lens and repeller electrode, after which an electrostatic einzel lens and deflection plates form an image at the variable-width entrance slit of the mass spectrometer. The mass spectrometer is double-focussing with Mattauch-Herzog geometry; mass resolution (M/ΔM) is adjustable from ∿ 200 to greater than 5000. Secondary ion intensities are recorded by ion counting using a 20-stage electron multiplier and a 100 MHz scale system coupled to an on-line computer. The ion probe is shown schematically in figure 2.

Isotopic data were collected by step scanning the spectrometer magnet through the mass range 27 to 20 using a computer-controlled, Hall-effect peak switching system. Isotope ratios were calculated after each scan from the secondary ion intensities integrated over the central 22 percent of each peak; 100 to 500 scans were averaged in a typical analysis and errors quoted are two standard deviations of the mean. The background between peaks was always less than 0.1 counts/second and no corrections for counting system dead-time losses were necessary.

The generation of large numbers of complex, poly-atomic and multiply charged ions in the sputtering process [15,16,17] creates potentially severe problems for low resolution secondary ion isotopic analysis. To minimize the formation of hydride and hydrocarbon secondary ions (e.g., $^{24}MgH^+$ and $^{12}C_2H_2^+$, respectively) the sample chamber is maintained at a pressure less than ∿ 1 x 10^{-7} torr and the ion extraction system is cooled to liquid nitrogen temperature. In addition, the secondary ion mass spectrum was frequently scanned at high mass resolution (M/ΔM > 3000) over the mass interval 28 to 22 to check for possible interfering species. The only interference detected at a level greater than 0.1 percent of the $^{25}Mg^+$ signal in anorthite was $^{48}Ca^{++}$ at mass 24. Figure 3 shows a high resolution scan of the Mg isotopes. Isotopic data were collected at a mass resolution of ∿ 300 for which $^{48}Ca^{++}$ and $^{24}Mg^+$ are not resolved; the $^{48}Ca^{++}$ correction was calculated by monitoring $^{40}Ca^{++}$ at mass 20, using $^{48}Ca/^{40}Ca = 1.9078$ x 10^{-3} [18]. The maximum correction to the $^{24}Mg^+$ intensity was ∿ 3.3 percent.

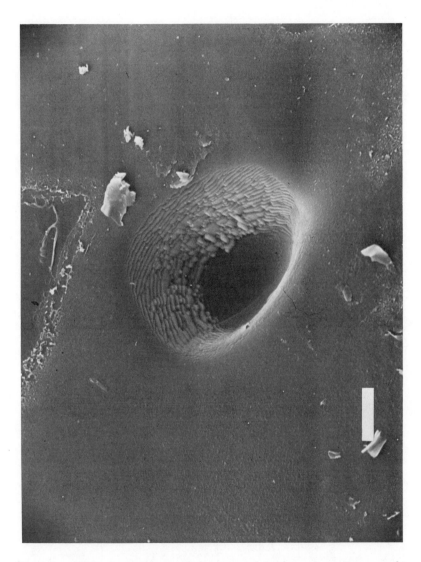

Figure 1.   SEM micrograph of a hole excavated in the surface of an anorthite crystal by the primary ion beam during approximately 4 h of isotopic analysis. The scale bar is 5 μm in length.

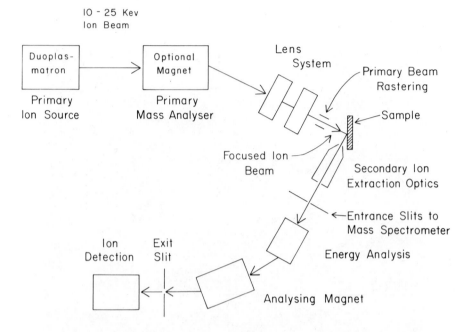

Figure 2. Schematic drawing of the AEI IM-20 ion microprobe. The magnetic analyzer and ion counting system are linked to an on-line computer for automated analysis.

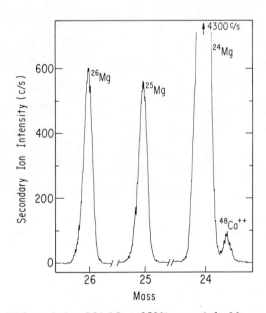

Figure 3. High-resolution (M/ΔM ~ 3500) scan of the Mg mass spectrum for a terrestrial anorthite. The scan is not continuous and shows only the region around each Mg isotope. The absence of any interferences other than $^{48}Ca^{2+}$ at mass 24 and the 0.1 count/s background between peaks are evident.

Analysis of terrestrial samples demonstrated that the tuning of the ion extraction system, and consequently the amount of instrumental mass fractionation, was very sensitive to charge build-up on the sample, the position of the primary beam relative to the spectrometer optic axis, and the position of the sample relative to the extraction lens. To optimize reproducible tuning of the extraction system from sample to sample, we developed the following criteria: (1) resistance of the sample (Au-coated) to ground less than 10^6 Ω; (2) alignment of the primary beam to within 10 μm using an optical microscope; (3) maximum $^{24}Mg^+$ secondary ion intensity; (4) flat-topped peaks ($\pm1\%$) at a mass resolution of $\sim$ 300; (5) uniform tuning (no changes in lens voltages to maintain maximum intensity) over the mass range 48 to 20. By following this procedure it was possible to control the magnitude of instrumental mass fractionation by adjusting only the deflection voltages to maintain maximum signal voltages (lens voltages constant) when moving from one sample to another. The operating conditions remained relatively unchanged throughout the majority of the analyses discussed below.

All of the samples analyzed were standard one-inch diameter polished thin sections. Whenever feasible the samples received a final, cleansing polish with 1 μm diamond compound made from commercial graded diamonds embedded in "vaseline". Commercial diamond paste has proved unsatisfactory due to high levels of K, Na, Cl, Si, F, and Ca. Samples are then cleaned with carbon tetrachloride, rinsed in ethanol, and coated with $\sim$ 100 Å of gold in a vacuum evaporator. This sample preparation technique was developed during our studies of minor elements [16,17] and has proved to produce consistently contamination-free samples.

Meteoritic Samples

Initially, our ion probe study concentrated on Type B inclusions from the Allende meteorite, since previous studies [6] had shown that Mg isotopic effects were largest in anorthite crystals from these inclusions. As discussed extensively by Grossman [1,19], Allende Type B inclusions are subspherical, centimeter-sized objects composed predominantly of coarse-grained Ti-Al-pyroxene, spinel, melilite, and anorthite. Characteristic of many Type B inclusions is a thick ($\sim$ 1 mm wide) outer mantle of gehlenitic melilite surrounding the 4-phase core (see, e.g., figure 4 and [20]). Wark and Lovering [21] subdivided Type B inclusions into B1 and B2 based on the presence or absence of this melilite mantle. We examined six Allende Type B inclusions, of which four are B1's (TS-23, TS-33, TS-34, and Al 3529G) and two are B2's (TS-8 and TS-21). (The nomenclature used is given in [19]. Inclusions are referenced by thin section number except Al 3529G and Al 3529-45, which were provided by Mason [22].) Anorthite crystals are large (up to 500 μm in length) and relatively

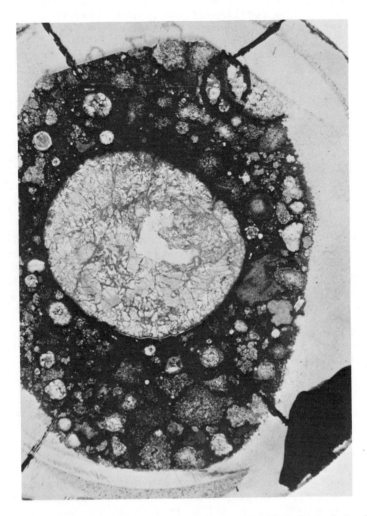

*Figure 4.   Photomicrograph of TS-23, a coarse-grained Type B1 inclusion from
Allende.*

*The outer 1.5 mm is a mantle of polycrystalline melilite whose Mg content increases
radially inwards, as does the abundance of included spinel. The interior of the inclusion
is predominantly coarsely crystalline Ti-Al-pyroxene, melilite, and anorthite, all containing euhedral crystals of spinel. The entire inclusion is bounded by a fine-grained rim of
complex mineralogy. The inclusion is ~ 1.5 cm in diameter.*

abundant (5 to 25% modal abundance) in Type B inclusions but characteristically are intergrown with high-Mg phases making separation of high purity, low-Mg phases for isotopic analysis by conventional techniques difficult. Type B anorthites also exhibit large correlated variations of their Mg and Na contents ($\sim$ 100 to > 1000 ppmw Mg [12]) such that the isotopic analysis of even a single crystal [23] averages over several distinct compositional zones (see following discussion). The oxygen isotopic compositions of inclusions TS-8, TS-23, and TS-34, measured by Clayton et al. [2 and 42], exhibit ^{16}O excesses typical of Allende coarse-grained inclusions.

We also analyzed one Allende Type A inclusion, an irregularly shaped object consisting predominantly of melilite (åkermanite content 0 to 33%) and spinel. The inclusion studied, Al 3529-45, was particularly chosen because it also contains hibonite ($CaAl_{12}O_{19}$), a highly refractory mineral, which on thermodynamic grounds, is expected to be one of the first major-element bearing phases to condense in the solar nebula [1]. The final inclusion examined in this study was Al 3510, a small ($\sim$ 3 mm diameter) circular object with an igneous-like texture composed predominantly of intergrown laths of anorthite (An 95) and olivine (Fo 88). Relative to most Allende coarse-grained inclusions, Al 3510 is enriched in Na, Mg, Si, and Fe and depleted in the refractory elements Al, Ca, and Ti.

All of the Type A and B inclusions studied are surrounded by a layered rim sequence of complex mineralogy [21] which clearly defines the inclusion-matrix boundary. Secondary alteration phases (grossular and nepheline, especially) are also a common feature of these inclusions, suggesting that vapor phase reactions with a relatively cool nebula occurred after formation of inclusions. Anorthite, in particular, is usually one of the most heavily altered phases; the relationship between Mg isotopic composition and alteration is discussed below. (See [12] for striking cathodoluminesce photographs of typical Allende alteration mineralogy.) Inclusion Al 3510 does not fit the normal pattern as it has no Wark-rim and does not contain the usual array of secondary minerals.

Terrestrial Samples and Standardization

Since the IM-20 had not previously been used for high precision isotopic measurements and since previous isotopic measurements with other ion microprobes [24,25] were characterized by percent level relative errors, we carried out an extensive series of Mg isotopic analyses of terrestrial samples. Terrestrial standards were especially emphasized prior to the study of meteoritic samples, but were also periodically interspersed with later meteoritic analyses as a check on the performance of the ion probe. Terrestrial standards included Ceylon spinel ($MgAl_2O_4$), Madagascar hibonite ($CaAl_{12}O_{19}$), a suite of olivines (Fo 100 to Fo 5) used in

a previous ion probe study [26], and numerous plagioclase feldspars (An 98 to An 50) whose Mg concentrations overlapped those of
anorthite from Allende inclusions. Analyses of terrestrial
plagioclase were particularly important for establishing the
validity of the $^{48}Ca^{++}$ correction, for testing the limits of
precision of Mg isotopic measurements in samples with low Mg
levels (< 300 ppmw), and for calibrating ion probe $^{27}Al^{+}/^{24}Mg^{+}$
ratios against Al/Mg elemental ratios measured with the electron
probe.
     A unique feature of the ion microprobe is the potential to
measure both elemental concentrations and isotopic ratios in the
same spot. This capability was particularly valuable in the
present study since the possible correlations between the
$^{26}Mg/^{24}Mg$ and Al/Mg ratios are central to the interpretation of Mg
isotopic anomalies. The Al/Mg ratio is calculated from the
$^{27}Al^{+}/^{24}Mg^{+}$ secondary ion ratio measured every twenty scans. The
wide range in sensitivity for the detection of different elements
with the ion probe is well documented (e.g., [16,27,28]), requiring calibration of secondary ion intensities against electron probe
measurements. Figure 5 shows the data for Al and Mg in plagioclase. The excellent linear correlation indicates that the yield
of Mg^{+} is independent both of Mg content and of plagioclase
mineral chemistry, at least over the range An 100 to An 50. The
slope of the Williamson least squares fit [29], 0.81 ± 0.05,
reflects the higher yield of $^{24}Mg^{+}$ relative to $^{27}Al^{+}$. For a more
complete discussion of quantitative elemental analysis with the
ion probe and ion probe-electron probe intercalibrations see
[16,17].
     A representative sample of the isotopic data from terrestrial
standards is given in Table 1 and plotted in figure 6 together
with data from Mg-rich minerals (spinel and Ti-pyroxene) from
Allende inclusions. Raw isotopic data are presented, corrected
only for any $^{48}Ca^{++}$ interference. Figure 6 utilizes δ-notation
which expresses the deviation of an isotope ratio from the standard
ratio in parts-per-thousand; i.e.,

$$\delta^m Mg = \left[\frac{(^m Mg/^{24}Mg)_{OBS}}{(^m Mg/^{24}Mg)_{STD}} - 1 \right] \times 1000.$$

It is clear from figure 6 that the terrestrial data do not cluster
about a single point but instead lie along a line of slope ∿ 0.5
on the three-isotope diagram, indicating isotopic variation due to
mass-dependent fractionation. Since mass fractionation effects in
Mg have not been observed in terrestrial materials [30,31], this
distribution of observed isotope ratios must be due to fractionation in the ion probe. The physical process which produces the

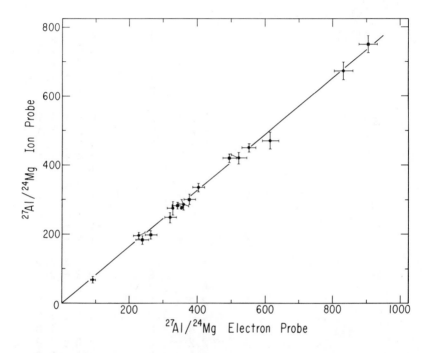

Figure 5.   Correlation of $^{27}Al^+/^{24}Mg^+$ secondary ion intensity ratios vs. $^{27}Al/^{24}Mg$ elemental ratio measured with the electron probe for Allende anorthite (●) and terrestrial plagioclase (■).

The fit of the data to the correlation line indicates that the relative yields of $^{27}Al^+$ and $^{24}Mg^+$ secondary ions are constant over the compositional range anorthite 100 to anorthite 50. The slope of the line, 0.81, reflects the higher sensitivity of the ion probe for Mg, and is used to convert measured secondary ion ratios to elemental ratios.

Table 1. Unnormalized Mg Isotope Data from Terrestrial Samples
and Mg-rich Minerals from Allende

Sample	$\delta^{25}Mg$ (permil)	$\delta^{26}Mg$ (permil)
Terrestrial Minerals		
Olivine	2.1 ± 1.5	4.8 ± 2.0
	-4.3 ± 2.5	-8.0 ± 3.0
	-3.6 ± 1.5	-3.2 ± 2.2
	-3.5 ± 1.3	-8.0 ± 1.8
	-5.4 ± 1.8	-10.4 ± 2.0
Spinel	3.8 ± 1.2	6.6 ± 1.9
	0.2 ± 1.0	0.8 ± 1.4
	-2.4 ± 1.1	-4.5 ± 1.7
	1.9 ± 1.3	3.5 ± 1.3
	0.7 ± 1.5	1.8 ± 1.0
Pyroxene	-1.5 ± 1.2	-2.7 ± 1.3
	3.4 ± 1.1	6.4 ± 1.7
	-2.4 ± 1.3	-4.1 ± 1.5
	-0.6 ± 1.0	-1.4 ± 1.2
Plagioclase[a]	1.2 ± 2.0	0.5 ± 3.0
	0 ± 1.8	-0.3 ± 2.0
	0.5 ± 2.0	-0.5 ± 2.2
	-2.3 ± 2.2	-5.2 ± 2.5
	2.6 ± 1.8	4.8 ± 2.0
	-2.4 ± 1.8	-4.7 ± 1.8
	-2.2 ± 2.0	-4.0 ± 1.8
	2.1 ± 1.5	1.5 ± 2.0
Allende Minerals		
TS-21 pyroxene	2.4 ± 1.0	4.8 ± 1.2
	-1.3 ± 1.2	-3.0 ± 1.2
	-0.1 ± 1.0	1.0 ± 1.0
	4.4 ± 1.4	9.5 ± 1.6
	2.9 ± 1.0	6.0 ± 1.0
	0.2 ± 1.2	0.6 ± 1.4
	1.3 ± 1.0	2.7 ± 1.0

Table 1.   (continued)

Sample	$\delta^{25}Mg$	$\delta^{26}Mg$
TS-23 pyroxene	-5.2 ± 1.4	-9.9 ± 1.5
	2.8 ± 1.0	5.3 ± 1.0
	3.3 ± 1.2	6.0 ± 1.3
	2.1 ± 1.0	4.0 ± 1.0
	0.3 ± 1.0	-0.3 ± 1.2
	0.1 ± 1.2	-0.4 ± 1.4
3529G pyroxene	-2.2 ± 1.2	-4.8 ± 1.5
	-2.6 ± 1.3	-4.7 ± 1.5
	0.3 ± 1.0	-0.7 ± 1.5
3529-45 spinel	-0.8 ± 1.0	-0.4 ± 1.0
	-3.3 ± 1.2	-6.5 ± 1.2
3510 olivine	-3.6 ± 1.0	-7.2 ± 1.2
	-4.8 ± 1.2	-10.5 ± 1.4
	-1.0 ± 1.0	-2.5 ± 1.5
TS-34 spinel	-1.6 ± 1.2	-3.6 ± 1.5
	0.6 ± 1.0	1.6 ± 1.4
TS-34 pyroxene	1.0 ± 1.2	1.9 ± 1.4
	1.7 ± 1.2	3.3 ± 1.5
TS-8 pyroxene	0 ± 1.0	1.2 ± 1.4
	1.3 ± 1.0	3.0 ± 1.2
	4.0 ± 1.0	7.9 ± 1.1

[a]Corrected for $^{48}Ca^{++}$ interference.

fractionation is poorly understood and fractionation may occur either during sputtering (generation of secondary ions) or during transfer of secondary ions from the vicinity of the sample to the mass spectrometer. The range of fractionation, ± 7 permil/amu, reflects the sensitivity of the ion probe to small changes in extraction conditions and it was only by rigidly adhering to the tuning criteria discussed above that we limited the fractionation within this range. Similar instrumental mass fractionation has also been reported in other ion probe studies [24] and in solid-source mass spectrometric measurements using the direct loading technique [32]. The slope of the fractionation line on the $\delta^{25}Mg$ versus $\delta^{26}Mg$ diagram in figure 6 is 0.51 ± 0.03, consistent with the slope expected for simple mass-dependent fractionation.

It is important to recognize that all of the terrestrial isotopic data lie along the best-fit line (nearly always within 2σ) independent of sample mineralogy and Mg content. We observed no systematic deviations from this line due to molecular

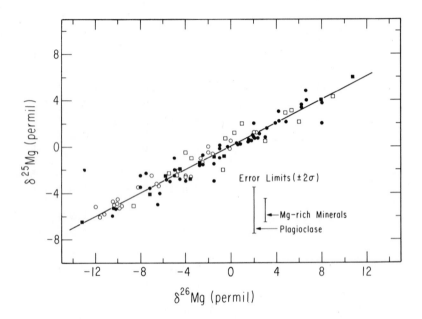

Figure 6.   Three-isotope correlation diagram for Mg using δ-notation (see text) for unnormalized data from terrestrial samples and Mg-rich Allende minerals.

The data plot along a line of slope ∼ 0.5, reflecting instrumental mass fractionation of ±7%/amu. The proper correction for $^{48}Ca^{2+}$ is indicated by the terrestrial plagioclase data which, if uncorrected, would plot far below the mass fractionation line off the diagram. Data from Allende Mg-rich minerals overlap those of terrestrial samples, denoting the absence of fractionation indigenous to the Allende material. Key: ○, terrestrial olivine + pyroxene; □, terrestrial plagioclase; ●, Allende Ti-pyroxene; and ■, Allende spinel.

interferences or unspecified instrumental effects of any kind. Data from some mineral groups clustered in certain areas along the line rather than being uniformly distributed; data from olivine and hibonite, for example, typically had $\delta^{26}Mg$ less than -4 permil. Terrestrial plagioclase data all lie within 2σ of the line, demonstrating a valid correction for $^{48}Ca^{++}$. Without this correction these data would be displaced from the fractionation line by up to nearly 30 permil along a line of slope one on the three-isotope plot.

The ion probe isotopic data were corrected for fractionation using the measured slope of the $^{25}Mg/^{24}Mg$ versus $^{26}Mg/^{24}Mg$ correlation line from terrestrial samples, 0.454 ± 0.022, after normalizing the observed $^{25}Mg/^{24}Mg$ ratios to the NBS standard $^{25}Mg/^{24}Mg$ ratio, 0.12663 [33]:

$$(^{26}Mg/^{24}Mg)_N = (^{26}Mg/^{24}Mg)_{OBS} + \frac{1}{0.454} \left[0.12663 - (^{25}Mg/^{24}Mg)_{OBS} \right].$$

With this normalization the terrestrial Mg isotopic data define a "normal" $^{26}Mg/^{24}Mg$ ratio of 0.13938 (corresponding to $^{25}Mg/^{24}Mg$ = 0.12663). Variations in the ^{26}Mg content of samples were then calculated as $\delta^{26}Mg$ in permil ($^o/oo$) relative to the measured normal value $(^{26}Mg/^{24}Mg)_o = 0.13938$:

$$\delta_N^{26}Mg = \left[\frac{(^{26}Mg/^{24}Mg)_N}{0.13938} - 1 \right] \times 1000 .$$

All of the terrestrial samples yielded $\delta^{26}Mg = 0$ within $2\sigma_{mean}$, where σ_{mean} is calculated from the standard deviation of the mean of the isotope ratio measurement, propagated to account for the fractionation correction. This fractionation correction removes the effect of mass-dependent fractionation both in the ion probe and in nature. The corrected values ($\delta_N^{26}Mg$) thus reflect only isotopic effects other than those due to mass-dependent fractionation.

Results

Mg Isotopic Composition of Inclusions

The Mg isotopic data for the Allende inclusions are summarized in Table 2 where we present the raw, unnormalized isotope ratios, the ^{26}Mg excess expressed as $\delta_N^{26}Mg$ (permil), and the $^{27}Al/^{24}Mg$ ratio for each sample. Data for Allende Ti-Al-pyroxene and spinel fall within the range of the terrestrial standards: $\delta^{25}Mg = 0 \pm 7$ permil. $\delta^{26}Mg = 0 \pm 14$ permil. It is important to recognize that the spread in the data for Allende Mg-rich minerals due to instrumental fractionation is the same as that observed for

Table 2. Mg Isotopic Composition of Mg-poor Minerals from Allende

Sample	$^{25}Mg/^{24}Mg$	$^{26}Mg/^{24}Mg$	$\delta^{26}Mg$ (‰) N	$^{27}Al/^{24}Mg$
B1 Anorthite				
TS-23				
An E-1	0.12721 ± 22	0.15095 ± 30	74 ± 4	229
E-2	0.12706 ± 25	0.15692 ± 33	120 ± 5	376
An 7	0.12676 ± 22	0.15400 ± 24	103 ± 3	318
An 8-1	0.12628 ± 30	0.15071 ± 32	86 ± 7	255
8-2	0.12725 ± 28	0.15560 ± 30	108 ± 5	353
An 20	0.12575 ± 30	0.14834 ± 33	77 ± 7	244
An 23	0.12661 ± 27	0.15161 ± 34	88 ± 5	269
TS-33				
An 1	0.12552 ± 28	0.15252 ± 33	112 ± 4	374
An 4-1	0.12674 ± 25	0.15672 ± 35	121 ± 4	366
4-2	0.12681 ± 32	0.15214 ± 28	89 ± 5	282
TS-34				
An 3	0.12643 ± 30	0.15887 ± 35	143 ± 6	425
3529 G				
An 1	0.12685 ± 25	0.14944 ± 32	69 ± 4	231
An 2	0.12622 ± 30	0.14955 ± 35	79 ± 5	250
B2 Anorthite				
TS-8				
An A	0.12564 ± 30	0.14096 ± 30	19 ± 5	251
An 1-1	0.12750 ± 35	0.15866 ± 30	126 ± 5	839
1-2	0.12690 ± 40	0.16266 ± 45	163 ± 7	964

Table 2. Mg Isotopic Composition of Mg-poor Minerals from Allende (continued)

Sample	$^{25}Mg/^{24}Mg$	$^{26}Mg/^{24}Mg$	$\delta^{26}Mg$ (‰)	$^{27}Al/^{24}Mg$
An 2	0.12672 ± 27	0.14094 ± 30	10 ± 4	465
An 3	0.12673 ± 28	0.14414 ± 32	33 ± 4	344
An 4	0.12625 ± 28	0.14255 ± 31	43 ± 3	354
An 5	0.12647 ± 32	0.14430 ± 35	48 ± 5	260
TS-21				
An F	0.12593 ± 35	0.14883 ± 40	78 ± 6	344
An S	0.12782 ± 20	0.15603 ± 26	102 ± 3	344
An 1-1	0.12773 ± 35	0.16570 ± 40	172 ± 6	559
An 1-3	0.12703 ± 30	0.16457 ± 26	175 ± 4	508
An 1-4	0.12677 ± 30	0.16650 ± 34	193 ± 4	572
An 1-5	0.12696 ± 23	0.15384 ± 30	99 ± 4	321
An 3	0.12688 ± 29	0.15197 ± 32	87 ± 4	362
An 11	0.12600 ± 30	0.15643 ± 36	132 ± 5	412
An 19-1	0.12736 ± 32	0.14926 ± 44	60 ± 6	311
An 19-2	0.12757 ± 32	0.15156 ± 40	73 ± 5	375
An 19-5	0.12738 ± 32	0.15426 ± 36	97 ± 5	550
An 21	0.12677 ± 40	0.16865 ± 40	66 ± 6	836
An 26	0.12663 ± 29	0.14754 ± 33	59 ± 4	350
An 28-2	0.12540 ± 32	0.13714 ± 36	2 ± 5	124
An 28-4	0.12677 ± 29	0.13960 ± 34	0 ± 4	155
An 31	0.12641 ± 32	0.15471 ± 40	113 ± 5	683
An 32	0.12675 ± 35	0.16304 ± 50	170 ± 6	531
Type A Hibonite				
3529-45				
Hb 1	0.12622 ± 20	0.14054 ± 26	15 ± 3	28
Hb 2	0.12567 ± 22	0.13966 ± 28	16 ± 4	30

Table 2. Mg Isotopic Composition of Mg-poor Minerals from Allende (continued)

Sample	$^{25}Mg/^{24}Mg$	$^{26}Mg/^{24}Mg$	$\delta^{26}_{N}Mg$ (‰)	$^{27}Al/^{24}Mg$
Hb-4	0.12652 ± 16	0.13959 ± 20	3 ± 3	30
Hb-5	0.12632 ± 22	0.14071 ± 30	14 ± 4	27
Type A Melilite				
3529-45				
Mℓ 1	0.12600 ± 24	0.13830 ± 23	2 ± 3	16
Mℓ 3	0.12662 ± 16	0.13980 ± 20	3 ± 3	22
Mℓ 5	0.12582 ± 26	0.14034 ± 32	18 ± 5	90
Mℓ 7	0.12680 ± 23	0.14061 ± 28	6 ± 4	29
Mℓ 11	0.12654 ± 31	0.13977 ± 36	4 ± 5	25
Mℓ 12	0.12631 ± 25	0.14024 ± 27	10 ± 4	50
Mℓ 13	0.12705 ± 32	0.14204 ± 32	13 ± 4	71
Mℓ 16	0.12620 ± 19	0.13930 ± 20	6 ± 3	36
Mℓ 17	0.12630 ± 35	0.14193 ± 37	23 ± 6	106
3510 Anorthite				
An 1	0.12629 ± 19	0.13900 ± 27	5 ± 3	98
An 2-1	0.12680 ± 28	0.14002 ± 28	6 ± 4	128
2-2	0.12638 ± 20	0.13978 ± 20	7 ± 3	103
2-3	0.12645 ± 25	0.13938 ± 31	3 ± 4	86
An 5	0.12646 ± 32	0.13928 ± 35	2 ± 4	114

$^{25}Mg/^{24}Mg$ and $^{26}Mg/^{24}Mg$ ratios corrected for $^{48}Ca^{++}$ interference.

terrestrial minerals. We believe that, using the tuning criteria discussed above, instrumental fractionation is limited to this range and that any measured fractionation greater than ± 10 permil/amu would be intrinsic to the sample. The fact that the raw $^{25}Mg/^{24}Mg$ ratios for all our Allende samples are within 7 permil of normal clearly indicates that none of the inclusions studied contains highly fractionated Mg such as found in Allende FUN inclusions [34] or in hibonite from the Murchison carbonaceous chondrite [35].

Unnormalized $^{25}Mg/^{24}Mg$ ratios from anorthite in Allende Type B inclusions also fall within the range of data from terrestrial samples but $^{26}Mg/^{24}Mg$ ratios show large excesses of ^{26}Mg (up to nearly 20%) and plot far to the right of the fractionation line on the three-isotope diagram (figure 7). Anorthite #28 from the B2 inclusion TS-21 is an important exception; see the following discussion. Increased $^{26}Mg/^{24}Mg$ ratios coupled with normal $^{25}Mg/^{24}Mg$ ratios represent enrichments in ^{26}Mg due to a nuclear isotope effect much greater than any possible instrumental effect. Isotope data from melilite and hibonite in the Type A inclusion, Al 3529-45, show smaller but clearly resolved ^{26}Mg excesses, while data from anorthite in the igneous inclusion, Al 3510, show only a very small enrichment in ^{26}Mg. After normalizing (assuming $^{25}Mg/^{24}Mg$ ratios are precisely normal) Allende Mg-rich minerals all yielded $\delta^{26}Mg = 0$ within ±2σ. All Allende anorthite, excluding 28-2 and 28-4 in TS-21, yielded positive $\delta^{26}Mg$, as did melilite and hibonite in Al 3529-45.

In the following discussion the degree of correlation between ^{26}Mg excesses ($\delta^{26}Mg$) and $^{27}Al/^{24}Mg$ ratios will be examined closely. Inclusions were first grouped according to the presence or absence of a unique linear correlation but, as will become clear, the Mg isotopic characteristics of the inclusions could have been distinguished equally well on the basis of mineralogy and petrology.

Allende B1 Inclusions

The data in Table 2 show that $^{26}Mg/^{24}Mg$ ratios from anorthites in all four Allende B1 inclusions are strongly correlated with their respective $^{27}Al/^{24}Mg$ ratios measured at the same point of analysis. This correlation defines a good linear array on the Al-Mg evolution diagram (figure 8) with a Williamson best fit slope of the isochron corresponding to $(^{26}Al/^{27}Al)_0 = (4.6 \pm 0.3) \times 10^{-5}$ and an intercept at $^{27}Al/^{24}Mg = 0$ of $(^{26}Mg/^{24}Mg)_0 = 0.13945 \pm 15$ ($\delta^{26}Mg = 0.5 \pm 1.1‰$). Note that the slope of the isochron is ~ 8 percent greater here than in our previous publications (excepting [12]) due to a recently discovered error in the ion probe calibration of Mg abundance sensitivity. All of the isochrons discussed in this paper reflect this change. Similar data showing the same linear correlation for other Allende inclusions have been reported in [6,23,24,25].

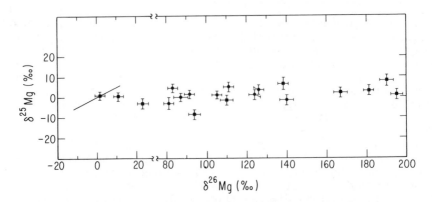

Figure 7.   Three-isotope correlation diagram for Mg using δ-notation for unnor-
malized data from Allende anorthite.

The range of normal Mg analysis (Figure 6) is shown by the solid line. Allende anor-
thites have normal $^{25}Mg/^{24}Mg$ ratios but greatly increased $^{26}Mg/^{24}Mg$ ratios, and plot
far to the right of normal Mg samples, indicating a nuclear isotopic effect in ^{26}Mg. Key:
●, B1 inclusions; and ■, B2 inclusions.

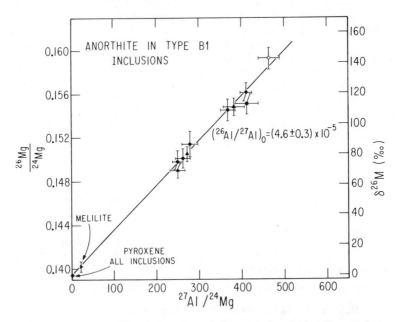

Figure 8.   The ^{26}Al–^{26}Mg evolution diagram for four Allende B1 inclusions.

The linear correlation between $^{26}Mg/^{24}Mg$ and $^{27}Al/^{24}Mg$ ratios measured in anorthite,
spinel, and pyroxene provides strong evidence for the in situ decay of ^{26}Al, and the slope
of the isochron indicates the abundance of ^{26}Al at the time the inclusions formed. Since
data from four inclusions plot along a single isochron, these inclusions must have formed
contemporaneously from a common reservoir. Key: ●, TS-23; ■, TS-33; ▲, 3629-G;
▶ TS-12; and ○, TS-34.

Allende B2 Inclusions

Most anorthites from the Allende B2 inclusions are also characterized by a positive correlation between excess ^{26}Mg and the $^{27}Al/^{24}Mg$ ratio but, as indicated both by the data in Table 2 and by the Al-Mg isochron plot for B2 inclusions (figure 9), the data fail to define the unique linear array found for B1 inclusions (cf. figure 8). The striking difference in Mg isotopic composition between B1's and B2's is well illustrated by data from inclusion TS-21 (B2). This inclusion was extensively studied utilizing the unique capabilities of the ion probe to combine petrographic and isotopic measurements on the same anorthite crystals. The isotopic data show that TS-21 can be divided into two sections with distinctly different Mg isotopic compositions. With one exception (An F) all of the interior anorthites (arbitrarily defined as those located more than $\sim$ 200 μm from the rim of the inclusion) show large ^{26}Mg excesses, ranging up to nearly 20 percent, and precisely the same linear correlation between $\delta^{26}Mg$ and $^{27}Al/^{24}Mg$ as found for the B1 inclusions. Exterior anorthites (those located within $\sim$ 300 μm of the rim), on the other hand, generally contain excess ^{26}Mg but without a unique correlation between $\delta^{26}Mg$ and $^{27}Al/^{24}Mg$. Data from exterior anorthites #19, 26, and 31 define a linear array with slope corresponding to $(^{26}Al/^{27}Al)_0 = (2.3 \pm 0.2) \times 10^{-5}$, about one-half the slope of the isochron defined by TS-21 interior and all B1 anorthites. Two analyses of anorthite #28, a crystal located within 50 μm of the rim, show that it contains no excess ^{26}Mg, $\delta^{26}Mg = 0$, making it unique among all Allende anorthites analyzed in this study.

Anorthites from TS-8, a second B2 inclusion, have Mg isotopic compositions similar to those of TS-21 exterior anorthites; no data lie on the standard B1 isochron. Anorthites 1 and 5 lie on the isochron defined by TS-21 exterior anorthites, while anorthites A, 3, and 4 from TS-8 together with anorthite #21 from TS-21 define a third isochron with slope corresponding to $(^{26}Al/^{27}Al)_0 = (1.2 \pm 0.3) \times 10^{-5}$, approximately one-fourth the slope of the standard isochron. TS-8 anorthite 2 has only a small ^{26}Mg excess, $\delta^{26}Mg \sim$ 10 permil and lies below all of the isochrons on figure 9.

Single Crystal Isochrons

A common feature of anorthite crystals in Allende Type B inclusions is large variations (up to a factor of $\sim$ 5) in Mg content on a scale of 10 to 50 μm. (Mg variability and its correlation with Na content and cathodoluminescence color are discussed extensively in [12].) Guided by cathodoluminescence micrographs, we used the microscopic spatial resolution of the ion-probe to measure the Mg isotopic composition at several points with distinct $^{27}Al/^{24}Mg$ ratios within individual crystals from TS-21 and TS-23. Data from each individual crystal define an

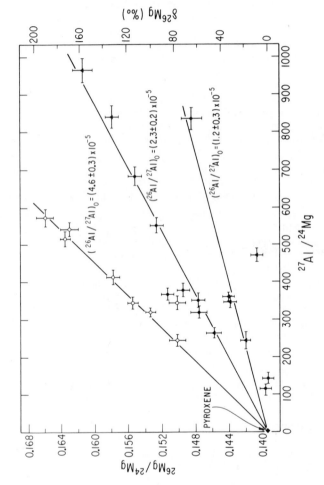

Figure 9. The $^{26}Al–^{26}Mg$ evolution diagram for two Allende B2 inclusions.

In contrast to Allende B1 inclusions (Figure 8), ^{26}Mg excesses in B2 anorthites do not exhibit a unique correlation with Al/Mg but, as discussed in the text, several correlation lines can be drawn through the data. Interior anorthites from TS-21 (○) plot along an isochron with the same slope as the B1 isochron, while TS-21 exterior anorthites (●) and all TS-8 anorthites (▲) plot below this isochron. Two new iso-chrons are defined by these data. The isotopic heterogeneity of B2 inclusions is consistent with either differences in formation times, isotopic variation due to secondary alteration, or a nonuniform distri-bution of ^{26}Al.

"internal" isochron. As shown in figure 10, the single crystal isochrons from two TS-23 anorthites and one TS-21 interior anorthite have the same slope as the standard B1 isochron (figure 8). Data not plotted but contained in Table 2 from one TS-21 exterior anorthite defines an internal isochron with slope equal to the TS-21 exterior anorthite isochron in figure 9, $(^{26}Al/^{27}Al)_o$ = 2.3 x 10^{-5}.

Allende Type A Inclusions

Most of the melilite in Type B inclusions is too Mg-rich for precise measurement of ^{26}Mg excesses using current ion probe techniques. Melilite in Allende Type A inclusions, however, has considerably lower Mg content [19,37], making ion probe studies feasible. Data from nine melilite crystals from inclusion 3529-45, shown in figure 11, define a good linear correlation between $^{26}Mg/^{24}Mg$ and $^{27}Al/^{24}Mg$ with slope corresponding to $(^{26}Al/^{27}Al)_o$ = (2.7 ± 0.5) x 10^{-5}, roughly one-half the slope of the standard Type B1 isochron and similar to the slope of the isochron defined by TS-21 exterior anorthites. The large error in the isotopic measurements of melilites #5, 12, 13, and 17 reflects the limited number of scans due to sudden changes in the Mg^+ secondary ion signal. These changes most likely indicate variations in Mg content on a submicron scale sampled as the primary ion beam excavates a small volume of the sample. Hibonites #1, 2, and 5 show clear excesses of ^{26}Mg ($\delta^{26}Mg \sim 15\%$) but lie well above ($\sim 4\sigma$) the melilite isochron. These hibonites have nearly identical Mg contents and so do not define an isochron; a line drawn from normal Mg at Al/Mg = 0 through the hibonite data would have a slope of ~ 7.5 x 10^{-5}. Hibonite #4, located in an interior clast comprised of hibonite and melilite, has the same $^{27}Al/^{24}Mg$ ratios as other hibonites but a much smaller ^{26}Mg excess, $\delta^{26}Mg \sim$ 3 permil, and lies below the melilite isochron.

Allende Inclusion 3510

Anorthites from this igneous-appearing inclusion are relatively Mg-rich and contain only small enhancements in $^{26}Mg/^{24}Mg$, $\delta^{26}Mg \sim 4$ permil, which are correlated with $^{27}Al/^{24}Mg$ ratios. Data from olivine (normal Mg) and anorthite define an isochron with slope corresponding to $(^{26}Al/^{27}Al)_o$ = (6 ± 2) x 10^{-6} (figure 12), much less than the slope of the standard B1 isochron. Only data from anorthites #1 and 2 were clearly resolved from normal Mg at the 2σ level. No evidence for mass fractionated Mg such as in ophitic inclusions from Allende by [38] was observed.

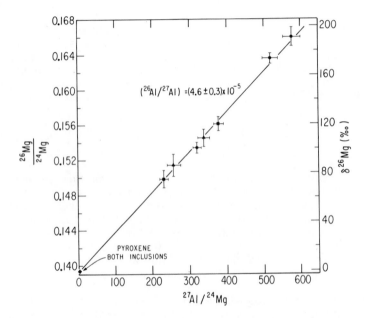

*Figure 10.   Single-crystal internal isochron plot of the $^{26}Al–^{26}Mg$ evolution dia-
gram for different areas with distinct $^{27}Al/^{24}Mg$ ratios within individual anorthite
crystals from TS-21 (●) and TS-23 (▲, ■).*

*The fit of the data to a single isochron indicates that individual anorthites from both B1
and B2 inclusions are internally isotopically homogeneous despite large internal variations in Mg and Na content and extensive mineralogical alteration. The slope of the
single-crystal isochron is the same as that of the B1 isochron.*

Discussion

Allende Inclusions

Two important conclusions can be drawn from the ion probe isotopic data from Allende Bl inclusions. The close agreement between the ion probe data and data collected by conventional solid source mass spectrometry on petrographically similar inclusions [5,6,10] together with ion probe data on terrestrial samples clearly establishes the capability of the Chicago ion microprobe for high quality isotopic studies of femtogram quantities of Mg. Secondly, the precise linear correlation between $^{26}Mg/^{24}Mg$ and $^{27}Al/^{24}Mg$ ratios exhibited by the ion probe data for anorthite from Bl inclusions provides strong evidence for the in situ decay of ^{26}Al in eight additional Allende inclusions. The ion probe data, in particular, are analyses of individual 5 to 10 μm diameter spots within single crystals and, unlike conventional mass spectrometry, elemental and isotopic ratios are measured on exactly the same material. Thus, there can be no question that the correlation is due to decay of ^{26}Al within the phases analyzed and not due to mixing of fossil ^{26}Mg produced elsewhere [8]. The close fit of data from four Bl inclusions suggests that Bl inclusions formed contemporaneously from a single nebular reservoir.

The addition of these ion probe data brings the number of Allende inclusions for which excesses of ^{26}Mg correlated with $^{27}Al/^{24}Mg$ ratios have been observed to fifteen, suggesting that ^{26}Al was reasonably abundant in the early solar system and could have provided an important heat source to produce differentiation in small planetary bodies [7,9]. The evidence for live ^{26}Al also changes the previously well-accepted view of solar system chronology. Until recently the interval between the end of galactic nucleosynthesis and the formation of solid bodies in the solar system was believed to be $\sim 10^8$ years based on studies of the decay products of ^{129}I ($T_{1/2} \sim 16 \times 10^6$ y) and ^{244}Pu ($T_{1/2} \sim 82 \times 10^6$ y) [39]. Short-lived ^{26}Al would, however, have been completely extinct after a decay interval of 10^8 years, requiring the addition of freshly synthesized nuclear material to the solar nebula on a time scale of a few million years. The discovery of excess ^{107}Ag, suggestive of the in situ decay of ^{107}Pd ($T_{1/2} \sim 6.5 \times 10^6$ y) is also compatible with a time scale of a few million years [40,41]. The conflict between the I-Pu and Al-Pd time scales strongly suggests that different events, in which the relative production rates for the various elements may have differed widely, contributed to the initial abundances of these radionuclides [9].

Alteration

^{26}Al has now totally decayed and the Al-Mg system cannot be used as a chronometer for early solar system events without knowledge of the initial $^{26}Al/^{27}Al$ ratio, corresponding to the first formation of solid bodies. The initial studies of Allende inclusions indicated $(^{26}Al/^{27}Al)_o \sim 5 \times 10^{-5}$, but additional studies have yielded $(^{26}Al/^{27}Al)_o$ ratios from $\sim 10^{-3}$ to $< 2 \times 10^{-7}$, thus inhibiting the use of the Al-Mg system as a timepiece. Since the majority of the discordant data lie below the standard B1 isochron (smaller $\delta^{26}Mg$ than expected for the $^{27}Al/^{24}Mg$), and since low-temperature secondary phases are ubiquitous in Allende inclusions, it was suggested that the Al-Mg system was disturbed in a back-retention with the nebula during which oxygen in anorthite and melilite was exchanged [42] or alkali- and halogen-rich material was introduced [34]. The effects of a back-reaction on the Mg isotopic composition of Allende inclusions can be examined by comparing the isotopic data from mineralogically altered and unaltered anorthites and data from inclusions which have experienced varying degrees of alteration. The single crystal isochron shown in figure 10 indicates that the linear correlation between $^{26}Mg/^{24}Mg$ and $^{27}Al/^{24}Mg$ ratios is maintained at all points throughout Allende anorthites despite large (up to $\sim 50\%$) variations in Mg content. The anorthites lying along this isochron differ widely in degree of mineralogical alteration — from undetectable to nearly complete replacement — yet exhibit no deviations from the isochron outside of analytical error. Anorthite #8 from TS-23 (squares in figure 10), for example, is more than 60 percent altered to grossular + monticellite + Na-rich plagioclase, yet isotopically appears identical to anorthite #1 from TS-21 (circles in figure 10) which shows no visible signs of alteration. The covariation of Na and Mg in these anorthites suggests that the zoning patterns were established during crystal growth [12] and are primary features. If, however, some of the complex, noncrystallographic patterns reflect subsequent redistribution of Mg accompanying mineralogical alteration, the high degree of linearity of the data requires that the Al-Mg clock must be been completely reset by this event. The observation that the single-crystal isochron in figure 10 has a slope $(^{26}Al/^{27}Al)_o = 4.6 \times 10^{-5}$ and that it intercepts the abscissa at normal Mg composition indicates that this event was not "late-stage" but must have occurred shortly after inclusion formation while ^{26}Al was still abundant. In this second senario the instability of secondary phases produced during alteration, such as grossular, at temperatures above 1000 C [43,44] requires rapid cooling of the nebula to temperatures below 1000 K within a few million years of the production of ^{26}Al.

The contrast between the extent of mineralogical alteration and isotopic behavior is emphasized by the data from B1 and B2 inclusions. B2 inclusions are on the whole more heavily altered,

as measured by the volume fraction of secondary phases, but
anorthite is generally not altered. In TS-8, for example, Ti-
pyroxene is commonly altered, while in TS-21 melilite is replaced.
In sharp contrast B1 anorthite is invariably altered to some
extent. A comparison of figures 8 and 9, however, clearly shows
the absence of any correlation between alteration of anorthites
and their Mg isotopic composition. As discussed above, all B1
anorthites, regardless of their state of preservation, lie
precisely along a single isochron. Mineralogically pristine B2
anorthites, on the other hand, fail to define a single isochron
and scatter widely on the Al-Mg evolution diagram (figure 9) as
discussed below. Ion probe analysis also failed to detect any
correlation between volatile element content and degree of alteration; Na, Cl, and K levels are extremely low in all Allende
anorthites. Extensive mineralogical alteration has also failed
to significantly disturb the oxygen isotopic composition of Ti-
pyroxene in TS-8 [2]. The majority of the evidence thus suggests
that the Mg isotopic composition of Allende anorthites was
unaffected by mineralogical alteration and the widespread formation of secondary phases. The possibility of additional events
during which Mg, ^{26}Al, and/or O were exchanged with a nebula
reservoir cannot be excluded. While the mineralogical evidence
suggests that the extensive zoning of Na and Mg in anorthites is
primary, the preservation of these features despite pervasive
oxygen isotope exhange [42] is poorly understood. Once again
there is no apparent connection between Mg and O isotopic effects.
The B2 inclusion TS-8 which has a complex Mg isotopic pattern
shows the same ^{16}O enrichment [2] as the B1 inclusion TS-34 which
has only a single Al-Mg isochron.

Allende B2 Inclusions

Data from the B2 inclusions are scattered over a large portion of the Al-Mg isochron plot (figure 9) but some systemmatic
trends can be recognized. Data from interior anorthites of TS-21
define an isochron whose slope is identical to that found for B1
inclusions. These isotopic data together with chemical data
showing that TS-21 interior anorthites exhibit the same linear
correlation between Na and Mg abundances as B1 anorthites [12]
suggest that the interior of TS-21 is, in fact, a B1 inclusion.
None of the data from either TS-21 exterior anorthites or TS-8
anorthites lie along the B1 isochron.

Three distinct nebular phenomena can in principle contribute
to the discordant Mg isotopic systematics: late formation of
individual anorthites followed by gentle compaction of B2 inclusions, selective secondary alteration of anorthites after
inclusion formation, and initial heterogeneities in ^{26}Al similar
to those observed in stable isotopes. The fact that data from
anorthites in both B2 inclusions are used to construct the two new
Al-Mg correlation lines suggests that whatever processes(es) or

environment(s) produced the Mg isotopic pattern must have been experienced by both B2 inclusions but not by any of the B1 inclusions.

In the first scenario (chronological interpretation; assumes an initially uniform distribution of ^{26}Al) the range in measured $(^{26}Al/^{27}Al)_o$ ratios for B2 inclusions, 4.6×10^{-5} to less than 5×10^{-6}, implies that chemically identical anorthite crystals were formed over an interval of at least 2.5×10^6 y. Individual anorthites must then have been incorporated into inclusions without disturbing their Mg isotope systematics. Each correlation line has approximately one-half the slope of the line above it, suggesting periodic formation of anorthite at $\sim 7 \times 10^5$ y intervals. All of the B2 anorthites except TS-21, 28 show clear excesses of ^{26}Mg and must have formed before the decay of ^{26}Al. Mg in TS-21, 28 is isotopically normal within errors, indicating that #28 formed more than 2.5 million years after most B1 anorthites. The similarity in chemical composition of all B2 and B1 anorthites suggests that conditions in the nebula were relatively stable during this period and the low abundance of Na indicates that the temperature remained above ~ 1200 K throughout this time [1].

Mg isotopic variations due to alteration have been discussed in a previous section; here we simply reiterate the fact that most B2 anorthites studied show no visible signs of chemical alteration. An alteration process which redistributed Mg was, therefore, most likely unconnected with the introduction of nonrefractory material into these inclusions and clearly had a much more profound effect on B2 anorthites than on their B1 counterparts. The heterogeneity of the Al isotopic distribution inferred from B2 anorthites indicates that alteration either did not affect all crystals to the same extent or was a recurrent phenomenon. Since variations in $\delta^{26}Mg$ within individual B2 anorthites are consistent with a single $(^{26}Al/^{27}Al)_o$ ratio, the alteration process must produce discordant neighboring grains while leaving individual grains internally concordant. In TS-21 we observed a correlation between Mg isotopic content and spatial position which in this model could be interpreted as preferential alteration of exterior grains. This correlation is, however, not perfect and was not observed at all in TS-8. Production of isotopic variations by alteration is thus severely constrained by the data.

The variations in the $(^{26}Al/^{27}Al)_o$ ratio may also reflect large non-uniformities in the distribution of ^{26}Al in the solar nebula, in which case the value of ^{26}Al as an easily applied and reliable chronometer is seriously compromised. The three Al-Mg correlation lines defined by the isotopic data indicate that Allende anorthites formed in reservoirs with distinct ^{26}Al contents but with otherwise similar chemical compositions and thermal conditions. Only one local reservoir was sampled by B1 anorthites while B2 anorthites were exposed to several; TS-8 anorthites did not sample the B1 reservoir, yet interior anorthites from TS-21

apparently did. The local reservoirs may have formed from a
single initial reservoir enriched in ^{26}Al by dilution with ^{26}Al-
free nebular gas. If hibonite and anorthite formed nearly
contemporaneously, the isochron through the hibonite data from
inclusion 3529-45 requires that the $^{26}Al/^{27}Al$ ratio was greater
than $\sim$ 7.5 x 10^{-5} in the initial reservoir and the wide distribution of Al isotopic compositions for B2 anorthite (figure 9)
requires at least four local reservoirs each with a different
$(^{26}Al/^{27}Al)_o$ ratio. Under these conditions the value of ^{26}Al as
a chronometer is uncertain and its usefulness as a heat source is
also diminished. Only in those reservoirs where the $^{26}Al/^{27}Al$
ratio is greater than $\sim$ 3 x 10^{-6} will the decay of ^{26}Al generate
sufficient heat to produce melting in chondritic bodies.

It is clear in figure 9 that not all of the B2 anorthites lie
within two sigma of any of the Al-Mg isochrons; more than one of
the processes discussed above may have been involved. Additional
data are surely needed to unravel the complex Mg isotopic pattern
of Allende B2 inclusions.

Allende Type A Inclusions

The evidence for live ^{26}Al and its in situ decay is the
precise correlation of ^{26}Mg excesses with Al/Mg ratios. The
observed variations in $(^{26}Al/^{27}Al)_o$ and large zoning of Mg,
together with the uncertainty over the location and behavior of
Mg in anorthite [12], have raised questions about the origins of
both normal and radiogenic Mg in Type B anorthite. Melilite
includes Mg in its crystal lattice and in Type A inclusions has a
sufficiently wide range in Al/Mg ratios to allow an independent
test of the correlation between $\delta^{26}Mg$ and Al/Mg. The quality of
the fit of the melilite data from 3529-45 to the isochron in
figure 11 is very similar to that of data from B1 anorthites
(figure 8), although the slope of this isochron is roughly one-
half that of the B1 isochron. These data show that the correlation between $^{26}Mg/^{24}Mg$ and $^{27}Al/^{24}Mg$ ratios established for
anorthite can also be found in a mineral which intrinsically
contains Mg in its crystal lattice, thus strengthening the argument for in situ decay of ^{26}Al. Two other features, the hibonite
data and the slope of the melilite isochron, however, make a
simple chronological interpretation difficult. The steeper slope,
$\sim$ 7.5 x 10^{-5}, of an isochron through the hibonite data could
indicate that the formation of hibonite preceded that of melilite
by about one million years. Earlier formation of hibonite would
be consistent with thermodynamic arguments which suggest that
hibonite should condense before melilite [43,44]. A one million
year interval between the formation of hibonite and melilite
corresponds to a cooling rate for the gas of $\sim$ 150 K/10^6 y (using
condensation temperatures for hibonite and gehlenite given in
[43].

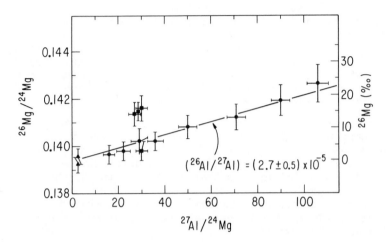

Figure 11.   The ^{26}Al–^{26}Mg evolution diagram for an Allende Type A inclusion 3529-45.

Data from spinel (▲) and melilite (●) define a good linear correlation between excess ^{26}Mg and $^{27}Al/^{24}Mg$ ratios with slope roughly one-half that of the B1 isochron. Data from hibonite (■), however, plot well above the isochron, suggesting either early formation of hibonite or heterogeneity in $^{26}Al/^{27}Al$.

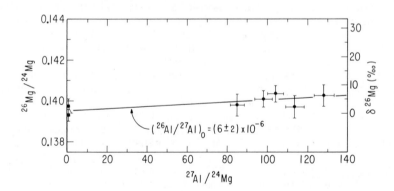

Figure 12.   The ^{26}Al–^{26}Mg evolution diagram for the igneous-textured Allende inclusion 3510.

Only a small ^{26}Mg excess is present in anorthite, and the slope of the olivine–anorthite isochron, $(6 ± 2) × 10^{-6}$, is correspondingly much less than that of the B1 isochron. The isotopic, mineralogical, and chemical data all suggest late formaton of this inclusion, about two million years after B1 inclusions. Key: ■, olivine; and ●, anorthite.

The data from 3529-45 can also be interpreted as reflecting an initially heterogeneous distribution of ^{26}Al. Melilite and hibonite may have formed from different reservoirs with enequal ^{26}Al contents in which case the different $(^{26}Al/^{27}Al)_o$ ratios for melilite and hibonite have no chronological significance. The clast hibonite (#4) has $\delta^{26}Mg$ about 7σ less than the other hibonites and in either scenario probably had its Mg isotopes re-equilibrated during clase formation. The slope of the melilite isochron is similar to one of the B2 anorthite isochrons (figure 9), suggesting either that 3529-45 melilite and some B2 anorthites are contemporaneous or that melilite and anorthite formed from the same nebular reservoir.

The Type A melilite data also reinforce earlier arguments about the absence of any connection between mineralogical alteration and Mg isotopic behavior. Melilites in 3529-45 are extensively altered, mainly to grossular and Na-rich plagioclase, yet show much less evidence for a disturbed Mg isotopic composition than relatively pristine melilites from B1 inclusions. B1 melilites exhibit much larger deviations from the standard isochron than anorthites (see also [1]).

Allende Inclusion 3510

Inclusion 3510 was studied isotopically because its mineralogy differs sharply from that of any other Allende inclusion studied here or elsewhere and because of the clear textural evidence that it had passed through a liquid stage. The data in figure 12 show that among Allende inclusion, 3510 also has a unique Mg isotopic composition. Once again the isotopic data alone are not sufficient to differentiate between late formation and heterogeneous $(^{26}Al/^{27}Al)_o$ ratios but in this case the combined evidence of mineralogy and chemical and isotopic compositions strongly favors the chronological interpretation. The shallow slope of the isochron suggests that 3510 formed about three ^{26}Al half-lives or $\sim$ 2 million years after Allende B1 inclusions but before ^{26}Al had totally decayed. Late formation in a cooler nebular environment is consistent with the abundance of olivine and the enhanced levels of Na and Fe; inclusion 3510 may represent the last glimmer of ^{26}Al activity. Assuming that the relatively high Na content of plagioclase in 3510 indicates a formation temperature below $\sim$ 1200 K [37], we can use the ^{26}Al decay interval to estimate a cooling rate for the nebula of $\sim$ 125 K/10^6 y between the condensation of diopside and the formation of 3510.

Conclusions

The isotopic studies of terrestrial minerals and Allende inclusions have clearly established the capabilities of the Chicago ion probe to perform precise isotopic measurements on femtogram

126   NUCLEAR AND CHEMICAL DATING TECHNIQUES

quantities of Mg. The measurements discussed in this paper represent the first extensive study of the distribution of Mg isotopes on a ten micron scale and have revealed large differences in Mg isotopic composition among petrologically distinct refractory inclusions from Allende. Large [26]Mg excesses in anorthites from four B1 inclusions show a common linear correlation with [27]Al/[24]Mg ratios, strengthening earlier arguments for the in situ decay of [26]Al [9]. Isotopic data from all of the anorthites analyzed from the four B1 inclusions fit closely to a single isochron, suggesting that these mineralogically and chemically similar inclusions [1] formed contemporaneously from a single nebular reservoir. The isotopic data are consistent with mineralogical and chemical evidence [20,45] suggesting that B1 inclusions were molten at one stage and allow these inclusions to differ in time of formation by no more than $\sim 10^5$ years.

The microscopic spatial resolution of the ion probe enabled us to measure the isotopic composition at several points within individual crystals and these ion probe data establish the first single crystal isochron for areas with distinct [27]Al/[24]Mg ratios. The Mg isotopic pattern within each B1 and B2 anorthite is individually consistent with a single $({}^{26}\mathrm{Al}/{}^{27}\mathrm{Al})_0$ ratio, suggesting that zoning of Mg and Na is a primary growth feature and indicating that any mobilization of Mg accompanying alteration occurred uniformly throughout these crystals.

The large heterogeneities in Mg isotopic composition found for the B2 and A inclusions confuse the simple chronological interpretation possible with the B1 data. The B2 and A data are generally consistent with three alternatives — actual differences in age, isotopic variation due to alteration, or heterogeneity in $({}^{26}\mathrm{Al}/{}^{27}\mathrm{Al})_0$ — which lead to different models for the early solar system. Only the data from Allende B1 inclusions provide compelling evidence for an initally uniform distribution of [26]Al and for the usefulness of [26]Al as a heat source and chronometer. This ion probe study emphasizes the complexity of the Mg isotopic record of Allende refractory inclusions and leads to one unmistakable conclusion: that as one in investigates the historical record stored in these relicts of the early solar system with ever-increasing spatial resolution, the mystery as to how they formed only deepens.

I thank my Chicago colleagues, R. N. Clayton, T. K. Mayeda, L. Grossman, and J. V. Smith for many helpful and stimulating discussions. I am particularly indebted to I. M. Steele for his invaluable help taming the beast. and I thank R. Draus for technical assistance, J. Eason for manuscript preparation, and L. Grossman and B. Mason for samples. This work was supported by NASA grant NGL 14-001-169.

Literature Cited

[1] Grossman, L., Ann. Rev. Earth Planet. Sci., 1980, 8, 559-608.
[2] Clayton, R. N., Grossman, L., Mayeda, T. K., Science, 1973,
     182, 485-488.
[3] Gray, C. M., Compston, W., Nature, 1974, 251, 495-497.
[4] Lee, T., Papanastassiou, D. A., Geophys. Res. Lett., 1974, 1,
     225-228.
[5] Lee, T., Papanastassiou, D. A., Wasserburg, G. J., Geophys.
     Res. Lett., 1976, 2, 109-112.
[6] Lee, T., Papanastassiou, D. A., Wasserburg, G. J., Astrophys.
     J. Lett., 1977, 211, L107-L110.
[7] Urey, H. C., Proc. Acad. Sci. U.S., 1955, 41, 127-144.
[8] Clayton, D. D., Icarus, 1977, 32, 255-269.
[9] Lee, T., Rev. Geophys. Space Phys., 1979, 17, 1591-1612.
[10] Esat, T. M., Papanastassiou, D. A., Wasserburg, G. J., Lunar
     Planet. Sci. X, 1979, 361-363.
[11] Hutcheon, I. D., Steele, I. M., Lunar Planet. Sci. XI, 1980,
     496-498.
[12] Hutcheon, I. D., Steele, I. M., Smith, J. V., Clayton, R. N.,
     Proc. Lunar Planet. Sci. Conf. 9th, 1978, 1345-1368.
[13] Banner, A. E., Stimpson, B. P., Vacuum, 1975, 24, 511-517.
[14] Steele, I. M., Hutcheon, I. D., Solberg, T. N., Smith, J. V.,
     Clayton, R. N., Int. J. Mass Spectrom. Ion Phys., 1977, 23,
     293-305.
[15] Bakale, D. K., Colby, B. N., Evans, C. A., Anal. Chem., 1975
     47, 1532-1536.
[16] Steele, I. M., Hutcheon, I. D., Smith, J. V., Proc. Lunar
     Planet. Sci. Conf. 11th, 1980, in press.
[17] Steele, I. M., Hervig, R. L., Smith, J. V., Hutcheon, I. D.,
     Amer. Mineralogist, 1980, in press.
[18] Russell, W. A., Papanastassiou, D. A., Tombrello, T. A.,
     Geochim. Cosmochim. Acta, 1978, 42, 1075-1090.
[19] Grossman, L., Geochim. Cosmochim. Acta, 1975, 39, 443-454.
[20] Steele, I. M., Hutcheon, I. D., Lunar Planet. Sci. X, 1979,
     1166-1168.
[21] Wark, D. A., Lovering, J. F., Proc. Lunar Sci. Conf. 8th,
     1977, 95-112.
[22] Mason, B., Martin, P. M., Smithsonian Contrib. Earth Sci.,
     1977, 19, 84-95.
[23] Lee, T., Papanastassiou, D. A., Wasserburg, G. J., Geochim.
     Cosmochim. Acta, 1977, 41, 1473-1485.
[24] Bradley, J. G., Huneke, J. C., Wasserburg, G. J., J. Geophys.
     Res., 1978, 83, 244-254.
[25] Lorin, J. -C., Christophe Michel-Levy, M., Fourth Int. Conf.
     on Geochron. Cosmochron. Isotope Geol., U.S.G.S. Open-File
     Rep. 78-701, 1978, 257-259.
[26] Steele, I. M., Hutcheon, I. D., Microbeam Analysis, 1979,
     1979, 338-340.

[27] Andersen, C. A., Hinthorne, J. R., Anal. Chem., 1973, 45, 1421-1438.
[28] McHugh, J. A., Secondary Ion Mass Spectrometry, A. W. Zanderna, ed., Elsevier, Amsterdam, 1975, pp. 223-278.
[29] Williamson, J. H., Can. J. of Phys., 1968, 46, 1845.
[30] Catanzaro, E. J., Murphy, T. J., J. Geophys. Res., 1966, 71, 1271-1274.
[31] Schramm, D. N., Tera, F., Wasserburg, G. J., Earth Planet. Sci. Lett., 1970, 10, 44-59.
[32] Esat, T. M., Brownlee, D. E., Papanastassiou, D. A., Wasserburg, G. J., Science, 1979, 206, 190-197.
[33] Catanzaro, E. J., Murphy, T. J., Garner, E. L., Shields, W. R., J. Res. Nat. Bur. Stand., 1966, 70A, 453-458.
[34] Wasserburg, G. J., Lee, T., Papanastassiou, D. A., Geophys. Res. Lett., 1977, 4, 299-302.
[35] Macdougall, J. D., Phinney, D., Geophys. Res. Lett., 1979, 6, 215-218.
[36] El Goresey, A., Nagel, K., Ramdohr, P., Proc. Lunar Planet. Sci. Conf. 9th, 1978, 1279-1303.
[37] Allen, J. M., Grossman, L., Davis, A. M., Hutcheon, I. D., Proc. Lunar Planet. Sci. Conf. 9th, 1978, 1209-1233.
[38] Lorin, J.-C., Christope Michel-Levy, M., Desnoyers, C., Meteroites, 1978, 13, 537-540.
[39] Schramm, D. N., Wasserburg, G. J., Astrophys. J., 1970, 162, 57-69.
[40] Kelly, W. R., Wasserburg, G. J., Geophys. Res. Lett., 1978, 5, 1079-1082.
[41] Kaiser, T., Kelly, W. R., Wasserburg, G. J., Geophys. Res. Lett., 1980, 7, 271-274.
[42] Clayton, R. N., Mayeda, T. K., Geophys. Res. Lett., 1977, 4, 295-298.
[43] Grossman, L., Geochim. Cosmochim. Acta, 1972, 37, 1119-1140.
[44] Blander, M., Fuchs, L. H., Geochim. Cosmochim. Acta, 1975, 39, 1605-1619.
[45] MacPherson, G. J., Grossman, L., Earth Planet. Sci. Lett., 1980, in press.

RECEIVED March 27, 1981.

Krypton-81–Krypton Dating by Mass Spectrometry

KURT MARTI

University of California—San Diego, Chemistry Department, B-017, La Jolla, CA 92093

The ^{81}Kr-Kr dating method is reviewed, the progress and experimental approaches are outlined, and some current applications and prospects are discussed.

The discovery of cosmic ray produced ^{81}Kr in meteorites [1][1] introduced a new method of high sensitivity measurements of ^{81}Kr concentrations and cosmic ray exposure dating. The method consists of a direct measurement of both radioactive ^{81}Kr atoms ($T_{1/2}$ = 2.13 x 10^5y, [2] and of stable spallation Kr atoms by a mass spectrometer and the calculation of cosmic ray exposure ages from measured Kr isotope ratios. The method is analagous to the ^{40}K-^{41}K method developed by Voshage [3] but provides much higher sensitivity, since static noble gas mass spectrometry is employed. Early measurements were carried out on meteorite samples corresponding to 10^6-10^7 atoms of ^{81}Kr [1,4]. The technique proved to be extremely useful in computing exposure ages of lunar samples returned by the Apollo program and it allowed the precise dating of lunar craters. Over the years, the sensitivity of the method has been improved and background interferences were reduced, and at present it is possible to measure samples corresponding to 10^5 atoms ^{81}Kr. On the other hand, attempts to detect ^{81}Kr by accelerator based high-energy mass spectrometry so far have not been successful. Progress has also been made in the understanding and calibration of cosmic ray induced reactions with the major target elements Sr, Y, and Zr in natural solid samples, and reasonably good agreement was obtained between predicted and observed production rates for the Kr isotopes in lunar samples [5]. ^{81}Kr measurements and ^{81}Kr-Ar ages provided useful tools in studies of

---

[1]Figures in brackets indicate the literature references at the end of this paper.

the cosmic ray exposure histories of chrondritic meteorites and elucidating the long-term average fluxes of solar and galactic cosmic rays [6,7].

Spallation Systematics

In natural silicate samples, cosmic ray spallation Kr is produced predominantly by high-energy reactions on Sr, Y, and Zr. Low-energy reactions on Rb often make contributions mainly to the heavier Kr isotopes, and slow neutrons with energies below 0.5 MeV produce ^{80}Kr and ^{82}Kr by neutron capture reactions on Br. Because of the low cosmic ray flux at the solid surface of the earth, studies so far have been restricted to extraterrestrial samples. The excitation functions for the production of spallation Kr on the above listed target elements were studied by Regnier et al., [5]. There is a systematic variation of the absolute and also of the relative production rates of the Kr isotopes with increasing degree of shielding due to the contribution by reactions of secondary cosmic ray particles.

In the ^{81}Kr-^{83}Kr dating method, which is discussed in the next paragraph, the production ratio of radioactive and stable Kr isotopes can be evaluated directly from the Kr spallation spectrum in a meteorite according to the relation [1]:

$$P_{81}/P_{83} + \frac{0.93}{2} [(^{80}Kr + {}^{82}Kr)/{}^{83}Kr] \qquad (1)$$

here the factor 0.95 ± 0.05 is the estimated isobaric fraction yield. Regnier [8] and Nishiizumi et al., [6] studied in detail measured P_{81} versus P_{81} (calc), as used in equation (1), and found that the factor 0.95 is appropriate for a variety of incident proton energies and chemical compositions, specifically also for chondritic meteorites. It appears that the production ratio P_{81}/P_{83} is well approximated by the interpolation used in equation (1). However, if specific neutron effects, due to neutron capture in Br, are present on the stable isotopes ^{80}Kr and ^{82}Kr, equation (1) can no longer be used and alternative treatments are required, such as interpolations between ^{78}Kr and ^{83}Kr [9,10]. Such alternatives, however, depend on the chemical composition of a sample and need to be adjusted accordingly.

The ^{81}Kr-Kr Method

The method assumes that the average cosmic ray flux has been constant over the mean-life of ^{81}Kr (τ_{81} + 3.07 x 10^5y) and over the overall exposure time of the sample cosmic rays. Furthermore, it is assumed that the exposure geometry has remained fixed over the period of irradiation, but no other assumptions are made regarding shielding.

Assuming constant production rates P_M for the Kr isotopes, the concentration of $[^{81}Kr]$ as a function of time (t) is given by

$$[^{81}Kr]_t = (P_{81}/\lambda_{81}) (1-e^{-\lambda_{81}t}), \qquad (2)$$

and the spallation yield of a stable isotope, e.g., ^{83}Kr, by

$$[^{83}Kr_s]_t = P_{83}t \qquad (3)$$

Therefore, at the time of fall, we expect a ratio

$$\left(\frac{^{83}Kr_s}{^{81}Kr}\right)_{T_e} = \lambda_{81} \frac{P_{83}}{P_{81}} \frac{T_e}{1-\exp(-\lambda_{81}T_e)}$$

and obtain

$$F(T_e) \equiv \frac{T_e}{1-\exp(-\lambda_{81}T_e)} = \frac{1}{\lambda_{81}} \frac{P_{81}}{P_{83}} \left(\frac{^{83}Kr_s}{^{81}Kr}\right)_{T_e} \qquad (4)$$

For cosmic ray exposure intervals $T_e \gg \frac{1}{\lambda_{81}} = 3.07 \times 10^5 \text{yr}$, the exponential term becomes negligible and equation (4) reduces to $F(T_e) \cong T_e$,

$$T_e \cong \frac{1}{\lambda_{81}} \frac{P_{81}}{P_{83}} \left(\frac{^{83}Kr_s}{^{81}Kr}\right)_{T_e} \qquad (5)$$

As discussed earlier, the production ratio P_{81}/P_{83} can, in general, reliably be obtained from equation (1) and, therefore, the ^{81}Kr-Kr method avoids much of the uncertainty arising from unknown production rates or production ratios. The method derives exposure ages from Kr isotopic ratios as obtained from mass spectrometry and does not require a knowledge of the concentration of spallation Kr, which makes the method inherently more precise.

Mass Spectrometry

Kr is extracted either by stepwise heating or melting of a sample followed by usually several gas clean-up steps and by separation of Kr from other noble gases by selective adsorption on charcoal at cryogenic temperatures [11]. Kr is analysed in ultra-clean static mass spectrometers which allows recycling of

non-implanted Kr atoms through the ion source. A clean separation
from abundant ^{40}Ar is generally required, since interferences from
charge exchange (Ar^{2+} → Ar^{+}) between ion source and magnetic
analyser may affect the Kr isotope at mass 80 in some mass
spectrometers. Interferences due to isobaric background can be
significantly reduced by rigorous bake-out techniques. Nevertheless, the remaining interferences may be significant, if ^{81}Kr
amounts equivalent to <10^{6} atoms are to be analysed. Two
different techniques have been used in our laboratory: (a) A high
mass resolution (about 500) which allows the separation of Kr
peaks from isobaric interfering background peaks, except for Br.
The Br isotopes are not resolved, but a possible interference at
mass 81 can be monitored at mass 79, since ^{79}Br is about equally
abundant as ^{81}Br (Br interferences may be time-dependent,
especially after bake-outs). Figure 1 shows, as an illustration,
the first measured cosmic ray produced ^{81}Kr which was found in the
Macibini meteorite. (b) Since the peak shape in method (a) may
not be suitable for automated peak-jump analysis, a method using
medium mass resolution (about 300) permits partial separation of
Kr peaks from isobaric interferences, again with the exception of
Br, as noted above. The measurement of background peaks, in
addition to the Kr mass spectrum, permits interference corrections
to be applied to the Kr isotopes, if the relative interferences
are carefully calibrated. A possible procedure is to pump out all
Kr at the end of the run and to determine the relative interferences at the Kr mass positions. Standards of atmospheric Kr are,
in general, used to determine mass discrimination corrections of a
mass spectrometer.

Data Analysis

Kr isotopic data derived from either stepwise or total sample
analysis represent generally mixtures of several components, but
equations (4) and (5) require the identification of the cosmic ray
produced spallation component. As long as spallation Kr is the
major component, the analysis of isotopic spectra is reasonably
straightforward. In many cases, however, spallation Kr may be
almost completely masked by abundant trapped Kr components, such
as solar wind Kr in the case of lunar soils. Identification of
spallation Kr in this case requires additional physical or
chemical methods for component separation (e.g. grain size or
mineral separations, etching of surface layers). In many cases,
it is important to know exactly the composition of trapped Kr
(terrestrial atmospheric, solar, meteoritic) and of fission Kr
before spectral decomposition is attempted. Analyses were
generally carried out by first partitioning ^{86}Kr into trapped,
fission and spallation components. ^{86}Kr is selected because the
relative spallation yield is very small. In fact, many authors
have adopted a zero spallation yield. Experimental data, however,
indicate that although this yield is small, it may not be

negligible. The smallest observed spallation ratios $(^{86}Kr/^{83}Kr)_s$ are $\sim$0.015 in lunar rocks and 0.008 ± 0.008 in phosphates of ADOR [12]; the latter exhibit a very low Zr/Sr ratio. Nuclear systematics indicate that spallation reactions on Rb and the heavier Zr isotopes or proton induced fission reactions may all make non-negligible contributions to ^{86}Kr [5]. Partitioning of ^{86}Kr can be achieved by adopting $(^{86}Kr/^{83}Kr)_s \equiv 0.015$, by computing the fission yield $^{86}Kr_F$ from measured fission components in the Xe mass spectrum, and by assigning the remaining $^{86}Kr_T$ to the trapped component. Of course, ^{81}Kr is not affected in this partitioning, since all ^{81}Kr belongs to the spallation component. The relevant data in equations (4) or (5), $(^{83}Kr_s/^{81}Kr)$ and P_{81}/P_{83}, are then directly obtained from the computed spallation components. A potential contribution to $^{80,82}Kr$ from neutron capture on Br has to be evaluated [13] (see also figure 1).

Present Applications and Prospects

The feasibility of ^{81}Kr-Kr dating of gram-size meteorite samples was demonstrated in several publications. Lunar rocks permit smaller sample sizes because of relatively high abundances of the target elements Sr, Y, Zr. Also, mineral separates thereof, corresponding to a few tens of milligrams, have been dated [5]. The ^{81}Kr-Kr method has provided reliable information on the exposure of lunar rocks to cosmic rays. It may provide important information in regard to the existence of groupings of exposure ages for certain meteorite classes and on the occurrence of multiple exposure stages of meteorites. The relatively long half-life of ^{81}Kr may prove to be useful in dating meteorites with rather long terrestrial ages such as the Antarctic meteorites. ^{81}Kr dating of polar ice, on the other hand, requires the measurement of extremely small ratios, say 1 atom of ^{81}Kr in about 10^{12} atoms of stable Kr [14]. A proposed solar neutrino experiment using the reaction $^{81}Br (\nu, e^-) ^{81}Kr$ may require detecting some 500 atoms of ^{81}Kr in 10^9 atoms of atmospheric Kr expected in the neutrino target tank [15].

Dating of Lunar Craters. ^{81}Kr-Kr exposure ages for lunar rocks were used to determine the ages of lunar impact craters. The basic assumptions in this method are that the studied rocks are true ejecta from the crater that is being dated, and that the rocks have been shielded from cosmic rays prior to the cratering event. Therefore, only when multiple samples which can be related to the same cratering event yield consistent ^{81}Kr-Kr ages, can the rock exposure age safely be equated with the crater age. Arvidson et al., [16] reviewed the information and concluded that the ages of three lunar craters are established: Cone Crater 26 My, North Ray Crater 50 My, South Ray Crater 2 My. More recent supplemental

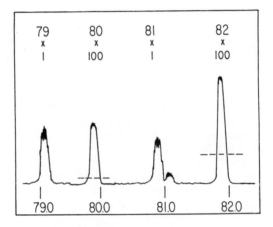

Figure 1.   A retraced portion of the (fading) 14-year-old Kr mass spectrum (79.0 to 82.0 amu) in the Macibini meteorite that established the presence of cosmic ray–produced ^{81}Kr (@ 80.917 amu).

The isobaric background peak is just resolved. Macibini also contains a trapped Kr component, which was evaluated according to the ^{86}Kr partitioning method explained in the text. The calculated amounts of trapped ^{80}Kr and ^{82}Kr are shown by the dashed lines, and the excesses are due to spallation Kr (1). (Note the scaling factors of 100 for ^{80}Kr and ^{82}Kr.)

information supports this.   In addition, downslope movements and emplacement times of lunar boulders can be determined.

Lunar and Planetary Surface Dynamics.   Two lunar rocks taken from opposite sides of a boulder during the Apollo 16 mission were studied by Drozd et al. [10], and the calculated ^{81}Kr-Kr ages were clearly incompatible with each other.   A detailed investigation of the spallation systematics, together with the exposure age record, revealed that the boulder was initially buried for about half of the time interval of exposure to cosmic rays, and was then transported to the surface and turned over to its present orientation at the time of the South Ray impact event.   Eugster et al. [17], in a detailed study of samples from a drive tube taken at the rim of Shorty Crater during the Apollo 17 mission, observed a dependence of inferred ^{81}Kr-Kr ages on the depth within the core from which samples have been recovered.   This information, again coupled with spallation systematics, reveals a multistep exposure of the studied sample, and, therefore, an irradiation prior to the impact that formed Shorty Crater.   Attempts by conventional methods to date the South Ray Crater event using returned soils from supposed ejecta layers failed, since the data indicated an age exceeding that of North Ray Crater, which is in conflict with the photogeological record.   Since ^{81}Kr-Kr ages obtained from a number of rocks from different stations revealed a 2 My exposure age, which is two orders of magnitude lower than that inferred from the soils, it can be established that the supposed South Ray soils actually contain very little material from this crater. These examples illustrate that discrepancies in the age record can be used to study surface dynamics, regolith evolution, and complex irradiation histories.   Such methods which have been developed in studies of lunar surface events and evolution ultimately can also be applied to returned samples from other solar system objects. In fact, this may provide an absolute time frame on which all photogeological chronology must be based.

Pre-Irradiation.   There has been considerable uncertainty regarding the absolute time scale for exposure ages of meteorites.   A number of different dating methods and varying production rates were proposed and used in the literature. ^{81}Kr-Kr dating, which so far was applied only to a limited number of chondritic meteorites, played an important role in recent attempts to resolve the problem.   Since this method is self-correcting for varying degrees of shielding, it assumes only a constant cosmic ray flux and single-stage irradiation.   Nishiizumi et al. [6] carried out a systematic calibration of the ^{21}Ne production rate by four independent methods and observed good agreement between three techniques, but disagreement with that obtained by the ^{26}Al method [18].   One possible explanation for this discrepancy suggested by Nishiizumi et al. [6] are excess concentrations of spallation Ne due to pre-irradiation either on the parent body or during early solar system history. Multi-stage irradiation histories are now documented for a number of meteorites, and their number appears to be increasing. A theoretical investigation by Wetherill [19] of the statistical expectation of multiple exposure, as a result of fragmentation, indicates that such effects should be rather common, if currently accepted models for the origin of meteorites are correct. It appears that meteorites will have to be studied for pre-irradiation effects on an individual basis and that ^{81}Kr-Kr studies, coupled with detailed analyses of spallation and neutron-capture effects, will play a key role.

Solar Wind and Solar Cosmic Rays. Recently, an "inverse" application of ^{81}Kr-Kr dating was used to correct measured "solar-type" isotopic Kr data in the Pesyanoe meteorite for a spallation component [20]. This work revealed discrepancies in the isotopic structure of Kr released at different temperatures. Studies of this type may provide more insight to Kr isotopic variations in the solar wind and in solar system reservoirs. Ongoing studies include the question of constancy of the solar cosmic ray flux on a time scale corresponding to the ^{81}Kr half-life. Spallation Kr is produced by solar cosmic rays in the surface layers of lunar rocks, and varying spallation ratios ^{81}Kr/^{83}Kr are expected to either document complex irradiation histories or possible changes in cosmic ray flux [9,7]. Studies of the latter record should provide information on the long-term solar flare activity.

———————

Over the years, the author has profited from collaborations and discussions with numerous colleagues. This work was in part supported by NASA Grant NGL 05-009-150.

Literature Cited

[1] Marti, Kurt, "Mass-spectrometric detection of cosmic-ray-produced ^{81}Kr in meteorites and the possibility of Kr-Kr dating", Phys. Rev. Lett., 1967, 18, (7), 264-266.

[2] Eastwood, T. A., Brown, F., and Crocker, I. H., "A krypton-81 half-life determination using a mass separator", Nucl. Phys., 1964, 58, 328.

[3] Voshage, Von H., "Bestrahlungsdter und Herkunft der Eisen-meteorite", Zeitschrift für Naturforschung, 1967, 22a, 477-506.

[4] Eugster, O., Eberhardt, P., and Geiss, J., "^{81}Kr in Meteorites and ^{81}Kr radiation ages", Earth Planet. Sci. Lett., 1967, 2, 77-82.

[5] Regnier, S., Hohenberg, C. M., Marti, K., and Reedy, R. C., "Predicted versus observed cosmic-ray-produced noble gases in lunar samples: Improved Kr production ratios", Proc. Lunar Planet. Sci. Conf., 10th, 1979, 1565-1586.

[6] Nishiizumi, K., Regnier, S., and Marti, K., "Cosmic ray exposure ages of chondrites, pre-irradiation and constancy of cosmic ray flux in the past, 1980, Earth Planet. Sci. Lett., 50, 156-170.

[7] Yaniv, A., Marti, K., and Reedy, R. C., "The solar cosmic-ray flux during the last two million years", Lunar and Planet. Sci. XI., Lunar and Planet. Inst., Houston, 1980, 1291-1293.

[8] Regnier, S., "Production of Kr isotopes by spallation on Y targets and implications for Kr-Kr dating", Lunar and Planet. Sci. X., Lunar and Planet. Inst., Houston, 1979, 1013-1015.

[9] Marti, K. and Lugmair, G. W., "^{81}Kr-Kr and K-^{40}Ar ages, cosmic-ray spallation products and neutron effects in lunar samples from Oceanus Procellarum", Proc. Second Lunar Sci. Conf., 1971, 2, 1591-1605.

[10] Drozd, R. J., Hohenberg, C. M., Morgan, C. J., and Ralston, C. E., "Cosmic-ray exposure history at the Apollo 16 and other lunar sites: lunar surface dynamics", Geochim Cosmochim. Acta, 1974, 38, 1625-1642.

[11] Lightner, B. D. and Marti, K., "Lunar trapped xenon", Proc. 5th Lunar Sci. Conf., Suppl. 5, Geochim Cosmochim. Acta, 1974, 2, 2023-2031.

[12] Lugmair, G. W. and Marti, K., "Sm-Nd-Pu timepieces in the Angra Dos Reis Meteorite", Earth Planet. Sci. Lett., 1977, 35, 273-284.

[13] Finkel, R. C., Kohl, C. P., Marti, K., and Martinek, G., "The Cosmic ray record in the San Juan Capistrano meteorite", Geochim. Cosmochim. Acta, 1978, 42, 241-250.

[14] Loosli, H. H. and Oeschger, H., "^{37}Ar and ^{81}Kr in the Atmosphere", Earth Planet. Sci. Lett., 1969, 7, 67.

[15] Hurst, G. S., Payne, M. G., Kramer, S. D. and Chen, C. H., "Counting the atoms", Physics Today, 1980, Sept. p. 24.

[16] Arvidson, R., Crozaz, G., Drozd, R. J., Hohenberg, C. M., and Morgan, C. J., "Cosmic ray exposure ages of features and events at the Apollo landing sites", 1975, The Moon, 13, 67-80.

[17] Eugster, O., Grögler, N., Eberhardt, P., and Geiss, J., "Double drive tube 74001/2: History of the black and orange glass; determination of a pre-exposure 3.7 AE ago by ^{136}Xe/^{235}U dating", Proc. Lunar Planet. Sci. Conf., 10th, 1979, 1351-1379.

[18] Herzog, G. F. and Anders, E., "Absolute scale for radiation ages of stony meteorites", Geochim. Cosmochim. Acta, 1971, 35, 605-611.

[19] Wetherill, G. W., Multiple Cosmic Ray Exposure Ages, paper presented at 43rd meeting of Meteoritical Society, La Jolla, Calif., 1980, Meteoritics (in press).

[20] Marti, K., "On Krypton isotopic abundances in the sun and in the solar wind", 1980 Proc. of the Conf. on the Ancient Sun (in press).

RECEIVED June 24, 1981.

Laser Microprobe Argon-39–Argon-40 Dating of Individual Mineral Grains

O. A. SCHAEFFER

State University of New York—Stony Brook, Department of Earth and Space
Sciences, Stony Brook, NY 11794

The use of a ruby laser to obtain K-Ar ages by
releasing the argon from selected sites, 50 μm in
diameter, representing about 0.2 μg of a mineral, is
described. The ages so obtained when used in conjunction with a conventional ^{39}Ar-^{40}Ar thermal release
study yield chronologically significant ages, often in
cases where the ^{39}Ar-^{40}Ar thermal release study is
disturbed. The method is illustrated by examples of
studies of meteorites, lunar rocks, and terrestrial
rocks.

The K-Ar isotopic system is useful in dating both terrestrial
and extraterrestrial samples. The strength of the method lies in
the widespread occurrence of K coupled with the fact that the
daughter, ^{40}Ar, is a rare gas and quite often absent in most mineral
systems. Because of the ability to detect extremely minute amounts
of argon, coupled with the long half-life of ^{40}K, the method gives
a wide age range of applicability from thousands to billions of
years. The K-Ar system has two problems: 1) the diffusive loss of
the daughter, ^{40}Ar, during metamorphic events, and 2) the presence
of extraneous ^{40}Ar not from the in situ decay of ^{40}K, especially in
the case of low-K minerals.
 The application of the ^{39}Ar-^{40}Ar thermal release method has
gone far to eliminate these difficulties. In this method, the
sample is irradiated by fast neutrons which produce ^{39}Ar from ^{39}K.
The age determination then reduces to an isotope ratio measurement
of the daughter to the parent: ^{40}Ar to ^{39}K, viz. ^{39}Ar [1][1].
By releasing the argon in temperature steps, and by determining
the ^{40}Ar/^{39}Ar ratio for each temperature step, it is possible to

[1]Figures in brackets indicate the literature references at the end
 of this paper.

0097-6156/82/0176-0139$05.00/0

obtain a so-called plateau age. That is, at the lower temperatures, the argon is released from mineral sites which have
suffered diffusive argon loss and show younger ages. After a high
enough temperature is reached, the argon is released only from
mineral sites which have not suffered diffusive argon losses, and
for subsequent temperatures the $^{40}Ar/^{39}Ar$ ratio remains constant
and is a measure of the age of the sample. In certain instances,
extraneous argon is released at the lower temperatures. This is
especially the case for implanted argon in lunar rocks or atmospheric argon in terrestrial samples. In other cases, however,
the temperature release pattern of $^{40}Ar/^{39}Ar$ values is not a
simple increase to a steady value but may show a drop at high
temperatures, show a saddle shape with intermediate temperature
releases giving lower ages, or give a more or less variable age
with temperature release. Samples which have suffered a metamorphic event or events often do not develop a well defined plateau
age. Non-ideal behavior is more evident in terrestrial samples
than lunar or meteorite samples; however, understanding of
non-ideal behavior is also important for the early chronology of
events on the lunar surface. In some cases, the complicated
release pattern is due to the existence of minerals whose ages
have been altered but which also release argon at all temperatures. In many cases, it is not possible to make a mineral
separation. The release of argon by a focused laser beam can be
of great assistance in interpreting the complicated release
patterns.

Laser Probe Mass Spectrometry

A focused laser beam of 0.2 joules per pulse is capable of
releasing rare gases from well defined 10-100 μm size spots on a
polished surface. As a result, it is possible to extend rare gas
mass spectrometry to the region of less than 1 μg samples. The
technique was first applied to the study of complex lunar samples
by George Megrue [2].
      If the samples are irradiated by fast neutrons, ^{39}Ar is pro-
duced from K by the reaction: ^{39}K (n, p) ^{39}Ar. The K can be
determined by neutron activation on the identical sample for which
the ^{40}Ar is determined, resulting in a K-Ar age.
      The principal advantages of the laser release over the
thermal $^{39}Ar-^{40}Ar$ age determination are: 1) The ability to
precisely define the minerals studied, i.e., it is possible to
avoid including material from grain boundaries or microscopic
inclusions within minerals grains. 2) The size of the sample
being investigated is approximately two orders of magnitude
smaller than that possible for a thermal release study. The lower
sample size is to a large extent possible because of the much
smaller values for the blanks. In the case of the thermal
release, the argon blank due to the heated materials surrounding
the sample is higher than the blank argon associated with a laser

study. The latter argon is only due to the mass spectrometer residual static vacuum. For a well baked system this can be <10^{-11} standard cm³/STP. 3) It is possible to obtain ages of various mineral grains from a single polished section. As a result, it is not necessary to make a mineral separation. 4) It is possible to obtain a laser microprobe age for different metamorphic grades of the same minerals. For example, due to collisional shock, tetonic folding, or a thermal event, quite often not all the mineral grains of a given mineral are affected. While it is relatively simple to distinguish the different metamorphic grades in a polished section, it is practically impossible to separate the individual mineral grains in a bulk crushed sample as is required for a thermal ^{39}Ar-^{40}Ar release study.

The principal disadvantages of the laser age method are: 1) The age obtained is a total K-Ar age, and as a result subject to the possibility of the loss of argon subsequent to the formation of the mineral grains: for example, as due to some metamorphic event. The error due to lost argon can be compensated for by heating the samples after neutron activation but before the laser study. In this way the argon is released from the low temperature sites, and does not contribute to the laser determined age. One essentially measures the integrated argon from the higher temperature sites. This is especially useful in the case of samples which show a well developed plateau with no complications at the higher temperatures. 2) For samples with low values of K, < 500 ppm, the laser method, because of its small sample size, becomes more uncertain than a thermal release study with a larger sample size. It is not possible to compensate in the laser probe method by multiple pulses above ~100 individual pulses which corresponds to evaporation of approximately 20 µg. It is conceivable that by completely redesigning a mass spectrometer with smaller volumes which is bakeable to higher temperatures to alleviate this problem. We have been able on occasion to work with samples as low as 100 ppm K but certainly not on a routine basis.

The use of a ruby laser coupled to a microscope which focuses on a polished surface is illustrated in figure 1. The samples are mounted inside a bakeable vacuum chamber with a pyrex viewing port attached to a standard vacuum chamber. The vacuum chamber is mounted onto the stage of a petrographic microscope, so that the sites which are degassed by the laser can be identified petrographically and mineralogically in reflected light. Oblique illumination is used in addition to measuring the thickness of the mineral crystals by focusing the microscope at different depths in order to make sure that the minerals selected are not penetrated by the laser pit. The laser, a Biolaser (Control Data Corp.) with an output of 0.2 joules in 150 µs, is mounted on the phototube of the microscope and focused by the microscope objective onto the specimen. The region melted by a single laser pulse is approximately a half-sphere whose diameter could be varied from ~100 µ to ~10 µ by the insertion of apertures into the laser beam.

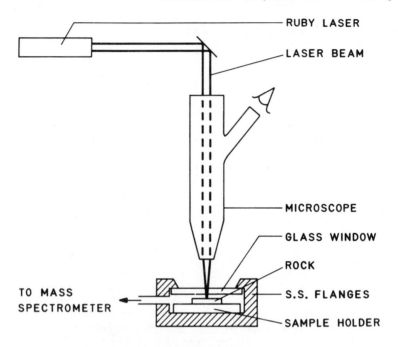

Figure 1.   Schematic laser probe.

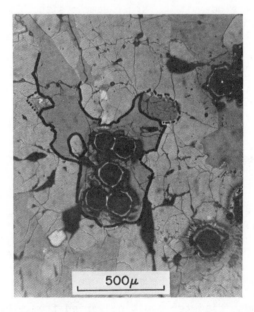

Figure 2.   Photomicrograph in reflected light of laser pits in a feldspar crystal of basalt 15607-4.   The solid line outlines the feldspar crystal, and the dotted line surrounds K-rich mesostatic material.

Figure 2 shows that it is possible to selectively degas single minerals in a polished section of a mineral assemblage by position-ing the laser pits within a single crystal. The amount of material melted by a laser pulse varies with the absorbency of tne material for the laser light, as can be seen in figure 3 where the pit on the left is in a colorless feldspar crystal whereas the pit on the right is in a brown pyroxene crystal. The amount of matter typi-cally melted by one laser pulse is ~0.2 µg for most silicates. As age measurements depend only on the ratios of the argon iso-topes, differences in the amounts of matter melted or degassed are of no importance. The specimen chamber is connected to the gas purification section of a Nuclide (6-60SGA) static mass spec-trometer. In a typical run, gas released by 1-100 laser pulses (depending on the K-content of the sample) is collected on activated charcoal at liquid nitrogen temperature while H_2 is gettered by a small Ti-Zr getter. The temperature of the charcoal is then increased to that of a dry ice-acetone mixture. At this temperature, Ar is released from the charcoal but most impurities are kept in the charcoal. The gas is then admitted to the mass spectrometer and the Ar isotopes are measured. The levels of the backgrounds depend somewhat on the amounts of gases released in previous experiments; therefore, a blank run without actual laser-ing but identical conditions is made before and after each experiment. The errors in the blanks are estimated from the daily reproducibility of the blanks. They are well within 10 percent except for mass 39 where the errors are governed mainly by statis-tics and are within 25 percent. The reproducibility of the peaks during a sample measurement is better than 2 percent, so that the total error is determined mainly by the amount of background correction necessary or, in other words, by the total amount of gas released. Typical blank levels for ^{40}Ar, ^{39}Ar, ^{38}Ar, ^{37}Ar, and ^{36}Ar are in units of 10^{-12} cm^3/STP respectively 10, 0.2, 0.5, 1.0, and 1.5. With these blank levels, it is possible to obtain a precise K-Ar age for a mineral which contains 1 percent K with one laser pulse.

We have an estimate of the spatial resolution of the laser pit--that is, the determination of the amount of material degassed compared to the volume of the material melted and evaporated. In the Allende meteorite, there are large melilite grains which contain smaller spinel inclusions. The melilite contains 40 percent CaO while the spinel contains 0.04 to 0.30 percent CaO as shown by electron microprobe analyses. The CaO in the melilite is thus at least 100 times greater than the CaO in the spinel. The grain size of the spinel is variable from somewhat lower than 10 µm to as much as 100 µm. We made individual analyses of the ^{37}Ar (produced from Ar by Ca (n, α) ^{37}Ar) released from 50 grains of diameter about 50 µm with laser pits of 30 µm. None of the pits touched the edge of the grains but in all cases came less than 10 µm of the edge of the grains. The average ratio as determined from the laser gas release of CaO (melilite)/CaO (spinel) was 50. It is clear that less than 1 percent of the

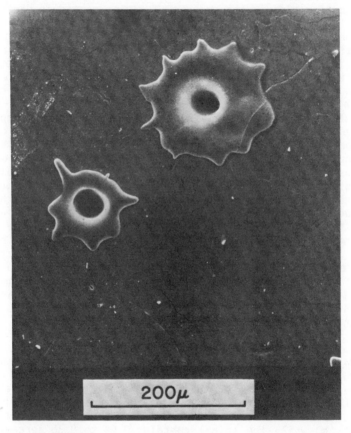

Figure 3. Scanning electron microscope picture of laser pits in different materials: (left) feldspar; (right) pyroxene. The difference in the amounts melted is caused by the different absorbencies of the minerals.

argon released came from the melilite. In other words, less than 1 percent of the gas released from a laser pit lies outside 10 microns of the physical size of the pit.

Surface Rare Gases

A study of the surface helium, neon, and argon in lunar rocks illustrates the sensitivity of the laser microprobe [3]. Lunar rock 12054 was collected from the surface of the moon with known orientation. This rock contains a glass coating which covers the face including a crack. As a result, part of the surface in the crack is never exposed directly to the sun whereas other parts are directly exposed. We made a traverse across the surface analyzing the gas from individual laser pits. Four types of sites were analyzed in this study.

1. Along two parallel <u>traverses</u> - Thirty-five sites were analyzed along a line from a well-shaded to a well-exposed portion of the sample.
2. Essentially <u>shaded surfaces</u> - Twenty laser pulses were applied to each of six areas in the dominantly shaded region. In addition, a freshly broken surface was analyzed in this way.
3. <u>Sunlit</u> surface - A total of 39 laser pulses were applied essentially at random to the sunlit host glass surface.
4. Microcrater <u>spall zones</u> - Ten laser pulses were applied to sites where surface material had been spalled away by nearby meteoroid impacts.

All laser spots were completely contained in the dark glass coating the sample.

It was found that the shaded side contains less ^{4}He, ^{20}Ne, and ^{36}Ar than the sunlit side, while the ^{40}Ar is generally the same on both the sunlit and shaded sides. The surface distribution of the rare gases observed can be understood as the ^{4}He, ^{20}Ne, and ^{36}Ar are primarily directly implanted from the solar wind and come from the direction of the sun while the ^{40}Ar represents argon degassing from the moon which is coming from an ambient atmosphere. The ^{40}Ar is probably implanted by some complicated process on the lunar surface and is not directly orientated toward the sun. The study on this essentially small scale has been possible because of the sensitivity of the laser method for analysis. The entire analysis could be made in the space of only a few days while the comparable study by conventional mass spectrometry would require an almost impossible painstaking removal and handling of sub-mg chips.

Lunar Ages

A laser probe study of Apollo 17 basalts [4] illustrates the use of the laser dating to interpret temperature release patterns. The laser $^{39}Ar-^{40}Ar$ method has been used to study three Apollo 17 basalts: 70215, a basalt with a normal well-behaved $^{39}Ar-^{40}Ar$ release pattern; 70017, a basalt with a disturbed $^{39}Ar-^{40}Ar$ release pattern, showing a limited intermediate age followed by a broad low age region; and 75035, a basalt with a pattern similar to 70017, followed by a high temperature maximum age. The laser study showed that all the mineral systems in 70215 had small and uniform temperature losses while for basalts 70017 and 75035 the laser ^{39}Ar-^{40}Ar ages were apparently mainly controlled by the minerals containing mesostasis inclusions, with the suggestion that the drop in ages observed by the conventional ^{39}Ar-^{40}Ar method was not only due to recoil of the ^{39}Ar during neutron irradiation but also due to gas loss from some minerals. From our measurements it appears that the plagioclases - at least for the intermediate temperature range between about 600 °C and 1000 °C - are the best minerals to use for a reliable age.

From the laser study of the Apollo 17 basalts, it is clear that by heating a sample after neutron irradiation but previous to the laser gas release it is possible to obtain a reliable and precise age for cases with good ^{39}Ar-^{40}Ar plateaus (for example, 70215). In addition, for samples with distributed plateaus (for example, 70017 and 75035), the laser study can elucidate the non-ideal ^{39}Ar-^{40}Ar release patterns. For samples as illustrated by the lunar basalts, the combination of a thermal release ^{39}Ar-^{40}Ar age and a laser study can yield reliable ages where such is not the case by the thermal release method alone.

As an illustration of the application of laser spot melting to a complicated lunar breccia, we studied breccia 73215. This rock has a fine-grained aphanitic matrix and contains numerous clasts of different types. Laser ^{39}Ar-^{40}Ar studies of a clast of ANT-suite anorthositic gabbro from breccia 73215 [5] show that the plagioclase in the rock has a wide range of K-Ar ages. These ages are correlated with the size and texture of the grains and the position of the material being studied within the grains. Material showing the oldest age (>4.11 G.y.) is at the centers of the largest grains, material showing the youngest age (3.81 - 3.88 G.y.) is near grain margins, and material showing intermediate age (3.99 - 4.05 G.y.) is in intermediate zones in the largest grains and at the centers of intermediate-sized grains. These results are confirmed by previous studies of another clast of similar anorthositic gabbro from 73215. The age pattern is interpreted as the result of partial outgassing of the clasts when they were incorporated in the 73215 breccia. The combined data for the two anorthositic gabbros set a lower limit of 4.26 G.y. on the date of an episode of high-temperature melting/recrystallization that affected the parent rocks of the clasts. Rb-Sr data (Compston et al., 1977) provide an upper limit of 4.45 G.y. on the date of this high-temperature event.

The laser ^{39}Ar-^{40}Ar results for a black aphanite clast from 73215 demonstrate that this rock is cogenetic with the aphanite that forms the matrix of the breccia. Ages determined for felsic glass fragments in the two types of aphanite, and for groundmass in the black aphanite, are identical within error with each other and with the data of the 73215 breccia-forming event, ∿3.87 G.y.

The time of crystallization of a lunar "granite" has also been determined by the laser method. This "granite" is the K- and Si-rich felsite that forms clasts in 73215, 73255, and Boulder 1 at Station 2. The laser results set a lower limit of 4.00 G.y. on crystallization of the parent body of this felsite, and the Rb-Sr data set an upper limit of ∿ 4.05 G.y. [6].

Meteorite Ages

The application of the laser probe to meteorite chronology is illustrated by a study of Ca-Al-rich inclusions from the Allende meteorite [7]. This study was able to show that the K in the inclusions studied mainly concentrated in veins and rims with very little, if any, K in the major minerals. The limit obtained is something of the order of 10 ppm. On the other hand, the major minerals do contain appreciable ^{40}Ar. Individual chondrules and the matrix were also studied in the Allende meteorite from places adjacent to the Ca-Al-rich inclusions. For these samples the ages varied from 3.3 to 4.4 G.y. There appears to be evidence that the Allende meteorite has been subjected to numerous metamorphic events, presumably of a collisional origin.

Terrestrial Ages

We have shown that for terrestrial samples with ideal plateaus it is possible by preheating the samples to obtain reliable and precise laser ages [8]. The study was made on granites and blue schists from the Alpine area for samples in the age 40 - 2000 M.y. For a Precambrian granite from the Ivory Coast, under the microscope, a polished section is seen to contain two kinds of biotites: dark biotites with an average size of 50 μm and light coarse biotites with a maximum length of 200 μm. For the analysis by ^{40}Ar-^{39}Ar stepwise heating, these two kinds of biotites were not separated due to a very similar density and the same magnetic susceptibility. The plateau age, resulting for the mixture of these two components was 1850 M.y. ± 80 M.y. The laser results show clearly two different ages: the dark biotites give ages of 2025 to 2100 ± 100 M.y. and the light ones 1550 and 1600 ± 100 M.y. The difference can be understood as related to a late thermal resetting event affecting only one kind of the biotites.

References

[1] Turner, G., Thermal histories of meteorities by the ^{39}Ar-^{40}Ar method. In: Meteorite Research, ed. P. M. Millman, Springer-Verlag, New York, 1969, p. 407-417.

[2] Megrue, G. H., Distribution and Origin of Helium, Neon, and
    Argon Isotopes in Apollo 12 Samples by In Situ Analysis with
    a Laser-Probe Mass Spectrometer, J. Geophys. Res. 76, 4956
    (1971).
[3] Hartung, J. B., Plieninger, T., Müller, H. W., and Schaeffer,
    O. A., Helium, Neon, and Argon on Sunlit and Shaded Surfaces
    of Lunar Rock 12054, Proc. Lunar Sci. Conf. 8th, 1977,
    p. 865.
[4] Schaeffer, O. A., Müller, H. W., and Grove, T. L., Laser
    ^{39}Ar-^{40}Ar Study of Apollo 17 Basalts, Proc. Lunar Sci. Conf.
    8th, 1977, p. 1489.
[5] Eichhorn, G., James, O. B., Schaeffer, O. A., and Müller,
    H. W., Laser ^{39}Ar-^{40}Ar Dating of Two Clasts From Consortium
    Breccia 73215, Proc. Lunar Planet. Conf. 9th, 1978, p. 855.
[6] Compston, W., Foster, J. J., and Gray, C. M., Rb-Sr Syste-
    matics in Clasts and Aphanites from Consortium Breccia
    73215, Proc. Lunar Sci. Conf. 8th, 1977, p. 2525.
[7] Herzog, G. F., Bence, A. E., Bender, J., Eichhorn, G.,
    Maluski, H., and Schaeffer, O. A., ^{39}Ar/^{40}Ar Systematics of
    Allende Inclusions, Proc. Lunar Planet, Sci. Conf. 11th,
    1980, p. 959.
[8] Maluski, M. and Schaeffer, O. A., manuscript in preparation.

RECEIVED July 28, 1981.

Resonance Ionization Spectroscopy for Low-Level Counting

G. S. HURST, S. D. KRAMER, and B. E. LEHMANN[1]

Oak Ridge National Laboratory, Chemical Physics Section, Health & Safety Research Division, Oak Ridge, TN 37830

Resonance Ionization Spectroscopy (RIS) can, in principle, be used to remove one electron from each atom (of a given type) in a pulsed laser beam. Thus, with time and space resolution it would be possible to detect each daughter atom emitted from the nuclear decay of a parent atom. However, in many cases the daughter atom will be thermalized into a +1 charge state. Such atoms must be neutralized before detection with the RIS process. Presented here is research in progress which is aimed at the detection of lithium in the electron capture decay of ^7Be (for solar neutrino measurements) and the detection of potassium in the beta decay of ^{39}Ar (for oceanography research). The technique could also be used in many other applications of similar nature, such as geochronology.

A variety of techniques for the detection of extremely low levels of stable or radioactive substance at the one-atom level are being developed at the Oak Ridge National Laboratory (ORNL) [1][2]. Such methods are in great demand for research on the sun, the ocean, and the earth. To meet these demands for ultrasensitive detection, ORNL is now engaged in an effort that represents substantial involvement with other laboratories. Some of these

---

[1] Postdoctoral Research Appointment through Western Kentucky University; supported in part by the Swiss National Foundation for Scientific Research.

[2] Figures in brackets refer to the literature references at the end of this paper.

0097-6156/82/0176-0149$05.00/0

are (a) the detection of solar neutrinos (with Brookhaven National Laboratory), (b) the study of ocean water circulation (with the Scripps Institution of Oceanography), and (c) the detection of plutonium (with EG&G, Santa Barbara Operations). Resonance ionization spectroscopy (RIS) [2] has already been used to detect single stable atoms and individual daughter atoms in time coincidence with the decay of parent atoms. At the present time we are attempting to detect single daughter atoms even when the decay of the parent leaves the daughter with a deficiency of one or more electrons. If this can be accomplished, we believe it possible to develop an ultralow-level counter having backgrounds on the order of one count per week. Another objective of the ORNL effort is the counting of daughter atoms which accumulate over a long period of time as a consequence of slowly decaying parent atoms, see figure 1. These counting methods would have application to solar neutrino flux measurements, to tracer studies of oceanic circulation, to geology, to the age of the universe, and to a variety of weak interaction problems in nuclear physics.

Resonance Ionization Spectroscopy

Resonance ionization spectroscopy is a photophysical process in which one electron can be removed from each of the atoms of a selected type. Since the saturated RIS process can be carried out with a pulsed laser beam, the method has both time and space resolution along with excellent (spectroscopic) selectivity. In a recent article [2] we showed, for example, that all of the elements except helium, neon, argon, and fluorine can be detected with the RIS technique. However, with commercial lasers, improved in the last year, argon and fluorine can be added to the RIS periodic table (see figure 2).

Detection of Individual Daughter Atoms

The first demonstration [3] that single atoms could be detected involved the use of RIS and a proportional counter to detect one atom of stable cesium. Later it was shown [4] that individual daughter atoms could be detected in time coincidence with the decay of parent atoms. Thus, single cesium atoms resulting from spontaneous fission of ^{252}Cf nuclei were detected by the RIS method. The ^{252}Cf nuclei were ion implanted in a nickel foil mounted in front of an apertured surface-barrier detector. A signal from this detector was used to indicate that a cesium fragment had been injected into a sample region and stopped between two plates of an ionization chamber. The electrons generated by the energy dissipation of the cesium fission fragment in the gas mixture of the sample region were drifted and collected on one plate of the ionization chamber and thus eliminated from the detector region. Shortly after collection of the fission recoil ionization which contained about one million electrons, a tuned

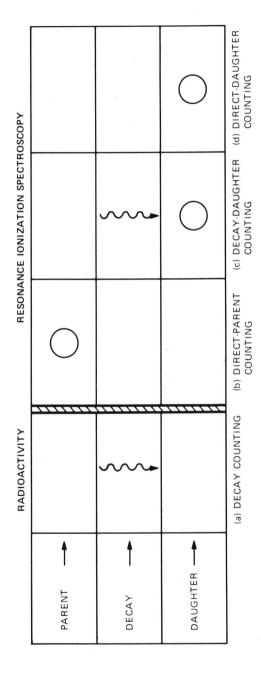

Figure 1. A perspective on atom counting.

Traditional decay counting is compared with new counting methods made possible by the tunable pulsed laser. (a) Instrumentation for decay counting includes ionization chambers, proportional counters, Geiger–Mueller counters, scintillation counters, particle detectors, and the like. Counting facilities with low backgrounds are known as low-level counters. (b) The use of the RIS laser technique makes possible the direct counting of individual parent atoms whether radioactive or stable. Instrumentation involves one or more pulsed lasers and a detector of single electrons or positive ions. When a buffer gas is needed, proportional counters or Geiger–Muller counters are used as ionization detectors. When a vacuum is desired, electron multipliers are used to detect single electrons or ions. Mass spectrometers can also be included to reject backgrounds or to provide isotopic selectivity. (c) RIS techniques using pulsed lasers make it possible to detect daughter atoms in time coincidence with the decay of a parent atom. Such coincidence techniques, especially when used with position-sensitive proportional counters could further reduce backgrounds in low-level counting facilities. (d) Another RIS technique under development at ORNL involves the accumulation of daughter atoms for subsequent counting. Instrumentation requirements are an extension of those for direct-parent counting.

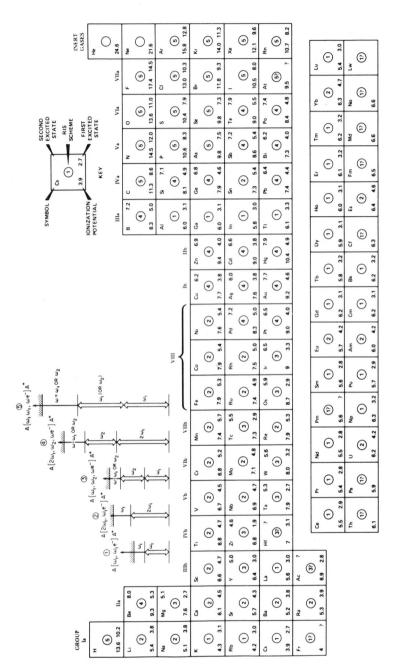

Figure 2. RIS schemes for the periodic table. The inset defines RIS of five different types. For more information on the RIS process, see Ref. 2. The schemes for F and Ar have been added since publication of that reference.

laser was triggered to ionize the neutralized cesium atom. The one electron from this ionization of the one cesium atom was drifted through the opposite field plate into a proportional counter volume where it was detected. Nearly all of the cesium atoms were neutralized and detected.

Solar Neutrino Detection

The major U.S. effort on solar neutrino experiments is concentrated at Brookhaven National Laboratory under the direction of Ray Davis [5]. This group relies primarily on radiochemical methods to detect a few atoms of a certain kind that are generated by prolonged exposure of very large samples (e.g., 400,000 liters) to the sun. A famous experiment by the Davis group involved filling the tank with C_2Cl_4 so that neutrinos could induce ^{37}Ar via $^{37}C(\nu,e^-)^{37}Ar$. Recovery of argon from the Homestake (South Dakota) tank and subsequent counting for ^{37}Ar at BNL showed that the neutrino flux was lower than that predicted by the best solar models by a factor of three. To check this result, Davis and John Bahcall [6] suggested other radiochemical detection schemes — in particular, $^7Li(\nu,e^-)^7Be$. However, the detection of ^7Be is nearly impossible with the standard radiochemical approach since the decay of ^7Be leads only to a low-energy Auger electron that "looks like" noise in a proportional counter. With RIS we can also detect the lithium daughter atom in time coincidence with the decay of the ^7Be parent. We expect that backgrounds could be reduced about 1000 times, and thus a few (e.g., 10-100) atoms of ^7Be can be counted as they decay back to lithium with the 53-day half-life. In figure 3 we show a concept for ultralow-level counting in general [7]. For an analysis of the background levels to be expected in such an apparatus, see reference [8].

Ocean Water Circulation

Another possible application of a RIS-based, low-level counter is for the study of ocean water circulation. We are working with Prof. Harmon Craig (Scripps Institution of Oceanography at LaJolla, California) and R. D. Willis (currently on assignment from Scripps to ORNL) to develop an ^{39}Ar detector for oceanographic research. Leading oceanographers believe that ^{39}Ar could be an excellent dating isotope for ocean water. The tracer is generated by cosmic rays and comes into equilibrium with surface water. At the greater depths in the ocean the water has been out of contact with the atmosphere on the order of 1000 years; thus, the ^{39}Ar half-life of 270 years is nearly ideal. Again, ^{39}Ar is difficult to count at low concentration because its radiation signature is a β^- interaction with the detector. The use of the decay counting technique [9] for radiochemical analysis of ^{39}Ar requires taking 2000-liter samples of water at various depths in the ocean. But with ultralow-level counting (ULLC) we should be

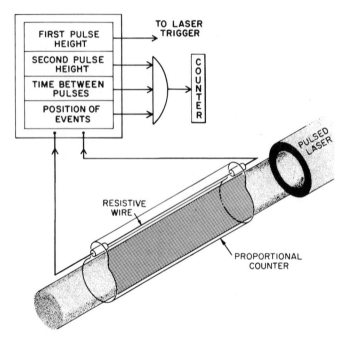

Figure 3.   Concept for the ultralow-level counter. Low-level counting facilities could be improved by using pulsed lasers to ionize daughter atoms selectively (when neutral) by using a RIS scheme. Shown here is the electronics logic that could be used to further reduce background; for more details, see Ref. 8.

able to reduce the required sample to one Gerard bottle (270 liters).

Charge Neutralization and Resonance Ionization Spectroscopy with Amplification (RISA)

The success of the ULLC facility proposed above for [7]Be and [39]Ar counting depends on a deeper scientific question — namely, are daughter atoms electrically neutral? Consideration of this question has led to development of a new method of RIS which we have called resonance ionization spectroscopy with amplification (RISA) [10]. The method uses tunable laser energy to repeatedly ionize neutral daughter atoms, the ions thus created being repeatedly neutralized in a gaseous environment. This cyclic process can be repeated many times during a laser pulse, so that more than one free electron can be generated from just one of the spectroscopically selected sample atoms. If a daughter atom is born as a positive ion, neutralization would occur automatically as a part of a RISA cycle. Four possible schemes for accomplishing RISA are shown in figure 4, while in figure 5 we show details on the ion molecule reaction scheme by using NO as an example. In figure 6 we show a laser experiment at ORNL which has been set up to demonstrate the scheme of figure 5.

Conclusion

The emphasis in this article has been the RIS detection of daughter atoms in time coincidence with the decay of parent atoms. Returning briefly to the perspective of figure 1, it is worthwhile to emphasize that we have discussed only one type of application of RIS. The order given in figure 1 is chronological. Considerable work has already been accomplished on the "direct-parent counting" applications. These include the following studies: (a) cross section for dissociation of alkali-halide molecules [11], (b) diffusion of free atoms [2], (c) chemical reaction of free atoms [2], (d) fluctuation phenomena [2] and tests of statistical mechanics [2], (e) semiconductor impurities [12], and (f) selective sources for mass spectrometers [13]. We are now concentrating on the decay-daughter atom application as we discussed here. Finally, through a joint effort involving ORNL with Harmon Craig and R. D. Willis of the Scripps Institution of Oceanography, the direct-daughter counting is receiving attention as another means of counting [39]Ar for the oceanographic application.

The capability of neutralizing daughter ions emitted by a parent atom in nuclear decay would result in practical realization of ULLC for solar neutrino detection, weak interaction physics, cosmochronology, geophysics, environmental research, and other important applications.

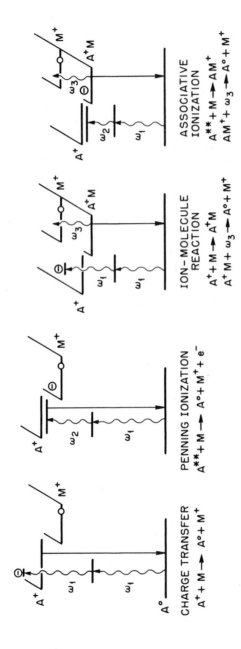

Figure 4. A family of four schemes has been considered for the RISA process.

The simplest of these, the charge transfer process, requires having molecules with ionization potentials less than that of the atom—a difficult condition when detecting alkali daughter atoms. Thus, the ion-molecule reaction may be the most useful since lasers producing photons at frequency ω_3 can add energy to an ion–molecule complex. Penning ionization and associative ionization are variations that take advantage of highly excited (Rydberg) states so that ionization can occur by collisions.

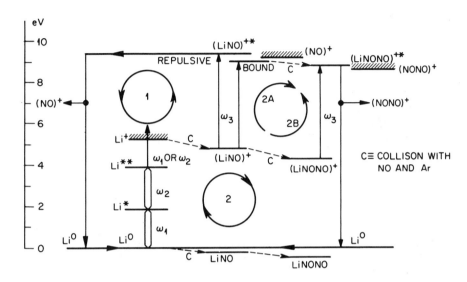

Figure 5. Example of the ion–molecule RISA cycle using NO. Experiments using this process are in progress at ORNL.

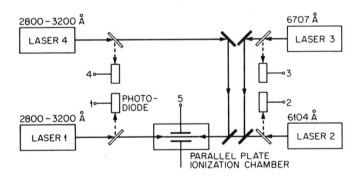

Figure 6. A laser scheme for the NO experiment suggested in Figure 5.

Laser 1 prepares a Li atom (by photodissociation of LiI—see Ref. 11 for a recent measurement of the cross section), while Lasers 2 and 3 ionize Li according to $Li(\omega_1,\omega_2,\omega e^-)Li^+$. Laser 4 is intended as a means of photodissociating $(LiNO)^+$ to complete the RISA cycle.

Acknowledgment

The research was sponsored by the Office of Health and Environmental Research, U.S. Department of Energy under Contract W-7405-eng-26 with the Union Carbide Corporation.

Literature Cited

[1] Health and Safety Research Division Progress Report for the Period May 1, 1978 – September 30, 1979, Oak Ridge National Laboratory Report ORNL-5595, pp. 29-35.
[2] Hurst, G. S., Payne, M. G., Kramer, S. D., and Young, J. P., Rev. Mod. Phys., 1979, 51, 767.
[3] Hurst, G. S., Nayfeh, M. H., and Young, J. P., Appl. Phys. Lett., 1977, 30, 229.
[4] Kramer, S. D., Bemis, C. E., Jr., Young, J. P., and Hurst, G. S., Optics Lett., 1978, 3, 16.
[5] Davis, R., Jr., Harmer, D. S., and Hoffman, K. C., Phys. Rev. Lett., 1968, 20, 1205.
[6] Bahcall, J. N., Rev. Mod. Phys., 1978, 50, 881.
[7] Hurst, G. S., Kramer, S. D., Payne, M. G., and Young, J. P., IEEE Trans. Nucl. Sci., 1979, NS-26, 133.
[8] Kramer, S. D., Hurst, G. S., Young, J. P., Payne, M. G., Kopp, M. K., Callcott, T. A., Arakawa, E. T., and Beekman, D. W., Radiocarbon, 1980, 22, 428.
[9] Oeschger, H. and Wahlen, M., Ann. Rev. Nucl. Sci., 1975, 25, 423.
[10] Hurst, G. S., Payne, M. G., Kramer, S. D., and Young, J. P., Chem. Phys. Lett., 1979, 63, 1.
[11] Lehmann, B. E., Kramer, S. D., Allman, S. L., Hurst, G. S., and Payne, M. G., Chem. Phys. Lett., 1980, 71, 91.
[12] Private communication with Santos Mayo, National Bureau of Standards. See also reference [13].
[13] Beekman, D. W., Callcott, T. A., Arakawa, E. T., Kramer, S. D., Hurst, G. S., and Nussbaum, E., Int. J. Mass Spectrosc. and Ion Phys., 1980, 34, 89.

RECEIVED April 27, 1981.

Counters, Accelerators, and Chemistry

L. A. CURRIE and G. A. KLOUDA

National Bureau of Standards, Center for Analytical Chemistry, Washington, D.C. 20234

Important advances in nuclear dating techniques and microchemical characterization are making major contributions to our ability to extract reliable information from archaeological and environmental samples. The capability of measuring very much smaller and older samples promises major progress in our understanding of both anthropogenic and natural processes because of the tremendous increase in information content which comes about through multidimensional chemical/physical sample characterization and the isotopic analysis of individual chemical fractions. Following a brief discussion of the comparative performance of small sample liquid scintillation counting, gas proportional counting and direct atom (accelerator) counting (with respect to precision, sample size and destruction, and non-Poisson error components), we shall illustrate the critical role that serial and parallel chemical information has played in the modeling and interpretation of environmental radiocarbon data. In our investigation of sources of carbonaceous gases and particles, serial data have included selective sampling (< 10 mg-C samples) followed by the determination of ^{14}C in specific classes of compounds and particle size fractions; parallel (multidimensional) data have included isotopic (^{13}C), elementary and organic composition.

During the past several years exciting advances have taken place in radioactive dating techniques--advances which have made it possible to determine unusually small isotope ratios and to determine the radioisotopic composition of individual chemical fractions and particle size fractions of extremely small samples.

The increased power to extract information from such samples, through combined microchemical and isotopic analysis, is enormous; and the ability to work with tiny samples allows us to address numerous problems which were previously beyond our reach. These problems extend beyond simple dating. They include, as noted in the Keynote Address [1][2], understanding and characterizing the environmental system. The importance of this has been manifest in recent societal concerns relating to the coupling of energy choices to environmental and climatic consequences. One phase of this question has been the focus of the recent work in our laboratory. That is, we have been attempting to learn more about the impact of human activities on carbonaceous gases and particles in the atmosphere through the use of receptor modeling together with chemical, physical, and isotopic (^{13}C, ^{14}C) characterization of selected samples. Following a comparative review of the characteristics of three advanced techniques for the measurement of radiocarbon in small samples, we shall summarize some of our recent observations related to sources of carbonaceous pollutants.

THE REVOLUTION IN RADIOCARBON MEASUREMENT: COMPARISON OF METHODS

The conventional approach to radiocarbon dating utilizes large samples (5 g to 50 g-carbon) which are measured in gas proportional or liquid scintillation counters, capable of yielding excellent precision (0.2% to 0.5% RSD) [2]. Recent developments in both of these low-level counting techniques (gas, liquid) have made it feasible to perform reliable measurements of natural radiocarbon in small (10 mg to 100 mg-carbon) samples [3]. Still smaller samples (< 100 µg-carbon) may now be assayed by direct atom counting with a cyclotron [4] or tandem accelerator [5]. At this point in time it appears that the three small sample techniques are somewhat complementary. Direct atom counting is in a much earlier stage of development--in terms of stability, sample preparation, sources of contamination--than the other two, but it is evolving rapidly. Besides significant differences in minimum sample size, the alternative techniques differ greatly in instrument availability and capital expenditure. (Any attempt at economic comparison in this paper would be presumptuous. Although the capital cost of a suitable tandem facility exceeds that of the other two by more than a factor of ten, labor costs, shared uses and "market" forces may tend to equalize charges per sample [6].)
Principal characteristics of small sample liquid scintillation counting (lsc), gas proportional low-level counting (llc) and atom counting by accelerator mass spectrometry (AMS) are summarized in Table 1, and systems we have used are shown in figure 1. The most important differences (apart from cost and availability)

---

[2]Figures in brackets indicate the literature references at the end of this paper.

c

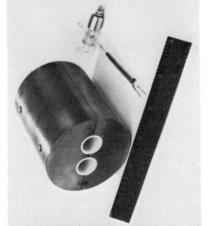

b

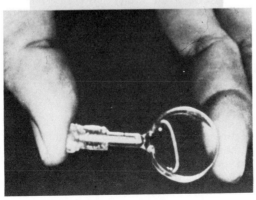

a

Figure 1. Small sample counting apparatus.

(a) Liquid scintillation vial containing 1 mL of benzene; (b) quartz 5-mL gas proportional counter and intercounter Cu shield; (c) main acceleration tank of 3-MV Tandem van de Graaff used in atom-counting experiment of 40 µg C shown loaded in 0.5-mm cup in 12-sample wheel.(See text for references.)

Table 1. Alternative Methods for Small Samples[a] (radiocarbon).

		lsc	llc	AMS
Samples	non. destr.	(C-compound)	element	destruction
	ideal form	C_6H_6	CO_2	$C[CO_2]$
Mass-C		> 100 mg	> 10 mg	> 10 µg
Counting		serial	parallel ($\stackrel{\sim}{>}$ 10 counters)	serial
Time[b]		1.8 hours	0.9 days/10	3^+ minutes
Modern/Background		∿3.	∿2.	∿200.
Stability	Effic.	∿0.5%	∿0.5%	few %
	BG	<2.5%	∿2.%	?
Contamination	Activity	^3H	Rn	^{14}C
	Chemical	quenching	electronegative	H-, O-, Li_2

[a] Column headings represent small sample liquid scintillation counting (lsc), low-level gas counting (llc), and accelerator mass spectrometry (AMS).

[b] Assumes modern carbon, Poisson statistics (10% RSD for lsc and llc, 4% for AMS), a $^{12}C^-$ ion current of 0.5 µA, and (Effic., BG) = (90%, 0.42 cpm)$_{lsc}$, (90%, 0.07 cpm)$_{llc}$, and (0.1%, 1. cpm)$_{AMS}$.

follow. First, it is possible in certain cases to preserve the chemical identity of a sample when using liquid scintillation counting. Non-destructive assay may be possible by reversible addition of (or solution in) a scintillant. Both liquid scintillation and gas proportional counting preserve the carbon sample (for subsequent reanalysis or archiving), whereas AMS consumes a portion of the sample. Sample preparation is most straightforward (combustion only, to CO_2) for llc and cyclotron-AMS, but the optimal chemical forms for the other techniques require reduction (to benzene, for lsc; to elemental carbon, for tandem-AMS). Regarding sample size: it was shown in reference 3 that the choice passes from lsc to llc for samples smaller than about 100 mg; when less than 5 mg to 10 mg-carbon is available, adequate (Poisson) precision may be obtained only by AMS. (It would seem that there is an inverse correlation between sample size and detector size-see figure 1.) Counting times, for 10 percent Poisson precision are seen to range from minutes to about a day. Thus, at this level of precision and for the sample sizes indicated, counting time is not out of proportion to other parts of the measurement process (sample preparation, accelerator tuning, etc.). Time compression exists with 11c, however, in that several counters are operated in parallel.

Because of the relatively low background, AMS has an over-whelming advantage for very old samples. If machine stability can be improved, it holds great promise also for high precision measurements of small samples. (The best precision to be demon-strated for 11c, with ~10 mg-carbon, is about one percent-- obtained by means of three months of counting [7].) As implied above, ultimate precision depends on stability of the overall system efficiency and background. Efforts at reducing the effects of the latter are underway (for 11c and AMS) by means of isotopic enrichment (vide infra). Finally, as indicated in the bottom line of Table 1, peculiar chemical and radioactivity contamination effects are frequently observed with each of the three modes of measurement.

Linked Systems: Isotopic Enrichment. The power of advanced measurement techniques can often be extended by linking them with other techniques. Such is the case for the detection of trace quantities of natural radionuclides and isotope enrichment. We have already found this extremely valuable for gas proportional counting of ^{37}Ar and accelerator atom counting of ^{14}C [8,9]. The first nuclide was enriched by means of thermal diffusion (Ar); the second, by electromagnetic isotope separation (CO^+). The major benefit of isotopic enrichment, of course, is the increase in signal-to-noise, where the "noise" in this case consists of both the Poisson component of the background error and more importantly the instability of the background. This latter point, which will be discussed below, cannot be overemphasized. Signals which are significantly smaller than the background (or blank) simply cannot be trusted in the absence of an inordinate amount of testing.

Practicable isotopic enrichment has the following prerequi-sites: adequately short time for the enrichment process, acceptable asymptotic enrichment factor, and adequate accuracy for the estimation of the enrichment factor. (When total activity, rather than specific activity, is limiting, one must also pay attention to losses during enrichment.) For the argon and carbon enrichments referred to above, enrichment factors of about 100 and 500 were obtained within a week and a few hours, respectively; and enrichment factors were deduced from direct observations of adjacent, stable isotopes. The ^{14}C enrichment process provided extra dividends for AMS measurement: the sample was implanted in an ideal form for the accelerator ion source, and it was spatially localized (depth) which gave added signal-to-noise enhancement.

Ironically, our current plans call for the reverse linkage of the above enrichment procedures. That is, we shall use an electromagnetic isotope separator to enrich argon isotopes for a mass spectrometry experiment, and we shall enrich radiocarbon via thermal diffusion for improved mini-gas proportional counting.

Some Precautions and Pitfalls Associated with Non-Counting Errors

The average efficiencies, backgrounds, and Poisson statistics determine the <u>best</u> results which may be obtained with the alternative measurement techniques. The actual reliability is limited also by non-Poisson variability (instability), contamination, spurious signals, and (hardware, software, sample preparation) blunders. Detection of such additional sources of error is difficult if they are random in origin, and close to impossible if they are erratic or unsuspected. The purpose of the following text is simply to catalogue some such pitfalls we have encountered and to note controls that we have incorporated to improve reliability. It is perhaps obvious that, in the absence of exhaustive attention to such control of quality, counting data may be totally misleading.

One common characteristic of many advanced scientific techniques, as indicated in Table 2, is that they are applied at the measurement frontier, where the net signal (S) is comparable to the residual background or blank (B) effect. The problem is compounded because (a) one or a few measurements are generally relied upon to estimate the blank--especially when samples are costly or difficult to obtain, and (b) the uncertainty associated with the observed blank is assumed normal and random and calculated either from counting statistics or replication with just a few degrees of freedom. (The disastrous consequences which may follow such naive faith in the stability of the blank are nowhere better illustrated than in trace chemical analysis, where $S \gtrless B$ is often the rule [10].) For radioactivity (or mass spectrometric) counting techniques it can be shown that the smallest detectable non-Poisson random error component is approximately $6\phi_p$, where ϕ_p represents the overall (Poisson) relative standard deviation (RSD) [11]. Thus, for our 5 mL quartz counter which has a background of about 0.04 cpm (in an underground laboratory [12]) the three-month detection limit for such non-counting error is equivalent to about 9 percent (RSD). And that assumes that the background distribution (mean, variance) is stationary over such a period of time. Similar considerations apply, of course, to the stability of counting efficiency, as well as to chemical (sample preparation) or sampling contamination. A quantitative treatment of the limitations of non-counting (random) error has been given in reference 13, where it is shown that if one wishes to quantitatively determine a sample whose activity is one-tenth of the Background Equivalent Activity (BEA), the non-counting RSD of the background must not exceed 0.7 percent, and that of the efficiency, 10 percent. Herein lies the real importance of isotopic enrichment, and the relative advantage of AMS where a modern 10 μg-C sample is 200 x BEA (See Table 1).

Table 2. Small Samples: Pitfalls and Control Measures

- Characteristics: S $\stackrel{\sim}{<}$ B; need long-term stability

- Sampling, Interpretation
  - non-representative, contamination
  - chemical (isotopic) heterogeneity
  - model validity (isotopic variations, receptor model)

- Sample Preparation
  - chemical fractionation (recovery)
  - contamination (chemical, radioactive)

- Measurement
  - new computer: software (B), hardware (Effic.)
  - fractionation (pure substance)
  - non-Poisson error (B, Effic.)
  - on-line control: meson (rate, spectrum); pulse-shape, time distribution

Some of the types of contamination that we have encountered include finding contemporary carbon on aerosol filter blanks [14], excess atmospheric ^{37}Ar arising from nuclear testing [8], intrinsic ^{14}C in a tandem accelerator [9], industrial (fossil) carbon in atmospheric methane [15], and the omnipresent radon in recently-prepared CO_2 counting samples [16]. Non-radioactive, chemical contamination also must be controlled, because of the well-known effect of traces of electronegative impurities on gas and liquid scintillation counting efficiency [13,17]. As with the problem of radon contamination, the chemical contamination problem may be quite sample specific. Because of the very complex and "unfriendly" chemical nature of the air particulate samples on which much of our environmental studies have focused, for example, decontamination requirements--from radon, halogens, nitrogen oxides, etc.--are both stringent and variable. Nor is the problem of chemical contamination limited to decay counting methods: with AMS, impurities (especially oxygen) may seriously interfere with the ionization process; and mixed carbonaceous species may suffer variable losses and/or ionization efficiencies in a sputter ion source.

Non-quantitative sample preparation, e.g., conversion to CO_2 provides two opportunities for bias. First, if the sample is a pure substance, such as methane or cellulose, isotopic fractionation can take place. Correction, using the stable isotope ratio $^{13}C/^{12}C$, is possible provided the initial ^{13}C concentration is

known. Second, for carbonaceous mixtures which typify archaeological and environmental samples, chemical fractionation accompanying poor recoveries can lead to severe information-loss and
bias, unless all components have the same "age". This latter
problem will be further discussed in the light of some of our
urban aerosol samples.

More insidious than contamination or fractionation problems
are erratic noise and "blunders" [18]. The former is illustrated
in figure 2 which depicts the "Saturday night excursion"--a host
of spurious pulses ranging up to thirty times the background
rate--which we occasionally encounter during weekend counting.
(The source(s) of the occasional excursions has not yet been
identified, but effective control measures will be described
below.) Hardware and software blunders, which accompanied the
installation of our new computer system, represented one of our
most difficult problems.[3] It is an instructive problem, however,
because it illustrates the statistical trap which is often
associated with attempts to validate a new or modified measurement system. In our case it was necessary to verify the
equivalence of the old minicomputer system with the new microprocessor system, where data were acquired from one to four
low-level counters simultaneously, and pulse rates varied from
0.07 to about 300 per minute. Unfortunately, a full parallel test
of the two systems was infeasible, and numerous intrinsic (timing,
bit manipulation, ...) differences existed. Although initial
results with a single intermediate size gas counter seemed
satisfactory, subsequent comparison with small counters and
multiple counters yielded results which were marginally different
(statistically). We seemed to be observing variable efficiency
and different backgrounds (doubled backgrounds for our smallest
counters).

The hardware and software experts at first responded that the
differences were indeed purely statistical in nature. Later, they
agreed the differences might be real, but that the results with
the new system rather than the old should be assumed correct.
(This was not a happy suggestion, for it carried the implication
that several years of data, acquired with the minicomputer, might
be invalid!) It was only through an exhaustive series of
carefully-designed experiments, combined with significant amount
of intuition to supplant the poor counting statistics, that it
was finally concluded (and later demonstrated) that there were
both hardware and software blunders with the new system. Not
surprisingly, one of the most critical supports for our intuition,
which led to an intensive re-examination of the new hardware and

---

[3]As noted in reference 18, the term "blunder" is not intended in
a pejorative sense. It refers, rather, to the class of non-
random errors or mistakes which are bound to occur in a new and
complex installation.

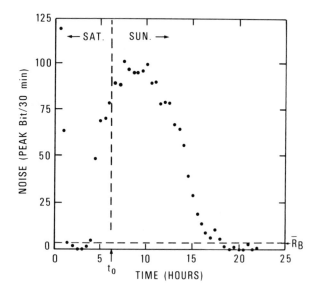

Figure 2.   Weekend noise.

Electronic noise, identified via rise-time characteristics, shown rising to 30 times back-ground rate for 15-mL gas proportional counter. A tiny fraction of such noise, not identified by means such as pulse shape analysis, could invalidate results which presume that the background is stable to < 10%.

software, were continuing series of observations of background
muons and a geologic (wood) reference sample. We shall conclude
this discussion with one example of each type of error: (a)
[Hardware] the new system read one counting channel at a time, in
contrast to the old system where all counter bits were read
simultaneously.   A hardware "lockout" held a second event in a
buffer only if it came from the same counter as the first event.
Since the overall processing time was increased from about 1 ms
to 10 ms, this lockout feature resulted in a variable deadtime
which gave the appearance of a change in counting efficiency
when more than one counter was connected to the system.   (b)
[Software] the background rate for our smallest counter appeared
to be severely increased whereas the rates for large counters
were scarcely affected as a result of the new data acquisition
software.   In this case, a bit which represented the scaled guard
counter pulses was mistakenly included among the net (anticoincidence) sample counter bits.   Since the scaling was 10^4 and
the guard rate was about 10^3 per minute, this represented an
apparent background increment (independent of counter size) of
0.1 cpm; and, of course, this increment was not distributed in a
Poisson manner.

Control Measures

     The first step in combating "chemical" problems (fractiona-
tion and impurities) is to assure quantitative yields and rigorous
purification, or to provide adequate time for decay and time-
scheduled counting in the case of radon.   Problems arising during
the counting process itself, however, are best treated by means
of on-line pulse selection and monitoring.   Referring again to
Table 2, we utilize the mu mesons (coincidence events) for
efficiency and overall gain monitoring.   The meson rate, corrected
for barometric pressure, gives a direct measure of efficiency
(therefore, gas purity) during counting; while the median point
of the meson spectrum provides an on-line measure of gain.   Pulse
shape discrimination allows us to reject electronic noise.
Finally, a direct evaluation of the time distribution of the
individual pulses permits a test of the Poisson assumption [19].
The importance of such on-line quality control cannot be overemphasized.   When one is working at extremely low-levels and
extended counting periods, control can be guaranteed only by
monitoring which takes place simultaneously with the sample
measurement.
     A complete monitoring program, as mentioned previously, must
include periodic replication for total precision assessment, plus
reference sample measurement to assure accuracy.   It was this
latter precaution which made it possible for us to infer the
reliability of the minicomputer system and the presence of
mistakes in the initial microprocessor system.

ATMOSPHERIC RADIOCHEMISTRY

The chemistry of carbon, and radiocarbon, in the atmosphere represents one of the most important areas of environmental research today. The primary practical reason for this is the increasing attention which must be paid to the critical balance between energy and the environment, especially from the viewpoint of man's perturbations of natural processes and his need to maintain control. Probably more than other species, carbonaceous molecules play a central role in this balance. Some of the deleterious effects of carbonaceous gases and particles in the atmosphere are set down in Table 3. The potential effects of increased local or global concentrations of these species on health and climate have led to renewed interest in the carbon cycle and the "CO_2 Problem". It should be evident from the table, however, that carbon dioxide is not the only problem. In fact, the so-called "trace gases and particles" in the atmosphere present an important challenge to our interpretation of the climatic effects of carbon dioxide, itself [20].

Table 3. Effects of Carbonaceous Species

Health

Tropospheric ozone production	(hydrocarbons)
Cancer	(PAH, respirable particles)

Climate

Greenhouse – effect	(CO_2, CH_4)
Albedo, cloud nucleation	(aerosols)

Stratospheric Ozone Reduction

Direct	(CCl_2F_2, CH_3CCl_3)
Indirect	(CO via OH)

Many fascinating questions in geochemistry and atmospheric chemistry are connected with the trace carbonaceous species. Apart from their various effects, certain of them are believed to be primarily anthropogenic (chlorofluorocarbons, methylchloroform), some primarily natural (terpenes, methane), and some of mixed origin (carbon monoxide, particles) [21]. Dispersion modeling, temporal and spatial patterns, and global/regional inventories have helped sort out sources, but many ambiguities remain. For these reasons, and in order to adopt meaningful

pollution controls, it has become interesting to employ chemical
and radiochemical "fingerprinting" techniques which will allow
one to identify pollutant sources from the observed characteristics of ambient samples.

The fingerprinting process--known more formally as Receptor
Modeling (RM) [22]--forms the basis of our recent work with
environmental radiocarbon. That is, we are engaged in "dating"
trace gases and particles in the atmosphere (excluding CO_2) in
order to draw inferences concerning their sources [23]. In this
case the radiocarbon dating is really radiocarbon isotope dilution, in which the $^{14}C/^{12}C$ ratio of specific samples is
interpreted in terms of the relative contributions from fossil
and biogenic sources rather than the age of the material. Just
as set forth by Professor Oeschger [1], we are using dating
techniques combined with chemical data and physical parameters, to
calibrate atmospheric models and to understand the environmental
system. The major difficulty in carrying out such measurements is
the small sample size. Except for carbon dioxide and methane, the
atmospheric concentrations range from roughly 10^{-5} to 10^{-1} mg-
carbon/m^3-air [21], so it is seldom feasible to collect more than
about 10 mg-carbon. Fortunately, development of the small sample
techniques discussed in the previous section has made it feasible
for us to pursue this form of dating; a review of some of our
recent results follows.

Contemporary Carbon Content of Environmental Samples

Carbon-14 was selected as a tracer for biogenic (fossil)
sources of atmospheric gases and particles because, unlike certain
alternative tracers, it is direct, unique, and robust. It is
direct in that carbon, rather than associated inorganic material
is measured; it is unique in that only the biogenic component
(like vegetation) contains ^{14}C; and it is robust in that the
isotopic ratio is essentially unchanged during chemical reactions
or atmospheric transport. Our recent results are summarized in
the following two tables. For each entry we have indicated our
primary objective, the observed fraction of contemporary carbon,
and concurrent chemical or (stable) isotopic data.

The first entry in Table 4 refers to the accelerator (AMS)
experiments noted earlier in which we investigated (a) the
feasibility of direct atom counting given only a few micrograms
of carbon (using the new international radiocarbon dating
standard), and (b) the applicability of an electromagnetic
isotope separator for the dual purposes of source preparation (by
implantation) and isotopic enrichment [9]. In this latter experiment the enrichment achieved, which was limited by the natural
isotopic abundance of $^{12}C^{18}O$, corresponded to an enrichment factor
of about 500. Experiments underway to investigate the global
methane cycle have begun with measurements of radiocarbon in
ambient (suburban) samples. Our initial measurements, though of

Table 4.   Recent Radiocarbon Results

Sample	Objective	Result $(f_c)^a$	Supporting Data	Ref.
^{14}C Standard[b] {direct, implant}	minimum size enrichment source preparation	0.95 ± 0.03 (^{14}C-pulse)	^{13}C	9
Atmospheric CH_4	CH_4 budget	0.98 ± 0.22	—	
Sediment {FA^c, PAH}	Source of PAH	0.86 ± 0.21 0.19 ± 0.21	^{13}C	24

[a] f_c = fraction of contemporary carbon. (Errors represent one (Poisson) standard deviation.) "Contemporary", as used here, refers to carbon from the living biosphere. Its ^{14}C concentration is currently about 30% higher than that of pre-nuclear era "modern" carbon, which is defined as 0.95 × NBS Oxalic Acid SRM-4990B [2].

[b] AMS, m_c < 40 μg; all others (11c) m_c = 5-10 mg.

[c] FA = fatty acid fraction; PAH = polycyclic aromatic hydrocarbons (recent sediment from Puget Sound, WA)

relatively poor precision, are consistent with a largely biogenic source. Prior measurements depended on air liquefaction plants for sampling, and they were consequently subject to possible industrial contamination [15]. Because atmospheric methane is a direct precursor of carbon monoxide, its isotopic composition is also important for inferences concerning the carbon monoxide budget.

Examination of carbon isotopes in individual organic species in different layers of sediment provides a critical tool for the evaluation of the historical record of man's influence on metabolites and toxic compounds. As shown in the final entry in Table 4, the polycyclic aromatic component in sediment (representing the last two decades) was essentially fossil in origin, unlike the fatty acid component [24]. This was a very interesting result, for there exist conflicting hypotheses suggesting that the PAH derives largely from forest fires [25], or from fossil fuel use [26]. The ^{14}C measurements were consistent with the latter hypothesis; the measurement of $^{13}C/^{12}C$ further ruled out residual oil as a principal source, in favor of coal or petroleum. The pattern of polycyclic compounds can sometimes assist in the source discrimination, but it does not approach the resolving power or reliability of the isotopic data. PAH's observed in this study were similar to what one finds associated with atmospheric particles coming from wood or fossil fuel combustion-- phenanthrene, pyrene, chrysene, benzopyrene, perylene, etc.-- several of which are direct or activable mutagens [27].

Table 5 summarizes our recent studies of carbonaceous particles. The urban samples, collected during the winter months in Denver, illustrate the importance of measuring radiocarbon in different chemical fractions. The sharp difference in isotopic composition between the volatile ("organic") and non-volatile ("elemental") carbon of course implies quite different sources. (This is rather analogous to the problem of non-contemporaneity in radiocarbon dating, where different chemical (or physical) fractions of a sample may derive from sources of different age; so separation is mandatory for the maximum, reliable information [28].) In the case of the Denver samples, the contemporary carbon, which dominated the volatile fraction presumably derived from the use of wood as a fuel, whereas the elemental carbon particles must arise from fossil fuel combustion sources, such as coal-fired plants and diesel vehicles. It is worth noting that the inorganic compositional data, when used alone for source apportionment left about 40 percent of the carbon unaccounted for; this, in turn, was ascribed to wood-burning [29].

Comparing rural with "typical" urban samples (2nd entry, Table 5), we find that fossil carbon predominates in the urban particulate matter, and the converse. A significant amount of biogenic carbon is found in the urban samples (even in the absence of wood burning), however; and this may be related to vegetative emissions. For example, besides fossil fuel indicators such as

Table 5.   Atmospheric Particles

Sample	Objective	Result $(f_c)^a$	Supporting Data	Ref.
Urban {volatile, non-volatile}	Natural Component (chem.)	0.69 ± 0.17, −0.06 ± 0.08[b]	Inorganic Composition	30
Regional {urban, rural}	Natural Component (locale)	0.23 ± 0.15, 0.88 ± 0.16	Organic Composition	31
Vegetative Burning (fine particles) {slash field, RWC[c]}	Urban Impact	0.83 ± 0.16, 1.07 ± 0.15, 0.57 ± 0.12	Particle Size and Inorganic Composition	33

a f_c = fraction of contemporary carbon. Errors represent one (Poisson) standard
    deviation.

b   Obtained by difference: total, less volatile.

c   Residential wood combustion.

branched cyclic hydrocarbons and dicarboxylic acids, vegetative
indicators such as predominantly odd carbon-numbered high-
molecular weight n-alkanes, and even carbon-numbered fatty acids
are found in samples from Salt Lake City and Los Angeles [31,
32].

Under certain circumstances, biogenic carbon may represent
the major component in urban particles. Such was the case for the
last study cited in Table 5, which was designed to evaluate the
impact of field, slash and residential (wood stove) vegetative
burning on the air quality of Portland, Oregon [33]. In this
case, ^{14}C was measured under different impacts (types of combustion known to be taking place) and in different size fractions.
Data for the respirable (< 2.5 μm) fraction, shown in the table,
demonstrated that field and slash burning could account for
practically all of the fine particle carbon in the city, and that
residential wood combustion could account for over half of this
carbon (in the winter time). Again, inorganic compositional data
were gathered for source reconciliation (chemical mass balance).
But these data were rather inadequate for the task of estimating
the biogenic (fossil) carbon source strength (uncertain up to
about 40%). This uncertainty is hardly surprising, because
assumed organic/inorganic emission ratios are not very reliable
even if all sources are known. With diminishing use of leaded-
gasoline, for example, it is becoming increasingly difficult to
estimate the total auto-exhaust carbon from the inorganic (Pb)
particulate data.

Higher Dimensions: Chemical and Physical Selectivity

All of the above particulate investigations were based on
mini-radiocarbon measurement techniques, with sample masses
typically in the range of 5-10 mg-carbon. This constituted a
major advantage, because it was practicable to select special
samples (given region, source impact, sediment depth) and to
further subject such samples to physical (size) or chemical
separation before ^{14}C measurement. This type of "serial selectivity" provides maximum information content about the samples;
and in fact it is essential when information is sought for the
sources or atmospheric distributions of pure chemical species,
such as methane or elemental carbon.

Complementing this is "parallel selectivity", where additional chemical or (stable) isotopic data are obtained in order
to provide the necessary degrees of freedom to estimate individual
source strengths--not just biogenic/fossil--through receptor
modeling and the use of the Chemical Mass Balance (CMB) [34]. The
power of the multidimensional approach is indicated in figure 3.
In this figure the dashed line represents the serial step--i.e.,
selection according to physical (particle size) and/or chemical
(volatility) characteristics which have special importance (for
source discriminating power, health effects, ...). The

three-dimensional projection, on the other hand, represents features which characterize the selected material in <u>parallel</u>, thus providing quantitative CMB input data.

The example shown, exhibiting an isotopic plane (two dimensions) plus one chemical (ratio) dimension, is appropriate for ambient carbonaceous particles. As shown on the abscissa, ^{13}C is effective for resolving carbonaceous material deriving from petroleum from that coming from methane or marine plants. Petroleum, coal, and (most) terrestrial vegetation overlap, however. The addition of the second isotopic dimension (ordinate) immediately adds to the resolving power of ^{13}C, and distinguishes biospheric from fossil source material. The ratio K/Fe, illustrating one out of many useful parallel chemical dimensions, complements the isotopic data and allows one to further resolve (imperfectly) field from slash burn vegetative material [35].

Note that the chemical dimensions do not yield such robust "markers", or tracers, as the isotopic dimensions. For a given combustion source, for example, one is apt to find chemical--but not isotopic--fractionation varying significantly with time, particle size, particle history (differential volatilization or reaction), etc. The ratio K/Fe, for example, was quite different in the fine and coarse fractions from a slash burn (1.5 <u>vs</u>. 0.3) [33]; and Pb/Br, which has been popular as an automobile exhaust tracer, varies with the "age" of urban particles [22].

This brings us to the question of modeling. In models which attempt to extract the anthropogenic component on the basis of temporal or spatial patterns of a given substance, source-sink-contamination pitfalls await. That is, unknown natural sources or sinks or baseline contamination may easily lead to false conclusions. This has been seen repeatedly in attempts to model the global cycles for CO and CO_2 [36,37], and it sparks a large part of the controversy over ambient atmospheric aerosols [38]. In the case of receptor modeling and CMB, source strengths are estimated from a linear model,

$$y_i = \Sigma A_{ij}x_j + e_i \qquad (1)$$

where the observed i^{th} characteristic (chemical, physical, isotopic) of the ambient sample (y_i) and of the jth source (A_{ij}) is utilized in a deconvolution process to estimate the strength of the jth source (x_j). Obviously, assumptions about the associated errors (e_i) and the identity and stability of the source vectors (A_{ij}) are critical for the success of CMB [22]. Here, again, resolving power or assumption validity may be inadequate to reliably estimate the anthropogenic component from chemical data alone. (A case in point is the controversy surrounding the sedimentary PAH "patterns" and the relative importance of forest fires and fossil fuel combustion [25,26].) The parallel use of isotopic information, as implied in figure 3, is therefore our goal, in order to generate redundant and assumption-resistant source identification and source strength estimates.

To illustrate the CMB method of "fingerprinting" carbonaceous particles, we shall examine data for one of the slash burn samples, using both direct (radiocarbon) and indirect (inorganic) methods of inference [33]. The sample in question (number-3, collected in downtown Portland in October, 1977) first underwent serial selection, according to particle size (< 2.5 μm). The indirect inference for biogenic carbon was then based on parallel observations of C, Si, Ni, and Pb--the latter three elements serving as orthogonal markers for "road dust", "residual oil", and "auto-exhaust", respectively. Five-dimensional patterns for the sample and the individual sources are shown in figure 4, together with the CMB matrix equation. (Ni serves as a tracer for only residual oil. Distillate oil, included in figure 4, derives from the emission inventory ratio.) It is clear by inspection that no one of the source vectors can adequately account for the observed ambient sample pattern. By combining all of the chemical and isotopic data, however, one can estimate contributions from each of the four sources of carbon, and have one degree of freedom left over for a consistency check. Alternatively, it is interesting to use the unique ^{14}C tracer to estimate the vegetative carbon, and the four chemical tracers to estimate all carbonaceous components. The results follow.

$$\text{Using } ^{14}C: \quad \hat{C}_V = 25.7 \pm 5.0\%$$

$$\text{Using C, Si, Ni, Pb:} \quad \hat{C}_V = 24.5\%, \quad \hat{C}_r = 0.9\%$$

$$\hat{C}_o = 0.7\%, \quad \hat{C}_a = 4.8\%$$

Thus, consistent results are obtained; about one-fourth of the fine particle mass, or 80 percent of the carbon is accounted for by vegetative-(slash) burn carbon. (The corresponding mass density of fine particles in downtown Portland from slash burning is 21 μg/m^3 [33].) The only other major source of carbon is seen to be auto exhaust.

The foregoing example illustrates the importance of combining independent chemical and isotopic data for maximum resolving power and reliability. Several cautions are evident, however. As indicated in figure 4 (matrix equation): all significant carbonaceous sources must be represented in the model; and uncertainties for both the sample (y_i) and the source matrix (A_{ij}) must be negligible, or at least estimable. These are non-trivial

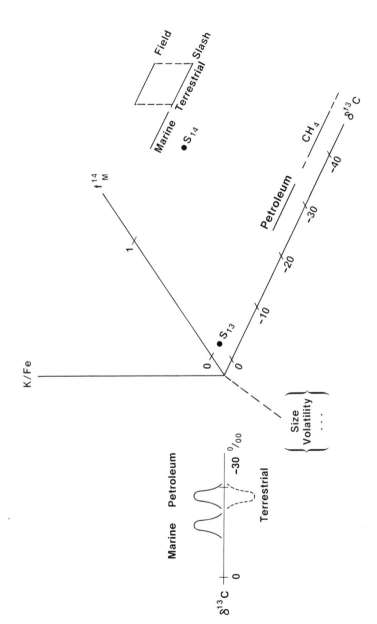

Figure 3. Parallel and serial selectivity.

Parallel (multidimensional) measurements of size- or chemically selected particles permit the simultaneous resolution of biogenic material from fossil sources (^{14}C), discrimination between certain fossil sources and between marine and terrestrial vegetation (^{13}C), and partial separation of agricultural burning sources (K/Fe). The term f_M^{14} represents the fraction of modern C based on standard S_{14}; $\delta^{13}C$ represents the deviation (per mil) of the $^{13}C/^{12}C$ ratio from standard S_{13} (2).

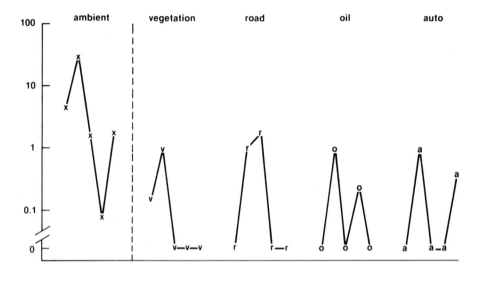

$$\begin{pmatrix} {}^{14}C \\ C \\ Si \\ Ni \\ Pb \end{pmatrix} = \begin{pmatrix} 4.4 \pm 0.8 \\ 31. \\ 1.54 \\ 0.083 \\ 1.94 \end{pmatrix} = \begin{pmatrix} 0.17 & 0 & 0 & 0 \\ 1 & 1 & 1 & 1 \\ 0 & 1.64 & 0 & 0 \\ 0 & 0 & 0.12 & 0 \\ 0 & 0 & 0 & 0.4 \end{pmatrix} \cdot \begin{pmatrix} C_v \\ C_r \\ C_o \\ C_a \end{pmatrix}$$

$$\hat{C}' = (24.5, 0.9, 0.7, 4.8)\%$$

Figure 4. Fine particle aerosol—individual source patterns and CMB equation.

Isotopic and chemical patterns are given for an ambient sample (downtown Portland) and four sources of carbonaceous particles. The source C contributions (C_j) are expressed as percent of total aerosol mass, and the subscripts refer to Vegetation (slash burning), Road dust, Oil (residual and distillate), and Auto exhaust. Units for the ordinate are also percent of total aerosol mass, except for ^{14}C which is expressed as dpm/g-aerosol. Abscissa points for each pattern are ordered as: ^{14}C, C, Si, Ni, and Pb.

questions for the Portland study, and they have been treated in detail elsewhere [39]. Two of the matrix elements, (A_{32}, A_{43}) for example, depend upon location or particle size, and a third (A_{11}) can be subject to uncertainties connected with vegetation age and isotopic heterogeneity.

These latter two problems must be faced both in environmental modeling and radioactive dating [40]. Radiocarbon variations and age of the source material in each case influence the $^{14}C/^{12}C$ ratio. For the vegetative burning aerosol, the ratio differed by about 13 percent between slash burning and residential wood combustion because of the more pronounced effect of "bomb carbon" during the more recent lifespan of the firewood [33]. With respect to isotopic heterogeneity, when the isotopic ratio differs among different physical or chemical fractions of the sample, at best information is lost if the individual fractions are not separately dated, and at worst the average result is biased if chemical fractionation occurs in connection with incomplete recoveries. Such was the case with the samples from Denver. Our initial experiments had non-quantitative recoveries (particulate carbon → purified carbon dioxide counting gas); and the organic (volatile) fraction was in fact substantially more biogenic than the elemental fraction. Such preferential loss of the volatile (more biogenic) component during sample preparation was equivalent to a bias toward fossil carbon for the overall particulate carbon. For example, the average fraction of contemporary carbon $\overline{f}_c$ for the urban sample in Table 5 (first entry) is 0.35, since it has approximately equal fractions of volatile and non-volatile components. If loss of volatiles led to a recovery of only 70 percent, then the apparent $\overline{f}_c$ would be but 0.20. A receptor model based on the average composition would thus yield a biased result. (This isotopic heterogeneity problem involving multiple organic forms in atmospheric particles is somewhat analogous to the situation with the ^{13}C content of lignin vs. cellulose in tree rings, and non-contemporaneity in radiocarbon dating.)

CONCLUSION

The advent of new techniques for measuring small samples of natural radionuclides is beginning to have important impacts in fields ranging from nuclear geophysics to radioactive dating to environmental chemistry. A comparative assessment of small sample liquid scintillation counting, small sample gas proportional counting and accelerator mass spectrometry suggests that the three techniques are complementary and each can yield important radiocarbon data in the mass range from 10 µg to 100 mg-carbon. Because of the different characteristics of the alternative techniques, individual problems involving non-counting error and sample preparation demand special attention. At the same time, there exist special opportunities ranging from non-destructive radioassay to linked systems for implantation and isotopic enrichment.

Given the present state-of-the-art, we conclude for the two decay-counting techniques that: they are well-suited for measurements of $\geq$ 100 mg-C (lsc) and 10 mg-C (llc) at moderate precision levels and ages; the measurement processes are sufficiently established that larger samples can be expected to yield improved precision (and accuracy) and older age limits; at the 10 mg–100 mg level, however, the fact that S $\sim$ B demands the use of sophisticated methods for on-line control of electronic noise, overall gain and counting efficiency. (It is interesting to note that on-line control in place of "statistical" correction, has long been the hallmark of low-level counting--viz., on-line control of background via. anticoincidence counting.) Atom-counting (AMS) is clearly superior for very small ($\lesssim$ 1 mg) or very old ($>$ 40,000 yr.) samples. For larger and younger samples, it is still attractive in that S $>>$ B, but its present reproducibility (a few percent) suggests that, for $\gtrsim$ 5 mg-carbon llc should not yet be discounted. Overall measurement times are comparable, considering (a) the fact that llc systems generally operate with several counters in parallel, and (b) the time required for sample preparation and purification. Comparative maintenance and personnel requirements, and real costs may be extremely important factors, but they are beyond the scope of this article. Perhaps the most important developments (already underway) with AMS will be its application to tiny isotopic ratios ($\gtrsim$ 10^{-15}) of longer-lived nuclides, such as ^{129}I [41]; the development of optimal, routine methods of sample preparation [42]; and the development of "dedicated" machines having low intrinsic blanks and improved precision through rapid switching [5] or on-line, parallel isotopic measurements [43]. Meanwhile, llc measurements are already practicable for environmental (this paper) and archaeological [7] radiocarbon measurements of 5-10 milligram samples.

One of the most exciting and most important opportunities provided by the new, small sample dating techniques is that of linking radioisotope measurement with detailed chemical and physical characterization of selected environmental species. With the possibility of working with as little as 10 μg of carbon it becomes practicable to assess anthropogenic contributions to several of the trace carbonaceous gases and particles--species which may have profound effects on health and future climate. Parallel or multidimensional chemical and nuclear measurements will become a major tool for extracting information about the environmental history of ambient samples; and the unique and robust character of the isotopic data has been demonstrated to be critical for reliable application of "chemical fingerprinting" (receptor modeling).

Recent societal changes, including changing energy patterns and major interest in man's perturbation of the environmental system, have made the evolution of the advanced radiocarbon techniques most timely. Preliminary results of our applying one of these techniques (llc) to single atmospheric species have shown that one (CH_4) is primarily natural (biogenic) in origin, and another (C-elemental) is largely fossil. By combining radiocarbon measurements with elemental analysis and size fractionation we have shown that significant portions of urban respirable particles are coming from residential wood-burning. The importance of this observation may be grasped when one considers that the sale of wood-burning stoves exceeds 10^6/year and wood already ranks as a major fuel in the U.S. (equivalent to hydroelectric and nuclear). The mutagens and carcinogens associated with wood-burning particles makes them especially noxious, and their emission is severe: 20-50 times that of oil or gas (per BTU) [33]. Other environmental questions which must soon be addressed with small-sample radiocarbon-chemical measurements include ice core/sediment studies of the carbonaceous pollutant record; radiocarbon "calibration" of the global CO budget (model) [36]; determination of the impact of diesel and unleaded gasoline emissions; and determination of biogenic and fossil contributions to trace atmospheric hydrocarbons. Although this discussion has emphasized environmental radiocarbon, there can be no doubt that similar attention to the selection of particular chemical and/or size fractions from archaeological samples will yield information of enhanced reliability [44]. That is, the capability of measuring µg-mg carbon samples makes possible the development of radiocarbon microchemical dating in which accurate, contamination-free dates may be guaranteed by the application of specific chemical knowledge rather than reliance on simple pretreatment.

The work reviewed in this chapter was made possible through the encouragement and cooperation of several colleagues. The most vital early encouragement came from W. F. Libby, to whom this article and this volume are dedicated. Major assistance with the early (1976) work on "miniradiocarbon counting" came from R. B. Murphy. Hardware and software support for the llc system came from J. R. DeVoe, F. Ruegg, and J. F. Barkley and the lsc data were obtained with the cooperation of J. E. Noakes. The Rochester (AMS) experiments took place through the hospitality of H. E. Gove with major assistance from D. Elmore. Others who provided samples and participated directly in the experiments cited in Tables 4 and 5 included J. R. Swanson, R. J. Countess, D. P. Stroup, S. M. Kunen, K. J. Voorhees, and J. A. Cooper. Partial support for this research was provided by the Office of Environmental Measurements, U. S. National Bureau of Standards, and the Energy-Environment Program (EPA-IAG-D6-E684), U. S. Environmental Protection Agency.

REFERENCES

[1] Oeschger, H., The Contribution of Radioactive and Chemical Dating to the Understanding of the Environmental System, Chapter 2 in this book.

[2] Currie, L. A., Polach, H. A., Exploratory Analysis of the International Radiocarbon Cross-Calibration Data: Consensus Values and Interlaboratory Error, Proceedings of the 10th International Radiocarbon Conference, Radiocarbon, 22, 933 (1980); and L. M. Cavallo and W. B. Mann, New National Bureau of Standards Contemporary Carbon-14 Standards, Radiocarbon, 22, 962 (1980).

[3] Currie, L. A., Noakes, J., Breiter, D., Measurment of Small Radiocarbon Samples: Power of Alternative Methods for Tracing Atmospheric Hydrocarbons, Ninth International Radiocarbon Conference, University of California, Los Angeles and San Diego, 1976.

[4] Mast, T. S., Muller, R. A., Tans, P. P., Radioisotope Detection with Accelerators, Proc. Conf. on the Ancient Sun, R. O. Pepin, J. A. Eddy, R. B. Merrill, eds., Geochim. et Cosmochim. Acta, 1980, Suppl. 13, 191.

[5] Purser, K. H., Russo, C. J., Liebert, R. B., Gove, H., Elmore, D., Ferraro, R., Litherland, A. E., Beukens, R., Chang, K., Kilius, L, Lee, H., The Application of Electrostatic Tandems to Ultra-sensitive Mass Spectrometry and Nuclear Dating, Chapter 3 in this book.

[6] Personal communication: G. Harbottle (1980), D. Donahue (1981).

[7] Sayre, E. V., Harbottle, G., Stoenner, R. W., Washburn, W., Olin, S., Fitzhugh, W., The Carbon 14 Dating of an Iron Bloom Associated with the Voyages of Sir Martin Frobisher, Chapter 22 in this book.

[8] Rutherford, W. M., Evans, J., Currie, L. A., The Application of Isotopic Enrichment and Pulse Shape Discrimination to the Measurement of Atmospheric ^{37}Ar, Anal. Chem., 48, 607 (1976).

[9] Currie, L. A., Klouda, G. A., Elmore, D., Ferraro, R., Gove, H., Accelerator Mass Spectrometry and Electromagnetic Isotope Separation for the Determination of Natural Radiocarbon at the Microgram Level (in preparation).

[10] Hume, D., Pitfalls in the Determination of Environmental Trace Metals, Progress in Analytical Chemistry, 5, Chemical Analysis of the Environment, Plenum Press, 1973.

[11] Currie, L. A., The Limit of Precision in Nuclear and Analytical Chemistry, Nucl. Instr. Meth., 100, 387-395 (1972).

[12] Measurements carried out at the University of Bern (1979), courtesy of H. Oeschger and H. H. Loosli.

[13] Currie, L. A., Accuracy and Merit in Liquid Scintillation Counting, Chapter 18, Liquid Scintillation Counting, M. Crook, ed., Heyden & Son, Ltd., London, 1976, pp. 219-242.

[14] Stevens, R. K., Currie, L. A., Dzubay, T. G., Mason, M., Particulate ^{14}C Measurements as Indicators of Biogenic and Fossil Carbon Sources, to be published (1981).

[15] Ehhalt, D. H., Methane in the Atmosphere, Carbon and the Biosphere, Woodwell and Pecan, eds., Conf. 720510, 1973, AEC, p. 144.

[16] The possibility of widely-varying radon levels from air filter samples can be a subtle trap. We were made aware of this when we were once under pressure to provide an immediate ("preliminary") result shortly after sample preparation.

[17] Brenninkmeijer, C. A. and Mook, W. G., The Effect of Electronegative Impurities on CO_2 Proportional Counting, Ninth International Radiocarbon Conference, University of California, Los Angeles and San Diego, 1976.

[18] Currie, L. A., Sources of Error and the Approach to Accuracy in Analytical Chemistry, Chapter 4, Vol. 1, Treatise on Analytical Chemistry, P. Elving and I. M. Kolthoff, ed., J. Wiley & Son, New York, 1978. (See especially "The Analysis of Blunders", pp. 12ff.)

[19] Currie, L. A., Lindstrom, R. M., The NBS Measurement System for Natural ^{37}Ar, in Proceedings of the Noble Gases Symposium, EPA and University of Nevada at Las Vegas, 1975.

[20] Geophysics Study Committee, Energy and Climate, NRC Geophysics Research Board, National Academy of Sciences, Washington, D.C., 1977.

[21] Covert, D. A., Charlson, R. J., Rasmussen, R., Harrison, H., Atmospheric Chemistry and Air Quality, Review of Geophysics and Space Physics, 13, 765 (1975).

[22] Watson, J. G., ed., Proc. Receptor Modeling Workshop, Quail Roost, N.C., Feb. 1980; Gordon, G. E., Receptor Models, Env. Sci & Tech., 14, 792 (1980).

[23] Currie, L. A., Murphy, R. B., Origin and residence times of atmospheric pollutants: application of ^{14}C, in Methods and Standards for Environmental Measurement, W. H. Kirchoff, ed., NBS Spec. Pub. 464, National Bureau of Standards, Washington, D.C., p. 439, Nov. 1977.

[24] Swanson, J. R., Carbon Isotope Analysis of Carbonaceous Compounds in Puget Sound and Lake Washington, PhD Thesis, University of Washington, 1980; and Swanson, J. R., Fairhall, A., Currie, L. A., Carbon Isotope Analysis of Sedimentary Polycyclic Aromatic Hydrocarbons (in preparation).

[25] Blumer, M., Youngblood, W. W., Polycyclic Aromatic Hydrocarbons in Soils and Recent Sediments, Science, 188, 53 (1975).

[26] Hites, R. A., Laflamme, R. E., Farrington, J. W., Sedimentary
     Polycyclic Aromatic Hydrocarbons: The Historical Record,
     Science, 198, 829 (1977).
[27] Kaden, D. A., Thilly, W. G., Mutagenic Activity of Fossil
     Fuel Combustion Products, Conference on Carbonaceous Parti-
     cles in the Atmosphere, T. Novakov, ed., University of
     California, Berkeley, 193, 1978.
[28] Schultz, H., Currie, L. A., Matson, R. R., Miller, W. W.,
     Pretreatment of Wood and Char Samples, Radiocarbon, 5,
     342 (1963).
[29] Wolff, G. T., Countess, R. J., Groblicki, P. J., Ferman,
     M. A., Cadle, S. H., Muhlbaier, J. L., Visibility-Reducing
     Species in the Denver 'Brown Cloud', Part III Sources and
     Temporal Patterns, GMR-3394, 1980.
[30] Currie, L. A., Countess, R. J., Klouda, G. A., Stroup, D.,
     The Contribution of Contemporary Carbon to the 'Denver Brown
     Cloud' (in preparation).
[31] Currie, L. A., Kunen, S. M., Voorhees, K. J., Murphy, R. B.,
     Koch, W. F., Analysis of Carbonaceous Particulates and
     Characterization of Their Sources by Low-Level Radiocarbon
     Counting and Pyrolysis/Gas Chromatography/Mass Spectrometry,
     Conference on Carbonaceous Particles in the Atmosphere,
     T. Novakov, ed., University of California, Berkeley, p. 36,
     1978.
[32] Kaplan, I. R., Currie, L. A., Klouda, G. A., Isotopic and
     Chemical Tracers for Organic Pollutants in the Southern
     California Air Basin (in preparation).
[33] Cooper, J. A., Currie, L. A., Klouda, G. A., Assessment of
     Contemporary Carbon Combustion Source Contributions to Urban
     Air Particulate Levels Using C-14 Measurements, to be
     published in Env. Sci. & Tech., 1981.
[34] Friedlander, S. K., Chemical Element Balances and Identifica-
     tion of Air Pollution Sources, Environ. Sci. Technol., 7,
     235 (1973)(See also ref. 22).
[35] Core, J. E., Terraglio, F. P., Field and Slash Burning Parti-
     culate Characterization: The Search for Unique Natural
     Tracers, Air Pollution Control Assn. Report, 1978.
[36] Pinto, J. P., Yung, Y. L., Rind, D., Russell, G. L, Lerner,
     J. A., Hansen, J. E., Hameed, S., A General Circulation Model
     Study of Atmospheric Carbon Monoxide, 1980.
[37] Bolin, B., Degens, E. T., Kempe, S., Ketner, P., The Global
     Carbon Cycle, John Wiley & Sons, New York.
[38] Kneip, T. J., Lioy, P. J., eds., Aerosols: anthropogenic
     and natural sources and transport, Annals N.Y. Acad. Sci.,
     338, 1-618 (1980).
[39] Watson, J. G., Chemical element balance receptor model
     methodology for assessing the sources of fine and total
     suspended particulate matter in Portland, Oregon, PhD Thesis,
     Oregon Graduate Center, Beaverton, Oregon, 1979.

[40] Neustupný, E., The Accuracy of Radiocarbon Dating, p. 23, in Olsson, I. U., ed., Radiocarbon Variations and Absolute Chronology, Proceedings of the 12th Nobel Symposium held at the Institute of Physics at Uppsala University, Wiley-Interscience, New York, 1970.
[41] Elmore, D., Gove, H. E., Ferraro, R., Kilius, L. R., Lee, H. W., Chang, K. H., Beukens, R. P., Litherland, A. E., Russo, C. J., Purser, K. H., Murrell, M. T., Finkel, R. C., Determination of ^{129}I Using Tandem Accelerator Mass Spectrometry, Nature, 286, 138 (1980).
[42] Rubin, M., Sample Preparation for Electrostatic Accelerator Dating of Radiocarbon, Chapter 5 in this book.
[43] Southon, J. R., Nelson, D. E., Korteling, R., Nowikow, I., Hammaren, E., McKay, J., Burke, D., Techniques for the Direct Measurement of Natural ^{10}Be and ^{14}C with a Tandem Accelerator, Chapter 4 in this book.
[44] Taylor, R. E., Problems in the Radiocarbon Dating of Bone, Chapter 23 in this book.

RECEIVED July 28, 1981.

Dating Groundwater

A Short Review

STANLEY N. DAVIS and HAROLD W. BENTLEY

University of Arizona, Department of Hydrology and Water Resources,
Tucson, AR 85721

The age of groundwater is the length of time the water
has been isolated from the atmosphere. Theoretically,
ages can be estimated by (1) the travel time of groundwater
from the point of recharge to the subsurface point of
interest as calculated by Darcy's law combined with an
equation of continuity, (2) the decay of radionuclides
which have entered the water from contact with the
atmosphere, (3) the accumulation of products of
radioactive reactions in the subsurface, (4) the degree of
disequilibrium between radionuclides and their radioactive
daughter products, (5) the time-dependent changes in the
molecular structure of compounds dissolved in water, (6)
the presence of man-made materials in groundwater, (7) the
correlation of paleoclimatic indicators in the water with
the known chronology of past climates, and (8) the
presence or absence of ions which can be related to past
geologic events that have been previously dated. Owing to
uncertainties in each of the methods, as many methods as
possible should be used in every field situation. Because
hydrodynamic dispersion and molecular diffusion always
take place, a single precise age for a given groundwater
sample does not exist. "Dating" the sample by various
methods, however, will help determine the extent of
dispersion and diffusion as well as mixing of water from
various aquifers which takes place within many wells. If
dispersion, diffusion, and cross-mixing are minimal, then
under ideal conditions ages can be determined for waters
less than about 30,000 years old, and rough approximations
of ages up to about one million years appear possible.

The age of groundwater is the length of time the water has
been isolated from the atmosphere. Although this definition is
useful for many purposes, it does not reflect the true complexity

0097-6156/82/0176-0187$09.00/0

of most circulation systems of groundwater. Only simple systems which can be compared with flow through a single long pipe can yield nearly homogeneous dates for water sampled from the same general part of an aquifer. Few natural systems approach this type of simple linear, non-mixing or "piston", flow. In addition, most groundwater is taken from wells which tap more than one restricted water-bearing zone. Consequently, a sample of water will be a mixture of waters of different ages even if the ground-water flow were to approach the idealized piston flow. Natural springs also are commonly connected in a complicated way to various water-bearing zones so that samples of spring water may represent mixtures of waters of vastly different ages.

Because of the complex history of most groundwater samples, the necessity of combining multiple dating methods with thorough regional and local hydrogeologic studies cannot be emphasized too strongly. Obviously, the larger the number of dating methods which are used, the more information can be obtained concerning the natural systems. The purpose of this present review is to emphasize the availability of many independent dating methods and to indicate the state of development of the methods. Although the best-known methods will be mentioned, an emphasis will be placed in this paper on some of the newer methods which are under development.

DARCY'S LAW

The oldest and most widely used method of estimating water age is the calculation of travel times using Darcy's law combined with an expression of continuity. If a field of steady-state, groundwater flow is subdivided into a two-dimensional flow net (figure 1), then Darcy's law can be written as:

$$Q = K \ m \ \Delta w \ \frac{\Delta h}{\Delta L} \tag{1}$$

in which

Q is the discharge per unit time,
K is the hydraulic conductivity, assumed to be isotropic,
m is the thickness of the flow field normal to the plane
   of the flow net,
Δw is the width of the stream tube, and
Δh is the change in hydraulic head over the incremental
   flow path, ΔL.

If the groundwater and aquifer are assumed to be incompressible, the continuity equation for water flow is:

$$Q = \frac{\Delta L}{\Delta t} \ n_e \ m \ \Delta w \tag{2}$$

in which

n_e is the effective porosity of the aquifer,

Δt is the time taken by the water to traverse the distance ΔL, and other symbols are as given above.

Combining equations 1 and 2 and solving for Δt yields:

$$\Delta t = \frac{n_e (\Delta L)^2}{k \Delta h} \tag{3}$$

The age of the groundwater is obtained by the summation of all Δt values along the flow path from the surface intake area to the point of interest in the subsurface. This has been done for the entire field of flow shown in figure 1. The resulting isochronal lines are shown in figure 2. Although figure 2 represents an aquifer of infinite depth and, therefore, does not represent a natural system, the fact that older water tends to rise along the axes of valleys containing perennial streams is widely recognized and has been documented by Carlston and others [1][1]. Figure 2 further emphasizes the fact that wells which are screened at various depths will produce water of mixed ages. A 200 m well near the river could have a mixture of water ranging in age from modern to 4,000 years old.

Figure 2 was constructed by contouring of values calculated from the flow net of figure 1. Much more complex diagrams are possible provided sufficient information concerning the aquifer and fluid-flow conditions can be obtained. For simple boundary conditions and homogeneous aquifers, a direct analytical solution for isochronal surfaces is available [2].

Unfortunately, despite the continued evolution of sophisticated numerical techniques which can be used to estimate water ages in idealized systems, the inability to define all the critical hydrogeologic details of aquifers will probably always leave large uncertainties in the estimation of groundwater ages by purely hydrodynamic methods. As Theis [3] has pointed out, detailed coring of many fluid-bearing zones which appear homogeneous has yielded samples with permeabilities ranging through at least two orders of magnitude. This variation of permeability will give rise to the phenomenon of megadispersion which, of course, gives rise in turn to water of mixed ages in any given zone in an aquifer. In fact, in many recent studies, radiometric methods of dating water have been used to help understand the possible extent of water mixing in aquifers [4-6].

Besides the problem of defining geologic details and the associated problems of megadispersion, the problem of defining ΔL and Δh values for past flow conditions should not be ignored. As

[1]Figures in brackets indicate the literature references at the end of this paper.

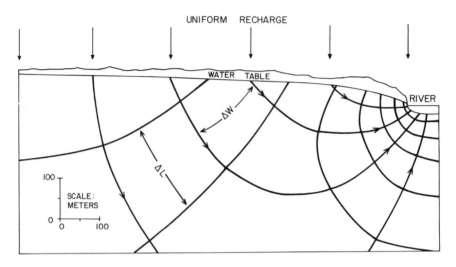

Figure 1.   Flow net representing groundwater circulation near a river which intercepts homogeneous and isotropic aquifer of infinite thickness.

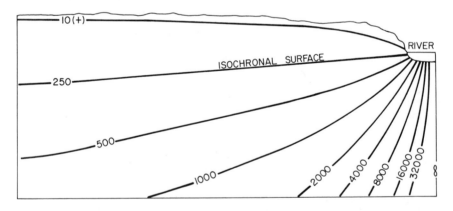

Figure 2.   Age of groundwater circulating in the flow system shown in Figure 1.
Isochronal numbers represent years. Ne is the effective porosity (Ne = 0.25; recharge (infiltration) = 12 cm/year), and K is the hydraulic conductivity (K = 2.50
cm/year = 8 × 10⁻⁶ cm/s (roughly equivalent to silt)).

can be seen from equation 3, the age calculations are particularly sensitive to the determination of ΔL because it is squared. Changes in past climates, tectonic movements, erosion of aquifers, pressures induced by continental glaciers, and changes in sea level may all possibly produce transient hydraulic heads which persist for thousands of years. If long-term transients exist as postulated by Tóth [7] and Kafri and Arad [8], then the determinations of long-term average values of ΔL and Δh for the purpose of water dating by hydrodynamic equations become very difficult indeed.

RADIONUCLIDES OF ATMOSPHERIC ORIGIN

A large number of radionuclides are produced continuously in the upper atmosphere through various interactions between gases and cosmic radiation [9,10]. Some of these radionuclides are produced also in the soil and bodies of surface water by cosmic radiation that penetrates the earth's atmosphere to interact with materials at the surface of the earth. Radionuclides which are of hydrologic interest and which are also produced primarily in the atmosphere (prior to 1945) are listed in Table 1.

Table 1.   Radionuclides of Atmospheric Origin and of Potential Use in Dating Ground Water.

Nuclide	Half-life (years)	Possible Initial Concentration in Rain Water (largely from Oeschger, [10]) (dpm/liter)
^{85}Kr	10.7	less than 10^{-8} dpm (before 1950)
^{3}H	12.26	3.6 (before 1954)
^{39}Ar	270	4×10^{-5}
^{32}Si	330^a	1×10^{-3}
^{14}C	5,730	2×10^{-1}
^{81}Kr	210,000	0.7×10^{-8}
^{36}Cl	301,000	1×10^{-5} to 2×10^{-4}

^{a}See discussion of half-life under ^{32}Si later in manuscript.

During the past 35 years, natural concentrations of ^{3}H and ^{85}Kr, and to a much lesser extent ^{14}C, have been masked by their man-made equivalents. In fact, natural concentrations of ^{85}Kr are completely masked at present by the vast amount of ^{85}Kr from artificial fission reactions. Owing to the relatively short

half-life and the very low natural production rate of [85]Kr, the
natural concentrations of [85]Kr in water which originated prior to
1945 will probably never be measured. Natural ^{3}H concentrations
were much larger than those of [85]Kr, so, at least theoretically,
traces of natural ^{3}H should still be barely detectable in groundwater between about 35 and 60 years old. On the other hand,
because of its longer half-life, ^{14}C concentrations from water
older than 35 years can be detected without difficulty.

The other radionuclides given in Table 1, namely [39]Ar, [32]Si,
[81]Kr, and [36]Cl, are not being produced at present in large amounts
by artificial means. However, nuclear detonations in salt and in
or near salt water have from time to time produced large amounts
of [36]Cl during the past three decades. Also, Dansgaard and others
[11] have presented evidence for a significant but short-term
pulse of bomb-produced [32]Si fallout following the weapons testing
of the early 1960's. The production of [39]Ar and [81]Kr by artificial means is probably small in comparison to natural production
(Oeschger, personal communication, 1978).

A number of general reviews of the use of atmospherically
produced radionuclides for dating groundwater have been written
[12-16]. Most of these reviews center on the use of ^{3}H and [14]C.

A fundamental assumption made for most dating with atmospheric radionuclides is that the cosmic radiation flux and hence,
the natural production of the radionuclides has been constant with
time. Various studies of this problem using ^{14}C and tree-ring
calibration have been made. Isotopic studies of meteorites have
also been useful [17]. Considering the probable lack of basic
accuracy of dating water, the problem of changes in cosmic ray
flux is not serious.

Carbon-14

K. O. Münnich [18] published the first description of the use
of ^{14}C to date groundwater. Since this pioneer paper, countless
studies have been made utilizing ^{14}C in conjunction with conventional hydrogeologic investigations in almost all parts of the
world [4,13,19-22]. One of the most extensive of these studies
was by Pearson who sampled the Carrizo aquifer in Texas and was
able to show a reasonable relationship between hydrodynamic and
[14]C ages of water over a wide region [23].

Despite the vast amount of work on ^{14}C dating which has
already been accomplished and despite the fact that it is the best
developed method available today, numerous difficulties still
exist with its application. First, carbonate geochemistry which
helped control ^{14}C concentrations in the past is not simple to
reconstruct. Carbonate minerals are commonly in a state of near
equilibrium with groundwater, and only slight changes in water
temperature or chemistry will promote either dissolution or precipitation of carbonate ions. In this way, the proportion of
modern carbon in the water can be changed and some isotope

fractionation could take place. Second, large amounts of organically derived carbon from ancient coal and lignite can enter into the groundwater by way of bicarbonate ions and this dead carbon can predominate even in relatively young groundwater [24]. Third, a minute but significant amount of ^{14}C is probably produced in the subsurface. Although not important in dating water less than a few thousand years old, ^{14}C produced in the subsurface may limit accurate dating to water which is 50,000 to 80,000 years old or less [25].

Under certain circumstances, the extent of dissolution of marine carbonate rocks can be estimated by using ^{13}C/^{12}C ratios which are much larger than carbon from terrestrial plants. Very roughly, biologically derived CO_2 in soil of the middle latitudes has a $\delta^{13}C$ value of -25 and marine carbonate rocks have $\delta^{13}C$ values close to 0.0 [26]. Because most ^{14}C enters groundwater through the dissolution of CO_2 which is, in turn, derived from terrestrial plants, a subsurface increase in ^{13}C/^{12}C ratios should reflect the dissolution of marine carbonate rocks which are assumed to be devoid of [14]C. The various steps for isotopic and geochemical corrections of ^{14}C dates have been reviewed by Fontes and Garnier [27]. They gave several examples of their method of correcting dates and compare their method with methods of several other authors. They pointed out that if the correct geochemical adjustments are not made, resulting dates can vary by more than 100 percent, thus, underscoring the necessity of using geochemically sound models to interpret the ^{14}C data.

Although not thoroughly documented, our judgment is that many ^{14}C dates of water should be considered only as order-of-magnitude estimates rather than "dates" in the usual meaning of the isotope geochemist. Where chemical complications are minimal, dates with ±20 percent accuracy may be possible. However, other published "dates" may be easily in error by more than ±100 percent. To the hydrogeologist, nevertheless, even an order-of-magnitude estimate of water age can be of great practical value in trying to decipher complex groundwater systems.

Hydrogen-3

Prior to 1952, most natural ^{3}H, or tritium, was derived from cosmic radiation interacting with the atmosphere. Historical concentrations in rain water in the middle latitudes prior to this time were on the order of 10 tritium units (TU), one tritium unit being equal to one ^{3}H atom per [10]18 atoms of stable hydrogen. The manufacture and testing of fusion devices have injected large amounts of tritium into the atmosphere during the past 28 years. Peak concentrations of more than 10,000 TU were measured in rain over Canada following massive weapons tests in the U.S.S.R. in the mid-1960's. Owing to the nature of atmospheric circulation patterns and the predominance of ocean surface which acts as a ^{3}H sink, precipitation in the southern hemisphere has roughly

one-tenth the tritium concentration of precipitation in the
northern hemisphere. Regional differences of tritium concentrations related to the distance from the coast and local climatic
controls on rainfall are also significant. Seasonal variations
are, in addition, large. In the northern hemisphere during the
1960's, summer maxima of tritium concentrations were ten times
the winter minima.

If tritium were evenly distributed in space and time within
the atmosphere, it would make an almost ideal radionuclide with
which to date very young groundwater [1,28]. Unfortunately, an
accurate historical reconstruction of the effective tritium concentration in past recharge water for a given aquifer is a
difficult task. Not only are original concentrations of tritium in precipitation at a given location poorly known, but
evapotranspiration and other natural phenomena related to local
weather, vegetation, and geology which will affect the tritium
concentrations in groundwater are largely unstudied. For example,
Ehhalt [29] has shown that microorganisms in the soil are able to
oxidize tritiated molecular hydrogen directly from the atmosphere. Inasmuch as the tritium content of the atmosphere may
reach 10^3 to 10^4 times the relative tritium concentrations of rain
water, the direct contribution of tritium to the groundwater
through soil bacteria may be as important under some circumstances
as tritium contributed from precipitation [29].

Owing to the complex problem of defining tritium concentrations at the time of groundwater recharge, most studies make only
a qualitative judgment of groundwater age based on tritium concentrations [5,30,31]. The Isotope Hydrology Section of IAEA [15]
recommended the following three-fold division of "tritium-ages":
1.   Water with concentrations less than 3 TU indicates ground-
      water ages in excess of 20 years.
2.   Water with concentrations between 3 and 20 TU indicates the
      presence of some tritium from testing of fusion devices and
      the water probably dates from the first testing period, that
      is between 1953 and 1961.
3.   Water with concentrations in excess of 20 TU would suggest
      water originating since 1961.

The relatively short half-life of tritium (12.26 years)
requires an appropriate modification of the above criteria for
tritium studies made at a date later than the date of publication
(1973). Also, the criteria are developed for the mid-latitudes
in the northern hemisphere and should not be applied elsewhere.

Tritium extracted from soil moisture in the unsaturated zone
at various depths below the surface has been used to infer the
progress of recharge of underlying aquifers. Studies of recharge
in arid and semiarid zones where water moves very slowly in a
downward direction have been particularly instructive [20,
32-35].

The short half-life of tritium imposes a time limit on the
usefulness of tritium dating. However, because tritium decays to

the stable helium isotope, ^{3}He, and because natural background concentrations of ^{3}He in water are so low, the original tritium concentration of groundwater can be determined theoretically by measuring the excess ^{3}He present [36,37]. Then, if the original tritium concentrations in the groundwater recharge can be determined as a function of time, dating of the water may be possible. To be useful for precise dating, the ^{3}H-^{3}He method would need to assume, first, that large seasonal fluctuations of original ^{3}H are "averaged" by some mixing process in the sub-surface and, second, that non-tritium sources of anomalously large ^{3}He concentrations are not present. Such sources, fortunately, are probably confined primarily to areas where deep thermal waters rise to the surface and would not be present in normal, near surface, groundwater. The matter of ^{3}He concentrations in water, nevertheless, needs further study. As is mentioned below, anomalous ^{3}He values could possibly be found associated with natural concentrations of lithium.

The theoretical aspects of subsurface production of tritium have been investigated. Normal aquifers should not have more than about 0.5 TU which originate in the subsurface primarily by natural fission of ^{238}U and by capture of thermal neutrons by ^{6}Li with a subsequent release of an alpha particle [38]. Unusually high concentrations of uranium and lithium, however, could give rise to perhaps as much as 1.5 TU through subsurface production. Such small concentrations are of little direct importance to normal tritium dating because the usual precision of tritium analyses is commonly about the same as the postulated background values produced by natural subsurface nuclear reactions. Never-theless, in view of the above discussion, trace amounts of tritium found in old groundwater should not be explained on the basis of sample contamination nor the mixing of small amounts of modern groundwater with predominantly old water unless the entire matter has received careful study.

Chlorine-36

The first analyses of ^{36}Cl in natural waters were reported by Schaeffer and others [39]. Based on this work, Davis suggested [40] that ^{36}Cl would be useful to date old groundwater because its half-life of 3.01 x 10^5 years is ideal for the range of 5 x 10^4 to 1 x 10^6 years which is beyond the normal range of ^{14}C dating. In addition, chloride in groundwater is neither derived normally from, nor reacts with, the solid matrix of the aquifer. Thus, the problems of geochemical interpretation are not as formidable with ^{36}Cl as with ^{14}C. Tamers and Ronzani [41] were the first to actually investigate dating of groundwater using ^{36}Cl. Unfortunately, they considered only cosmogenic ^{36}Cl production at the earth's surface and ignored the component of atmospheric origin, shown by Bentley [42] to be much more important.

Moreover, their conventional counting of the very low ^{36}Cl
activity required heroic analytical methods.

The development of mass spectrometric techniques for nuclide
identification using a tandem Van de Graaff accelerator at the
University of Rochester Nuclear Structure Laboratory by H. Gove,
K. Purser, A. Litherland, and numerous associates has provided an
excellent means for the precise measurement of ^{36}Cl concentrations
in natural water [43]. Thus far, about 40 groundwater related
samples which have been collected and purified chemically by
H. Bentley have been analyzed for ^{36}Cl by D. Elmore, H. Bentley,
and others using the University of Rochester machine. Some of
these samples are listed in Table 2.

Table 2.   Analyses of Chlorine-36 Using Accelerator at the University of Rochester Nuclear Structure Research Laboratory.

(Analyses are given as the ratio of ^{36}Cl nuclei to 10^{-15}
times the total number of chlorine nuclei)

Sample Number	^{36}Cl/Cl ($\times 10^{15}$)

Samples of water less than 20,000 years old

Tucson, Arizona, alluvial aquifer

| 1. | City well #B-18 . . . . | 365 ± 18 |
| 2. | City well #C-13 . . . . | 379 ± 22 |

Madrid, Spain, Tertiary alluvium

3.	Well 535-7-b	231 ± 21
4.	Well 535-7-a	295 ± 12
5.	Well 535-5-c	235 ± 7

Southern, Texas, Carrizo Sandstone

| 6. | Well . . . . | 32 ± 3 |
| 7. | Well . . . . | 64 ± 6 |

North Dakota, Fox Hills Sandstone

| 8. | Well, Bowman 131-102-14AAB . . . . . . | 258 ± 13 |

Samples of water between 10^5 and 10^7 years old

North Dakota, Fox Hills Sandstone

| 9. | Stanton, Well #400, 144-085-03DCD . . . . | 7.1 ± 2.3 |
| 10. | Mandan, Well #139-081-09 AAA1 . . . . . . | 10.2 ± 2 |

Table 2 continued

Sample Number	$^{36}Cl/Cl$ $(x10^{15})$

South Carolina, Metamorphic rocks below Coastal Plain sediments

    11.   Savannah River Plant, Test well #DRB6 . .   9.5 ± 1.0
    12.   Savannah River Plant, Test well #DRB11 .   6.9 ± .7

Samples of water likely to contain ^{36}Cl of bomb origin

Tucson, Arizona, alluvial aquifer

    13.   Well-Campbell Farms . . . . 1950 ± 150

Southern Texas, Carrizo Sandstone

    14.   Well A1-68-51-803 . . . .   90 ±   9

Ocean Water

    15.   Atlantic Ocean, surface water, AII85 . .   2 ±   2

Samples of solid material

    16.   Modern salt crust, Willcox Playa,
          Arizona . . . . 177 ± 10
    17.   Clear Fork, Texas, salt from salt dome .   1 ±   2
    18.   Thorium ore . . . . 256 ± 17

Variations in the $^{36}Cl/Clx10^{15}$ ratios of samples 1 through 12 shown in Table 2 are a function of numerous factors among which latitude, proximity to the coast, and age appear to be most important. Inasmuch as the average age of chloride in the oceans is probably in excess of 10^8 years, the concentration of ^{36}Cl per total chloride atoms is very low in ocean water (sample 15, Table 2). The effect of "dead" marine chloride near the coast is seen clearly in the contrast between young samples from Texas (Samples 6 and 7, Table 2) as compared with young samples from Madrid (Samples 3, 4, and 5), Tucson (Samples 1 and 2), and North Dakota (Sample 8). The $^{36}Cl/$ $Clx10^{15}$ ratios in all the young samples can be predicted [42] by considering the atmospheric cosmogenic ^{36}Cl which depends on latitude [44] and the average annual delivery of chloride to the sample area [45] which is largely a function of the distance to the coast.

Time, of course, allows disintegration of the ^{36}Cl once water enters the subsurface. Ideally, ^{36}Cl concentrations should decrease regularly downdip in an aquifer as water carries the chloride deeper into an aquifer. In the aquifer studied in most detail thus far, the Fox Hills Sandstone of North Dakota, a reasonable correlation exists between hydrodynamic ages and ^{36}Cl ages [46; Bentley, manuscript in preparation]. In the Carrizo Sandstone, however, the data obtained thus far (most of which are not shown in Table 2) are not as easily interpreted because there is an initial increase in ^{36}Cl concentrations as well as $^{36}Cl/Cl$ ratios in the uppermost part of the aquifer. Pearson's data on ^{14}C ages of Carrizo groundwater [23] indicate that these anomalous samples are all less than 30,000 years old. Induced upward migration of old water containing dead chloride from lower aquifers may accompany the development of some of the wells near the outcrop, thus lowering the $^{36}Cl/Cl$ ratios and causing water from these wells to appear older than the down-gradient waters where groundwater extraction is minimal. Also, lowering of the ocean levels with a retreat of the shoreline during the Pleistocene may have caused a reduction in dead chloride in the recharge areas of the Carrizo Sandstone during the period of approximately 15,000 to 70,000 years ago. This would make the Pleistocene waters which are now down-dip in the aquifer, appear younger than modern groundwater near the outcrop of the aquifer. Concentration increases of ^{36}Cl may have been caused by higher evapotranspiration of the older waters due to a more arid climate in the past. Another possible mechanism for ^{36}Cl concentration increase is ion concentration due to membrane effects of the clays and shales which confine the Carrizo aquifer.

Complications in ^{36}Cl dating of groundwater, which are also applicable to dating with other atmospheric radionuclides, are (1) possible isotope fractionation due to membrane effects of groundwater passing through silt and clay beds, (2) cross-formational flow of groundwater in aquifers which appear to be isolated by nonpermeable beds but actually are not isolated, (3) possible diffusion of dead chlorine from fluid inclusions in minerals within crystalline rocks, and (4) subsurface production of ^{36}Cl by the natural subsurface neutron flux. The actual effects of the items listed above are quite site specific and, therefore, need to be investigated on a case by case basis. Notwithstanding this site-specific nature, considerable general research is needed to help bound the problems. For example, Bentley [42] has calculated the normal ranges to be expected from subsurface production of ^{36}Cl and has concluded that it becomes significant after about two half-lives and may dominate the ^{36}Cl concentrations after four half-lives. The useful range of dating by ^{36}Cl is limited on the upper extreme at about one million years by effects of subsurface production. In the absence of significant neutron production in the subsurface, as would be expected in exceptionally pure deposits of limestone and halite, subsurface production should be very slight (Sample 17, Table 2). On the other hand, uranium or thorium ore should have a maximum subsurface production which would roughly be equivalent to the rate of atmospheric production (Sample 18, Table 2).

Whether or not cosmic radiation of the soil surface increases the available ^{36}Cl significantly is an open question. Certainly some ^{36}Cl is produced, but the effects are probably not dominant. Sample 16 (Table 2) is from a playa which has a thin salt crust. Even though the water seeping into the playa and the salt already there are both young and should contain relatively large amounts of ^{36}Cl, the $^{36}Cl/Cl$ ratio is still below that of shallow groundwater in the region (Samples 1 and 2). The extent of ^{36}Cl derived from cosmic radiation of soil is not particularly important to know, however, for most water dating projects. As we visualize the dating method, the initial ^{36}Cl concentrations will be established by sampling water which is near the aquifer intake area but which is still a few hundred to a few thousand years old. ^{36}Cl derived from rain, dry fallout, and soil leaching should be mixed together and roughly averaged over periods of several centuries. Because the ^{36}Cl method of dating may be sensitive to Pleistocene climatic fluctuations, extensive sampling in the aquifer of interest combined with other studies of radionuclides and paleotemperature indicators is advised.

Silicon-32

The fact that ^{32}Si has a half-life intermediate between ^3H and ^{14}C, the two radionuclides most commonly used for dating water, suggestes that it could be important for dating water which is generally between 50 and 1,000 years old [47]. Lal and his co-workers in India have been the most active in investigating the hydrogeologic applications of ^{32}Si dating [48].

Most ^{32}Si dates appear to be much younger than ^{14}C dates of the same water. This discordance may be explained in part by hydrodynamic mixing of waters in shallow aquifers [49]. A recent determination of the half-life of ^{32}Si, however, has suggested that the discordance in dating is even larger than formerly believed. Published values of the half-life vary from 101 years to 710 years. The smallest value is the latest determination [50,51] which, if adopted, will replace the previously accepted value of 330 years. Ages calculated using the 101-year half-life will be significantly smaller than any previously published dates.

In addition to the questions related to the half-life of ^{32}Si, many questions exist as to the details of the near-surface radiochemistry and geochemistry of silica. As already mentioned, there probably exists a significant but poorly known contribution of ^{32}Si from fallout from testing nuclear bombs [11]. The first few centimeters of soil are subject to nuclear reactions produced by cosmic radiation such as the possible spallation of ^{35}Cl to produce ^{32}Si. The extent of near-surface production is unknown. Even more serious is the very complex nature of silica geochemistry, particularly in the soil horizon. Significant amounts of silica accumulate in growing plant materials [52] and under some circumstances can also accumulate as phytoliths of opal [53] and possibly other more complex compounds in the upper part of the soil horizon. Inasmuch as most of the silica which is mixed with organic material in the soil is probably in a relatively soluble form or in a form which is easily desorbed, water which eventually becomes groundwater recharge could have initial ^{32}Si concentrations somewhat above those of the original precipitation.

At the present stage of development, dating with ^{32}Si is best applied to the establishment of relative ages of water in a single aquifer. The method is probably not a reliable means of establishing an absolute age.

Argon-39

The half-life of 270 years of ^{39}Ar makes it useful for dating materials in the 50 to 2,000-year range [54], which is in the age range between dating by ^3H and ^{14}C. Preliminary studies by Loosli and Oeschger [55] suggest that, indeed, ^{39}Ar will give good relative dates for different samples of groundwater. However, like ^{32}Si dates, the ^{39}Ar dates are generally much younger than dates of the same water obtained by using ^{14}C.

Loosli and Oeschger [55] considered three primary explanations for the ^{39}Ar-^{14}C dating discrepancies. First, and perhaps most important, the large natural abundance of potassium would make the reaction $^{39}K(n,p)^{39}Ar$ quite important. Thus, subsurface production of ^{39}Ar could produce apparent ages which are far too young. Loosli and Oeschger [55] as well as Zito (1980, personal communication) have estimated that, under certain circumstances, the subsurface production of ^{39}Ar could exceed the atmospheric production. However, data are lacking for reliable calculations of both the ^{39}K cross section for the capture of thermal neutrons and the rate of transfer to the water of the ^{39}Ar which is generated in solids. A second possible explanation for the discrepancies is that the ^{14}C dates are far too large because of dead carbon entering the system or a "chromatographic separation" of dissolved species containing ^{14}C takes place as water flows downgradient in the aquifer. Finally, Looslie and Oeschger [55] considered that a subsurface mixture of old and young waters can account for some of the different dates which are obtained by various methods.

Krypton-81

Dating groundwater with ^{81}Kr, if it ever proves feasible, would have several advantages. First, the long half-life of 210,000 years should allow dating of water beyond the range of ^{14}C. Second, the gas is inert which would simplify the problems of geochemical interpretation of analytical results. Third, the production of significant amounts of ^{81}Kr is probably confined to natural nuclear reactions in the atmosphere and shallow soil horizon induced by cosmic radiation. Natural subsurface production and artificial nuclear reactions should not complicate the interpretation of the results. Unfortunately, natural concentrations are very small and water sample sizes of more than 10^6 liters may be required to obtain measurable amounts of ^{81}Kr [10]. Such large sample sizes will require some type of gas separation system with a large through flowing system of water from the well or spring which is being studied. The potential problems of sample contamination with such a system will be difficult to handle. Thus far, water dating by ^{81}Kr has not been accomplished.

Krypton-85

The present atmospheric concentrations of ^{85}Kr are almost entirely from artificial nuclear fission. Its half-life of 10.7 years is nearly the same as ^{3}H, so the useful range of dating is similar. Unlike ^{3}H, however, concentrations of ^{85}Kr have been increasing at a more-or-less steady rate (figure 3) for the past 35 years. Therefore, the input function for ^{85}Kr in groundwater is much simpler than ^{3}H and, theoretically, the resulting dates should be more accurate. The primary difficulty with using ^{85}Kr for dating purposes is the fact that, even for modern water, the concentrations are very low, necessitating the separation of krypton gas from relatively large samples (120 to 360 liters) of water [56]. Although not widely used at present, ^{85}Kr dating could be superior to ^{3}H dating for groundwaters less than 30 years old.

ACCUMULATION OF PRODUCTS OF RADIOACTIVE DECAY

Introduction

As time passes, the direct or indirect products of various radioactive decay processes may accumulate. If these products tend to be formed in groundwater or tend to migrate into the groundwater and if the products move with a known relationship to the movement of the groundwater, then the concentration of the products in the water may indicate water age. A fundamental advantage of using decay products as a basis for dating is the fact that as time progresses, more products will be present and the analytical aspects of dating will become easier. This is in contrast with the use of atmospheric radionuclides which will become more difficult to detect as the age of the groundwater increases.

A number of decay products may be of interest ultimately as a basis of dating groundwater. At present, however, the accumulation of inert gases appears to offer the most significant possibilities for dating [19,36,58-60]. Some candidate gases are given in Table 3. Of those listed, ^{4}He will probably be the most useful because of its relatively rapid rate of production. As already mentioned, because it is the decay product of tritium, the other

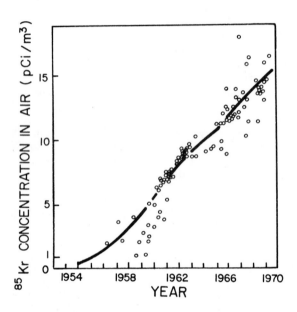

Figure 3.   Concentration of ⁸⁵*Kr measured in northern hemisphere air samples (57).*

Table 3. Noble Gases of Possible Use for Dating Groundwater.

Nuclide	Origin	Possible Range of Ages for Dating[a]
^{3}He	Stable end product of the decay of ^{3}H.	5 to 50 years
^{4}He	Neutralization of alpha particles.	10^4 to 10^7 years
^{21}Ne	Capture of alpha particles by ^{18}O.	10^6 to 10^7 years
^{40}Ar	Decay product of ^{40}K. (10.7% of ^{40}K decays to ^{40}Ar, the rest decays to ^{40}Ca).	10^5 to 10^7 years
^{81}Kr	Cosmic radiation interacting with the atmosphere.	10^4 to 10^6 years
^{85}Kr	Fission of ^{238}U for natural systems. Fission of ^{235}U and ^{239}Pu for artificial systems. Only anthropogenic ^{85}Kr is quantitatively important.	1 to 40 years
^{130}Xe	Natural fission of ^{238}U.	10^6 to 10^7 years
^{136}Xe	Natural fission of ^{238}U.	10^6 to 10^7 years
^{222}Rn	Decay product of ^{226}Ra in the ^{238}U decay series.	0.5 to 10 days

[a]Theoretically, the accumulation of noble gases could be used to date water older than 10^7 years; however, under most situations geologic considerations would cast doubts on the significance of such dates. For example, in the United States, more than half of the groundwater which is pumped comes from aquifers which did not exist 10^7 years ago.

isotope of helium, ^{3}He, may be also useful to date very young groundwater. The potential usefulness of only ^{4}He and ^{40}Ar will be discussed in this section. Not enough is known about the production rates and/or release rates to groundwater of the other gases.

Helium-4 Accumulation

The subsurface accumulation of ^{4}He is largely from the neutralization of alpha particles. Although alpha particles can originate in a variety of ways, most will come from the decay of heavy radionuclides such as ^{232}Th, ^{146}Sm, ^{190}Pt, and ^{174}Hf. Of the heavy radionuclides, ^{238}U is the most important source of alpha particles under normal circumstances, and ^{232}Th is next in importance. The third in importance is ^{235}U. Sources of ^{4}He other than uranium and thorium probably account for less than one percent of the total.

Owing to the fact that most ^{238}U and virtually all the ^{232}Th nuclides are bound in the structure of minerals and also to the fact that the mean path length of alpha radiation is very short, most newly formed ^{4}He resides originally in solid material. The rates with which ^{4}He diffuses from the solid rock matrix into adjacent water-filled pores is largely unknown. The fact that the ^{4}He is not retained perfectly in the lattice of the rock-forming minerals is well-known because of the early failures to date minerals by the "helium clock". The exact rate of escape of ^{4}He is undoubtedly a complex function of temperature, concentration gradients of ^{4}He, distribution of alpha-generating nuclides with respect to rock pores, original energy of the alpha particles, and types of minerals in contact with the groundwater. As a first approximation, the release of ^{4}He from solid material into the groundwater system might be considered as a steady-state process [60]. If so, then perfectly static groundwater would experience a nearly linear increase in ^{4}He with time. This linear relation is true because the half-life of most alpha-producing radionuclides is more than two orders of magnitude larger than the age of the oldest groundwater which might be dated.

Early suggestions to use helium to date groundwater were made by Savchemko in 1936 as quoted by Spiridonov and others [58] and Davis [40]. Serious attempts to date groundwater by helium accumulation measurements have been made recently by Spiridonov and others [58], Marine [60], Fritz and others [4], and Bath and others [19]. Of the various studies, those made by Bath and co-workers [19] have been most elaborate and shown most promise. Although ^{14}C and ^{4}He dating of the same groundwater samples yields a distinct correlation between the dates derived by the two separate methods, the correlation is not a direct one-to-one relationship. Almost all ^{4}He dates are distinctly older, some by a factor of 3 or more, than corresponding ^{14}C dates [19,4]. Some representative values from the literature are given in Table 4.

Table 4.   Comparison of Carbon-14 and Helium-4 Dates of Groundwater.

Location of Sample	^{14}C Age (years)	^{4}He Age (years)	Reference
Stripa, Sweden	30,630 (uncorrected)	900,000 (assuming 1 ppm U and 4 ppm Th)	Fritz and others, 1979 [4]
Savannah River, South Carolina	Less than 20,000 Possible sample contamination.	840,000	Marine, 1976 [60]
Eastern England, Halam	Between 1,500 and 4,300 years BP	27,000	Bath and others, 1979 [19]
Grove	Between 7,300 and 10,600 years BP	10,600	"
Egmanton	Between 4,000 and 7,300 years BP	17,600	"
Rampton	Between 33,000 and 36,100 years BP	42,100	"
Gainsborough	Between 24,900 and 28,200 years BP	65,600	"

The greatest obvious weakness in the [4]He method of dating is in the assumptions necessary to calculate the flux of [4]He into the groundwater. The exact distribution of U and Th in the rock mass is not generally measured but is, nevertheless, of critical importance. In particular, U is relatively mobile in water under oxidizing conditions and will commonly form fine-grained mineral coatings along water-filled fractures in otherwise solid rock. The opportunity for direct [4]He release to groundwater under these conditions is almost infinitely greater than [4]He release from larger, solid mineral grains imbedded in virtually nonpermeable rocks. The effects of microfractures might also be important as avenues of [4]He migration. The apertures of the microfractures will change with the stress conditions in the rock and will open or close in response to changes in surface loads imposed by glacial ice and fluctuations of nearby bodies of water. Thus, conflicting ^{14}C and [4]He dates from water from the Stripa Mine in Sweden [4], for example, may be caused by dilation of microfractures with accompanying release of accumulated [4]He. This might have taken place during the last period of deglaciation approximately 10,000 years ago. Heaton and Vogel [61] have postulated that the migration of methane could also act as a carrier gas for the migration of He. The matter of the flux of helium from intermediate and deep geologic sources is, moreover, an unsettled question. Relatively large helium fluxes have been measured for many years in areas of recent volcanism and geothermal activity [62]. Also, large helium concentrations have been found in young, nonthermal groundwater [55] and surface water [63]. The possibility exists that helium from deeper parts of the earth's crust or from the mantle can seep upward in significant quantities in areas outside of obvious geothermal activity. An index of this helium flux from deep sources might be the [3]He/[4]He ratio which would be much higher than values calculated from purely in situ generation of helium. The mere presence of [3]He, however, does not automatically indicate deep sources of the helium because of the significant production of [3]He from natural radioactivity [38].

Argon-40

Owing to the disintegration of [40]K, the amount of [40]Ar present in groundwater should increase slowly with time. Despite the natural abundance of [40]K, however, the volume of [40]Ar which is produced in the average rock is less per unit time than the volume of [4]He. If one kg of average igneous rock is assumed to have 1 mg U, 4 mg Th, and 25 g K, the production of gas at STP per kg of rock will be:

from ^{238}U, 1.160×10^{-10} mL/yr. of [4]He;
from ^{235}U, 4.73×10^{-11} mL/yr. of [4]He;
from ^{232}Th, 1.139×10^{-10} mL/yr. of [4]He;
and from ^{40}K, 9.49×10^{-11} mL/yr. of [40]Ar.

Thus, under the assumed conditions which were chosen to favor ^{40}Ar production, the ^{40}Ar production will be less than one half of the ^{4}He production. Furthermore, the diffusion of ^{40}Ar out of the production sites in the minerals will be less effective than the diffusion of ^{4}He. Even if all the locally generated ^{40}Ar were to enter the groundwater, the ^{40}Ar would be difficult, nevertheless, to differentiate from atmospherically derived ^{40}Ar. The ^{40}Ar/^{36}Ar ratio in groundwater should increase with time from the original atmospheric ratio of 295.5. However, the increase will be very slow. If one percent porosity and 2.8 g/mL density are assumed for the theoretical igneous rock discussed above and if the rock is saturated with water which has equilibrated with the atmosphere at 5 °C, then the ^{40}Ar/^{36}Ar ratio will only increase to 297 after 10^5 years provided all the ^{40}Ar which has been generated in the rock migrates into the water. After 10^6 years, the ratio will only increase to 313. Given certain errors related to sampling and analyses, it is doubtful that water less than 10^5 years can ever be dated by ^{40}Ar; this is particularly true if one considers the fact that most of the ^{40}Ar generated in the rock will probably be retained by minerals in the rock. Normal concentrations of ^{40}K dissolved in groundwater are at least an order of magnitude lower, and commonly four orders of magnitude lower, than ^{40}K in the solid rock, so direct generation of ^{40}Ar in the water would not provide significant amounts of ^{40}Ar for dating until after several million years.

URANIUM DISEQUILIBRIUM

If species with short half-lives are omitted, the first portion of the ^{238}U decay series can be written as

$$^{238}U \xrightarrow{4.47 \times 10^9 \text{ yr}} {}^{234}U \xrightarrow{2.45 \times 10^5 \text{ yr}} {}^{230}Th \xrightarrow{7.7 \times 10^4 \text{ yr}}$$

$$^{226}Ra \xrightarrow{1.60 \times 10^3 \text{ yr}} {}^{222}Rn.$$

Numbers of years refer to half-lives of the reactions. Because the half-life of ^{238}U is much longer than the other radionuclides in the series, the activities of all the radionuclides will be roughly equal after about 2×10^6 years provided the parent and daughter products are in a closed system. In natural systems which are exposed to circulating groundwater, however, the activity ratios of these species are rarely equal [64]. Theoretically, this disequilibrium of radionuclides in the water can be used to date the water provided enough is known about the local geochemistry of the dissolution, sorption, and precipitation processes affecting the radionuclides. In practice, however, such detailed information normally cannot be obtained, so major simplifying assumptions are commonly made. For some models, the geochemical system is assumed to be in a steady-state, thorium is assumed to be entirely in the solid form, uranium is assumed to go into solution in the near-surface portions of aquifers where oxidizing conditions prevail, uranium is assumed to precipitate in the solid form primarily in deeper portions of the aquifer where oxidizing conditions change to reducing conditions (figure 4), and lastly, deep within the aquifer where conditions are reducing, the uranium already in solution is assumed to remain in solution without significant effects of sorption on solid materials.

In dating groundwater with the $^{234}U/^{238}U$ activity ratio, an initial ratio must be estimated from field data. Most commonly, the initial ratios in water from shallow aquifers are greater than 1.0 with values higher than 10.0 encountered in some regions. High values are caused by direct recoil of alpha-emitting nuclei and the selective dissolution of ^{234}U from sites where the mineral structure has been damaged by alpha distintegration of ^{238}U [65]. If the only source of ^{234}U were direct alpha recoil of the nuclide into the water, the buildup of ^{234}U would be related primarily to time and to the microdistribution of the uranium in relation to the water-filled pores. If the geochemistry of the solid materials is uniform throughout the aquifer, then increases in ^{234}U might be used to deduce water ages up to about 1.5×10^6 years. However, selective dissolution of minerals along alpha-recoil tracks will also introduce ^{234}U into the water [65], so the buildup of ^{234}U is, in addition, some complex function of the hydrochemistry of the mineral-water system. Clearly, the buildup curves for $^{234}U/^{238}U$ must be interpreted with caution.

Another possibility for utilizing $^{234}U/^{238}U$ for dating is to measure the decrease of the $^{234}U/^{238}U$ ratio in the deeper part of the reducing zone of the aquifer as an index of relative water age (figures 4 and 5). Theoretically, as the water changes from an oxidizing environment to a reducing environment, most of the uranium will precipitate in this transition zone [12,66]. Groundwater flowing through the transition zone and continuing down-gradient should enter into a zone that is geochemically quite uniform and is quite close to being in chemical equilibrium. It has been postulated [12] that in this downgradient region the uranium which has entered into solution will tend to remain in solution and that additional uranium will not be dissolved. Thus, only radioactive decay will affect the $^{234}U/^{238}U$ ratios which then can be used directly for dating.

Most researchers studying uranium isotopes in water agree that only qualitative dating is possible at present. However, with more information concerning the entire geochemical system, the calculation of actual groundwater ages may be possible. Recent studies of uranium in ore deposits [67] and fractured source rocks as well as water-deposited calcite in caves (speleothems) have added useful background material [64, 68-71]. For example, under some circumstances the $^{234}U/^{238}U$ ratios in speleothems appear to be relatively constant over periods of several thousand years suggesting an average stability in that part of the geochemical system affecting uranium dissolution.

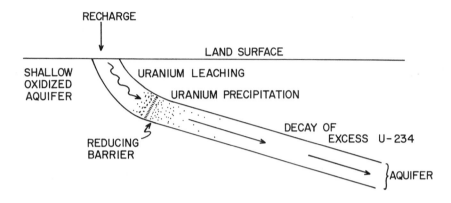

Figure 4.   *Simplified geochemical explanation of U migration in an aquifer.*

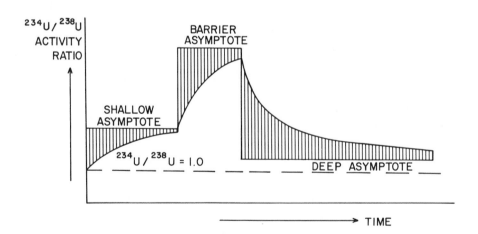

Figure 5.   *Changes in $^{234}U/^{238}U$ ratios in water as a function of time in the aquifer shown in Figure 4. Velocity of the water is assumed to be constant, so distance of travel is directly proportional to time (12).*

This in turn would lend credibility to the method of dating water by uranium disequilibrium. Additionally, the use of $^{230}Th/^{234}U$ ratios to date the solid parts of aquifers would give information on an upper age limit for the water saturating the aquifer. For example, if $^{230}Th/^{234}U$ dates indicate that minerals were being deposited in the aquifer 20,000 years ago, then the present water saturating the aquifer is probably less than 20,000 years.

CHEMICAL DISEQUILIBRIA

Under ideal circumstances, certain chemical processes which are relatively sluggish may possibly be used for water dating. Near-surface water which is low in dissolved silica, for example, might be undersaturated with respect to silica which in turn would suggest that the water is less than 10 years old and probably less than a few months old. Unfortunately, the large number of variables which control dissolution or precipitation of minerals in natural systems probably can never be defined with sufficient precision to enable more than the most general, qualitative dating.

Rather than considering mineral-water reactions as a basis for dating, the presence of certain metastable molecules which alter spontaneously with time may offer better opportunities for dating water. Some attention has been given to this possibility [12], but, to date, field-oriented data are lacking. The only chemical group mentioned thus far in the literature in connection with water dating has been the amino acids which undergo spontaneous changes with time. These changes, termed racemization by organic geochemists, are also temperature dependent [72,73]. To be useful for dating water, amino acids originating only at the surface must be identified, their rate of racemization determined, and the thermal history of the water after dissolution of the amino acids must be estimated. The presence of ancient organic materials, particularly buried soils, in the subsurface and the present-day activity of certain bacteria in the subsurface make the task of identifying unique amino acids for dating difficult. Moreover, many groundwater systems include deep circulation where temperatures may be more than 20 °C warmer than at the surface, so significant temperature-dependent effects could be encountered [73]. Although modeling the flow system may allow meaningful estimates of the temperature effects, dating with amino acids will probably always yield only qualitative results.

ANTHROPOGENIC CONSTITUENTS

Several anthropogenic constituents which are present in the atmosphere are potentially useful as an index of water age. Two radioactive gases from nuclear weapons and from power reactors, ^{3}H and ^{85}Kr, have been discussed already. Several other radionuclides of man-made origin are present in the atmosphere and in the soil in soluble form. Of these radionuclides, ^{90}Sr, ^{129}I, ^{99}Tc, and ^{106}Ru would have the longest half-lives combined with lesser tendency to be absorbed on solids than other man-man radionuclides such as ^{137}Cs. Even with the extensive testing of nuclear explosives in the atmosphere during the 1960's, environmental concentrations of most artificial radionuclides are at such low levels that analytical detection of their presence in recent groundwater would be difficult. Tritium and possibly ^{85}Kr will probably remain the only convenient radionuclides having man-made origins which can be used world-wide.

A family of anthropogenic chemicals called halocarbons offer some interesting possibilities for dating water which is less than about 40 years old [74,75]. Many of the compounds are easily detected in very small concentrations with a gas chromatograph having an electron capture detector. Thus far, trichlorofluoromethane (Freon-11) and dichlorodifluoromethane (Freon-12) have been investigated with some success. Two dating methods are used. One method is based on the fact that the atmospheric inventory of the compounds has increased dramatically during the past 40 years. Concentrations in rainwater and the resulting groundwater recharge water, consequently, also have increased systematically during this period. Because both compounds (Freon 11 and 12) are stable at ambient temperatures and neither compound adsorbs strongly on normal aquifer materials, the concentrations of the compounds in groundwater should correlate with water ages back to 30 or 40 years before the present [75]. The other method uses the fact that the two compounds have been introduced into the atmosphere at different rates [76,77]. Because of its early widespread use in refrigeration, Freon 12 concentrations increased in the atmosphere first. Later use of Freon 11 together with Freon 12 as aerosol propellents, foaming agents, cleaners, etc. has produced a changing Freon 11/Freon 12 ratio with time in the atmosphere as well as in newly recharged groundwater. This ratio, therefore, may be correlated with the age of the water.

Freon dating has advantages over tritium dating of lower potential cost and greater precision. The greater precision comes from a better knowledge of the initial input concentrations because the introduction of Freons into the atmosphere has been much more uniform than tritium in both space and time. Potential disadvantages come primarily from possible sorption of Freons, particularly on organic materials. Also, considerable development work with Freon dating is needed before concentrations in recent water can be correlated with confidence with local climatological and geographic factors even though the general world-wide buildup of the compounds in the atmosphere can be reconstructed with some confidence.

MATCHING PALEOCLIMATIC INDICATORS WITH WATER AGES

Several constituents in groundwater give either direct or indirect indications of original water temperatures or general climatic conditions during the time when the water infiltrated into the subsurface. At least three general approaches can be used to reconstruct these past climates. The most common method uses the $^2H/^1H$ and $^{18}O/^{16}O$ ratios in water to infer storm patterns, evapotranspiration, and past temperatures [78,79]. A second method uses the concentrations of noble gases in water also to infer paleotemperatures. A third method uses the chloride content of groundwater to interpret ancient rates of evapotranspiration and/or positions of ancient shorelines.

Once climatic trends as interpreted from the groundwater data are established, they can be matched with the known chronology of climatic fluctuations back to about 200,000 years before present. This method, of course, is not precise and may even be misleading unless abundant regional data are available. Notwithstanding many shortcomings of the method, the simple fact that groundwater may have recharged at a time when average surface temeratures were distinctly different than at present is a valuable bit of information which, when combined with other dating methods, may provide information on groundwater ages.

Hydrogen-2/Oxygen-18 Ratio

The stable isotopes of oxygen and of hydrogen fractionate in the atmosphere and at the earth's surface. In general, the lighter isotopes are associated with precipitation in cooler weather, at higher elevations, and at a great distance from the sea [80]. Because of several favorable geographic and climatological factors in Greenland, Dansgaard [81] was able to correlate concentrations of the stable isotopes of oxygen and hydrogen in ice with average temperatures on the ice cap. Other workers have attempted to extend the work of Dansgaard to the interpretation of paleotemperatures in old groundwater [78]. Nevertheless, this extension is open to question. Several physical factors must be, on the average, constant enough so that the temperature imprint on the isotope ratios can be detected. Some of the most important of these factors are:

1. Topography. The local elevation at the aquifer intake area as well as the configuration and elevations of surrounding mountains must be relatively constant.
2. Recharge inducing storms. The general nature of the storms and their trajectories over the aquifer intake area must be, on the average, constant. A change, for example, from dominantly local convective summer storms to dominantly frontal winter storms could account for changes in isotopic ratios in the water which neither reflect changes in average temperatures nor total annual precipitation.

3. <u>Evapotranspiration</u>. The total evapotranspiration which takes place prior to groundwater recharge should be constant.
4. <u>Position of the coastline</u>. The position of the coastline with reference to the position of the aquifer intake area should remain constant. This is particularly critical along coastlines with gently sloping topography and shallow water on the continental shelf. In such areas, lateral shifts of the shoreline of more than 200 km have been common during the past 20,000 years.
5. <u>Subsurface reactions</u>. The geochemical modifications of groundwater, particularly ion filtration and hydrothermal reactions, must not be of the type which would fractionate the isotopes being studied.

Fortunately, if enough samples are available from a given aquifer, the variations of ^{2}H and ^{18}O can be compared with Craig's empirical relationship, which for normal surface waters is $\delta^2H = 8\ \delta^{18}O + 10^o/_{oo}$. Large departures from Craig's correlation line could indicate the effects of evapotranspiration or of hydrothermal reactions.

At present, the viewpoints of hydrogeologists concerning the utility of ^{2}H and ^{18}O analyses to give paleoclimatic information vary from almost blind acceptance to rather general skepticism [82].

Noble Gases

The concentrations of noble gases in groundwater should reflect the surface temperature at the time of groundwater recharge, provided the recharge is rapid and goes directly into the aquifer [83-85]. The solubility of each noble gas in water is different, each with a unique relationship with temperature. At a fixed temperature, the heavier gases are more soluble, Xe being roughly 400 times more soluble (on the basis of mass ratios) than He at 20 °C. More importantly, the solubilities of the heavier gases are far more temperature sensitive than the lighter gases. For example, the solubility of He varies only six percent due to a temperature change from 5° to 20 °C; whereas, the same temperature change causes a 40 percent change in the solubility of Xe. Therefore, with a given sample of groundwater three independent paleotemperatures can be calculated, one for each of the heavier noble gases (Ar, Kr, and Xe).

Paleotemperatures derived from noble gas analyses are potentially more meaningful than those from oxygen-deuterium analyses because the noble gas content is a direct measure of the temperature of the water at the time of infiltration rather than a complex function of geographic and meteorological factors as is the case with ^{2}H and ^{18}O. Despite this potential superiority, few noble gas studies of water paleotemperatures have been published. Specifically, questions need to be answered relative

to possible subsurface changes due to re-equilibrium with soil
gases or gases generated in the aquifer such as methane. Also,
improvements are needed in techniques of field collection and
laboratory analyses.

Chloride

The chloride content of groundwater may be a sensitive indicator of either the distance between the intake area of the
aquifer and coast or the amount of evapotranspiration prior to
groundwater recharge. Because chloride is not normally derived
from dissolution of solid aquifer materials and it does not enter
into ion exchange reactions to any great extent, the chloride
content in shallow aquifers and aquifers isolated from sources
of connate water should reflect some of the original environmental factors of the outcrop area [19,86].
Within about 500 km of coastal areas, the chloride content
of precipitation is strongly related to the proximity of the
shoreline. The ocean-derived chloride in the precipitation may
commonly vary from 10 to 20 mg/L at the coast to less than 1 mg/L
at a distance of 200 km from the coast. Precise amounts are
related closely to climatological factors such as prevailing winds
and total precipitation. Local vegetation cover and topographic
effects may also be important, particularly in controlling dry
fallout of sea-spray particles within a few kilometers of the
coast.
If most of the climatological and topographic factors are
relatively constant with time, as may have been true along the
Gulf Coast of Texas, then groundwater which is a few thousand
years old may have the effects of a fluctuating coastline preserved in the form of bands of groundwater having differences in
their chloride content. Such appears to be the case in the
Carrizo aquifer of southern Texas.
The chloride content of groundwater in inland regions should
be more sensitive to concentration by evapotranspiration than to
fluctuations of the position of distant shorelines. A band of
lower chloride water in the Great Australian Artesian Basin has
been interpreted as groundwater originating during a period of
greater rainfall and/or less evapotranspiration [86]. This
interpretation is supported by independent calculations of the
hydrodynamic age of the water which suggest recharge during the
latest episode of glaciation.

GEOLOGIC RECONSTRUCTIONS

Standard methods of reconstructing geologic history are
essential to check other methods of water dating. Geologic
history is highly specific to each site of interest, so useful
generalizations are difficult to make. Nevertheless, the importance of general geologic reasoning cannot be emphasized too

strongly. For example, groundwater in a recharge zone composed of the latest Pleistocene outwash cannot be older than about 20,000 years which would represent the maximum age of the outwash. Another common situation would be water which has a chemical imprint of some known geologic event such as a volcanic eruption or an invasion of marine water along coastal areas of the world [87]. As important as such information may be, nevertheless, water ages in the normal sense can rarely be obtained. However, knowing upper or lower limits to water ages is a check of great importance on dates obtained by methods which yield specific numbers and appear to be more precise but commonly are not.

CURRENT PROBLEMS

Three primary problem areas exist in dating groundwater. These are: (1) Formulation of realistic geochemical-hydrodynamic models needed to interpret data which are generated by field and laboratory measurements, (2) development of sensitive and accurate analytical methods needed to measure trace amounts of various stable and unstable nuclides, and (3) theoretical and field oriented studies to determine with greater accuracy the extent and distribution of the subsurface production of radionuclides which are commonly assumed to originate only in the atmosphere.

Each dating method requires some type of model to aid in the interpretation of the data. The models may vary from so-called conceptual models which are universally required in geological investigations to very intricate, coupled geochemical-hydrodynamic models which are formalized in exceedingly complex computer programs. In general, the geologically oriented conceptual models are site specific and require extensive field work by experienced geologists. Such work is sophisticated at present and will increase in complexity with the continued development of the science as a whole. The development of purely hydrodynamic models is quite advanced. These models are generally adequate for hydrodynamic dating of water samples from sedimentary aquifers. For fractured igneous and metamorphic aquifers, however, hydrodynamic models which are sufficiently realistic for water dating are not available. Coupled geochemical-hydrodynamic models are in their infancy. Considerable development work is needed, although elementary chromatographic-like and mixing-cell models have been used with apparent success.

The extent to which molecular diffusion affects dating of fractured rock has yet to be evaluated thoroughly with proper models. Although diffusion is a slow process in dense crystalline rocks, it could still have an important influence on dates of very old groundwater. With atmospherically derived radionuclides, dates of water affected by this slow diffusion should appear too old. On the other hand, dating of water based on the accumulation of helium which diffuses out of solid rock into fissures should yield dates which are too young. Thus far, however, helium dates of water from fractured rocks appear to be older than ^{14}C dates of the same water, results that are opposite to those expected from dates influenced by slow diffusion.

Methods of sample preparation and nuclide analyses need to be improved materially if certain nuclides are to be used as a basis of dating water. Of particular interest is ^{81}Kr which has a low natural abundance and a low specific activity. Because it has a long half-life, is inert, and is probably produced exclusively in the atmosphere, it would be an ideal radionuclide for dating old groundwater. Other radionuclides of possible interest for dating water which might need the development of special analytical methods are ^{41}Ca and ^{79}Se. As yet, these radionuclides have not been reported from analyses of groundwater.

Theoretical studies [25,42] have shown that significant amounts of a number of radionuclides usually assumed to be derived only from the atmosphere may actually be produced in the subsurface, largely through interactions with secondary neutrons produced by alpha capture reactions. The alpha particles are derived mostly from normal decay of natural U and Th. Whether or not subsurface production of radionuclides can indeed influence dating has yet to be demonstrated by field and laboratory tests. The matter needs further study, particularly in relation to ^{14}C dating of water which is more than 40,000 years old.

Eugene S. Simpson, Glenn M. Thompson, Anthony Muller, Richard Zito, Juan Carlos Lerman and other associates at the University of Arizona have been most generous with their time and ideas. Many of the researchers cited in our review have also contributed to our study in ways too numerous to mention, to these and especially to Professor Hans Oeschger and Dr. David Elmore we owe our gratitude. The present study was funded by U. S. Nuclear Regulatory Commission Contract NRC-04-78-272.

References

[1] Carlston, C. W., Thatcher, L. L., Rhodehamel, E. C., Tritium as a hydrologic tool, the Wharton Tract study, Internat. Assoc. Sci. Hydrol. Publ. No. 52, 503-512 (1960).

[2] Nelson, R. W., Reisenauer, A. E., Application of radioactive tracers in scientific groundwater hydrology, Radioisotopes in Hydrology, Tokyo Symposium 1963, p. 207-230, Inter. Atomic Energy Agency, Vienna.

[3] Theis, C. V., Hydrologic phenomena affecting the use of tracers in timing groundwater flow, Radioisotopes in Hydrology, Tokyo Symposium 1963, p. 193-206, Inter. Atomic Energy Agency, Vienna.

[4] Fritz, P., Barker, J. F., Gale, J. E., Geochemistry and isotope hydrology of groundwater in the Stripa granite, results and preliminary interpretation, Lawrence Berkeley Laboratory Report LBL-8285, Berkeley, California, 135 p., 1979.

[5] Hobba, W. A., Jr., Fisher, D. W., Pearson, F. J., Jr., Chemerys, J. C., Hydrology and geochemistry of thermal springs of the Appalachians, U. S. Geol. Survey Prof. Paper 1044E, 36 p., 1979.

[6] Osmond, J. K., Kaufman, M. I., Cowart, J. B., Mixing volume calculations, sources and aging trends of Floridan aquifer water by uranium isotopic methods, Geochim. et Cosmochim. Acta, 38, 1083-1100 (1974).

[7] Tóth, J., Gravity-induced cross-formational flow of formation fluids, red earth region, Alberta Canada: Analysis, patterns, and evolution, Water Resour. Res., 14(5), 805-844 (1978).

[8] Kafri, U., Arad, A., Paleohydrology and migration of the groundwater divide in regions of tectonic instability in Israel, Geol. Soc. of America Bull., 89, 1723-1732 (1978).

[9] Lal, D., Peters, B., Cosmic-ray produced isotopes and their applications to problems in geophysics, In: Progress in Elementary Particle and Cosmic Ray Physics, North Holland Publishing Co., Amsterdam, Vol. 6, p. 1-74, 1962.

[10] Oeschger, H., Some cosmic ray produced radionuclides of interest in dating old groundwater, In: Davis, S. N., ed., Workshop on dating old ground water, Dept. Hydrology and Water Resources, University of Arizona, report on contract to Union Carbide Corp. in Oak Ridge (Y/OWI/SUB-78/55412), p. 129, 1978.

[11] Dansgaard, W., Clausen, H. B., Aarkrog, A., Evidence for bomb-produced silicon-32, J. Geophys. Res., 71(22), 5474-5477 (1966).

[12] Davis, S. N., ed., Workshop on dating old ground water, Subcontract 19Y-55412v, Report to Union Carbide Corp., Nuclear Division by Dept. of Hydrology and Water Resources, University of Arizona, Tucson, 138 p., 1978.

[13] Freeze, R. A., Cherry, J. A., Groundwater, Prentice-Hall, Inc., Englewood Cliffs, NJ, p. 134-139, 290-295, 1979.

[14] Gaspar, E., Onescu, M., Radioactive tracers in hydrology, Amsterdam, Elsevier Publishing Co., 342 p., 1972.

[15] Isotope Hydrology Section, International Atomic Energy Agency, Nuclear techniques in ground-water hydrology, In: Ground-water studies, UNESCO, Paris, Sections 10.1-10.4, 38 p., 1973.

[16] Plata Bedmar, A., Isotopos en Hidrología, Editorial Alhambra, S. A., Madrid, 328 p., 1972.

[17] Oeschger, H., Houtermans, J., Loosli, H., Wahlen, M., The constancy of cosmic radiation from isotope studies in meteorites and on the Earth, Nobel Symposium, Vol. 12, p. 471-498, 1970.

[18] Münnich, K. O., Messungen des [14]C Gehaltes vom hartem Grundwasser, Naturwiss., 44, 32 (1957).

[19] Bath, A. H., Edmunds, W. M., Andrews, J. N., Palaeoclimatic trends deduced from the hydrochemistry of a Triassic sandstone aquifer, United Kingdom, In: Isotope Hydrology 1978, Internat. Atomic Energy Agency, Vienna, Vol. 2, p. 545-568, 1979.

[20] Bergstrom, R. E., Aten, R. E., Natural recharge and localization of fresh water in Kuwait, J. Hydrology, 2(3), 213-231 (1965).

[21] Calf, G. E., The isotope hydrology of the Mereenie Sandstone aquifer, Alice Springs, Northern Territory, Australia, J. Hydrology, 38, 343-355 (1978).

[22] Grove, D. B., Rubin, M., Hanshaw, B. B., Beetem, W. A., Carbon-14 dates of ground water from a Paleozoic carbonate aquifer, southcentral Nevada, U. S. Geol. Survey Prof. Paper 650-C, p. 215-218, 1969.

[23] Pearson, F. J., Jr., White, D. E., Carbon-14 ages and flow rates of water in Carrizo Sand, Atascosa County, Texas, Water Resources Research, 3(1), 251-261 (1967).

[24] Winograd, I., and Farlekas, Problems in ^{14}C dating of water from aquifers of deltaic origin, Internat. Atomic Energy Agency, Vienna, Isotope Techniques in Groundwater Hydrology, Vol. II, p. 69-93, 1974.

[25] Zito, R., Donahue, D. J., Davis, S. N., Bentley, H. W., Fritz, P., Possible subsurface production of carbon-14, Geophys. Research Lett., 7(4), 235-238 (1980).

[26] Rightmire, C. T., Hanshaw, B. B., Relationship between the carbon isotope composition of soil CO_2 and dissolved carbonate species in groundwater, Water Resour. Research, 9(4), 958-567 (1973).

[27] Fontes, J. -C., Garnier, J. M., Determination of the initial ^{14}C activity of the total dissolved carbon, A review of the existing models and a new approach, Water Resour. Research, 15(2), 399-413 (1979).

[28] Libby, W. F., Tritium Geophysics, J. Geophys. Research, 66, 3767-3782 (1961).

[29] Ehhalt, D., On the uptake of tritium by soil water and groundwater, Water Resour. Research, 9(4), 1073-1074 (1973).

[30] Hufen, T. H., Buddemeier, R. W., Lau, L. S., Isotopic and chemical characteristics of high-level groundwaters on Oahu, Hawaii, Water Resour. Research, 10, 366-370 (1974).

[31] Poland, J. F., Stewart, G. T., New tritium data on movement of groundwater in western Fresno County, California, Water Resour. Research, 11, 716-724 (1975).

[32] Allison, G. B., Hughes, M. W., The use of environmental tritium to estimate recharge to a South-Australian aquifer, J. Hydrology, 26(3)(4), 245-254 (1975).

[33] Dincer, T., Al-Mugrin, A., Zimmermann, U., Study of the infiltration and recharge through the sand dunes in arid zones with special reference to the stable isotopes and thermonuclear tritium, J. Hydrology, 23, 79-109 (1974).

[34] Smith, D. B., Wearn, P. L., Richards, H. J., Rowe, P. C., Water movement in the unsaturated zone of high and low permeability strata by measuring natural lithium, In: Isotope Hydrology 1970, Vienna, Interna. Atomic Energy Assoc., p. 73-87, 1970.

[35] Vogel, J. C., Thilo, L., Van Dijken, M., Determination of groundwater recharge with tritium, J. Hydrology, 23, 131-140 (1974).

[36] Tolstikhin, I. N., Kamenskii, I. L., Determination of groundwater ages by the T-^{3}He method, Geochem. Int., 6, 810-811 (1969).

[37] Torgersen, T., Clarke, W. B., Jenkins, W. J., The tritium/helium-3 method in hydrology, In: Isotope Hydrology 1978, Interna. Atomic Energy Agency, Vienna, Vol. 2, p. 917-929, 1979.

[38] Zito, R., Davis, S. N., 1980 Subsurface production of the mirror isotopes ^{3}H and ^{3}He, unpublished manuscript, University of Arizona, Department of Hydrology and Water Resources, 24 p.

[39] Schaeffer, O. A., Thompson, S. O., Lark, N. L., Chlorine-36 radioactivity in rain, J. Geophys. Research, 65, 4013-4016 (1960).

[40] Davis, S. N., DeWeist, R. J. M., Hydrogeology, John Wiley & Sons, New York, 463 p., 1966.

[41] Tamers, M. A., Ronzani, C., Scharpenseel, H. W., Naturally occurring chlorine-36, Atompraxis, 15, 433-437 (1969).

[42] Bentley, H. W., Some comments on the use of chlorine-36 for dating very old ground water, In: Workshop on dating old ground water, S. N. Davis, ed., Subcontract 19Y-55412v, report to Union Carbide Corp., Nuclear Division, by Department of Hydrology and Water Resources, University of Arizona, 138 p., 1978.

[43] Elmore, D., Fulton, B. R., Clover, M. R., Marsden, J. R., Gove, H. E., Naylor, H., Purser, K. H., Kilius, L. R., Beukens, R. P., Litherland, A. E., Analysis of ^{36}Cl in environmental water samples using an electrostatic accelerator, Nature, 277, 22-25 (1979).

[44] Lal, D., Peters, B., Cosmic ray produced radioactivity on the earth, Handbuch der Physik, XLVI/2, 551-612 (1967).

[45] Eriksson, E., The yearly circulation of chloride and sulfur in nature, meteorological, geochemical, and pedological implications, Part II, Tellus, 12(1), 63-109 (1959).

[46] Bentley, H. W., Davis, S. N., Feasibility of ^{36}Cl-dating of very old ground water, EOS, American Geophysical Union Transactions, 61(17), 230 (1980).

[47] Kater, R., Development of a method for measuring natural ^{32}Si activities and aspects of its use in hydrogeological researches, Neue Bergbautechnik, 5, 941-943 (1975).

[48] Lal, D., Nijampurkar, V. N., Rama, S., Silicon-32 hydrology, Isotope Hydrology 1970, Inter. Atomic Energy Assoc., Vienna, p. 847-868 (1970).

[49] Gupta, S. K., Lal, D., Silicon-32, In: Workshop on dating old ground water, S. Davis, ed., Subcontract 19Y-55412v, report to Union Carbide Corp., Nuclear Division, by Department Hydrology and Water Resources, University of Arizona, Tucson, 131-138, 1978.

[50] Elmore, D., Anantaraman, N., Fulbright, H. W., Gove, H. E., Hans, H. S., Nishiizumi, K., Murrell, M. T., Honda, M., Half-life of ^{32}Si using tandem accelerator mass spectrometry, Nuclear Structure Research Laboratory, University of Rochester, NY, Publication UR-NSRL-220, 11 p., 1980.

[51] Kutschera, W., Henning, W., Paul, M., Smither, R. K., Stephenson, E. J., Yntema, J. L., Alburger, D. E., Cumming, J. B., Harbottle, G., Physical Review Lett., 45(8), 592-593 (1980).

[52] Lovering, T. S., Significance of accumulator plants in rock weathering, Geol. Soc. America Bull., 70, 781-800 (1959).

[53] Riquier, J., Les phytolithes de certains sols tropicaux el des podzols, Trans. 7th Internat. Congress Soil Sci., Madison, WI, Vol. 4, 1960.

[54] Oeschger, H., Gugelmann, L. H., Schotterer, U., Siegenfhaler, U., Wiest, A., Ar dating of groundwater, In: Isotope Techniques in Groundwater Hydrology 1974, Inter. Atomic Energy Agency, Vienna, p. 179-189, 1974.

[55] Loosli, H. H., Oeschger, H., Argon-39, carbon-14 and krypton-85 measurements in groundwater samples, In: Isotope Hydrology 1978, Internat. Atomic Energy Agency, Vienna, Vol. 2, p. 931-945, 1978.

[56] Rózański, K., Florkowski, T., Krypton-85 dating of groundwater, In: Isotope Hydrology 1978, Internat. Atomic Energy Agency, Vienna, Vol. 2, p. 949-959, 1979.

[57] National Council on Radiation Protection and Measurements, Krypton-85 in the atmosphere--accumulation, biological significance, and control technology, Nat. Council Rad. Protection and Meas., Washington, D.C., NCRP Report No. 44, 79 p., 1975.

[58] Spiridonov, A. I., Sultankhodzhayer, A. N., Beder, B. A., Taneyev, A. N., Tyminskiy, V. G., Some problems in the computation of the age of ground waters, Soviet Hydrology, selected papers, Issue No. 3, p. 265-267, 1973.

[59] Teitsma, A., Clarke, W. B., Fission xenon isotope dating, J. Geophys. Res., 83, 5443-5453 (1978).

[60] Marine, I. W., Geochemistry of ground water at the Savannah River Plant, Savannah River Laboratory Report DP-1356, Aiken, South Carolina, 102 p., 1976.

[61] Heaton, T. H. E., Vogel, J. C., Gas concentrations and ages of groundwaters in Beaufort Group Sediments, South Africa, Water S. A., (South Africa), 5(4), 160-170 (1979).

[62] Naughton, J. J., Lee, J. H., Keeling, D., Finlayson, J. B., Dority, Helium flux from the earth's mantle as estimated from Hawaiian fumarolic degassing, Science, 180, 55-57 (1973).

[63] Torgersen, T., Clarke, W. B., Excess helium-4 in Teggar Lake, Possibilities for a uranium ore body, Science, 199, 769-771 (1978).

[64] Rosholt, J. N., Doe, B. R., Tatsumoto, M., Evolution of the isotopic composition of uranium and thorium in soil profiles, Geol. Soc. America Bull., 77, 987-1004 (1966).

[65] Fleischer, R. L., Isotopic disequilibrium of uranium, alpha-recoil damage and preferential solution effects, Science, 207, 979-981 (1980).

[66] Cowart, J. B., Osmond, J. K., Oxidation/reduction in the Edwards Limestone aquifer as indicated by dissolved uranium isotopes, Geol. Soc. America, Abstracts with Programs, 9(7), 938 (1977).

[67] Airey, P. L., Australian Atomic Energy Research Establishment, personal communication, 1980.

[68] Harmon, R. S., Lively, R. S., ^{230}Th/^{234}U dating of Quaternary uranium deposits, Geol. Soc. America, Abstracts with Programs, 9, 1005 (1977).

[69] Lively, R. S., Alexander, E. C., Jr., ^{230}Th-^{234}U ages of speleothems from Mystery Cave, Minnesota; Abstract, National Speleological Society, Quarterly Journal, 42(2), 34 (1980).

[70] Thompson, P., Schwarcz, H. P., Ford, D. C., Continental Pleistocene climatic variations from speleothem age and isotopic data, Science, 184, 893-895 (1974).

[71] Thompson, G. M., Lumsden, D. N., Walker, R. L., Carter, J. A., Uranium series dating of stalagmites from Blanchard Springs Caverns, USA, Geochimica et Cosmochimica Acta, 39, 1211-1218 (1975).

[72] Bada, J. L., Shou, M., Effects of various environmental parameters on amino acid racemization rates in fossil bones, Geological Soc. America, Abstracts with Programs, 8, 762-763 (1976).

[73] Hare, P. E., Amino acid dating, limitations, and potential, Geological Soc. America, Abstracts with Programs, 9, 1004-1005 (1977).

[74] Davis, S. N., Thompson, G. M., Bentley, H. W., Stiles, G. K., Ground-water tracers--a short review, Ground Water, 18(1), 14-23 (1980).

[75] Thompson, G. M., Hayes, J. M., Trichlorofluoromethane in groundwater--A possible tracer and indicator of groundwater age, Water Resour Research, 15(3), 546-554 (1979).

[76] Hayes, J. M., Davis, S. N., Chlorofluoromethanes in ground-water, unpublished research proposal to U. S. Office Water Research and Tech., Indiana University, 19 p., February 21, 1975.

[77] Randall, J. H., Schultz, T. R., Davis, S. N., Suitability of fluorocarbons as tracers in ground water resource evaluation, unpublished Project Completion Report OWRT No. A-063, University of Arizona, 37 p., November, 1977.

[78] Gat, J. R., Comments on the stable isotope method in regional groundwater investigations, Water Resour. Research, 7, 980-993 (1971).

[79] Sonntag, C., Klitzsch, E., Löhnert, E. P., El-Shazly, E. M., Münich, K. O., Junghans, Ch., Thorweike, U., Weistroffer, K., Swailem, F. M., Palaeoclimatic information from deuterium and oxygen-18 in carbon-14 dated north Saharian groundwaters, In: Isotope Hydrology 1978, Inter. Atomic Energy Agency, Vienna, Vol. 2, p. 569-580, 1979.

[80] Dansgaard, W., Stable isotopes in precipitations, Tellus, 16 436-468 (1964).

[81] Dansgaard, W., Johnsen, S. J., Møller, J., Langway, C. C., One thousand centuries of climatic record from Camp Century on the Greenland ice sheet, Science, 166, 377-381 (1969).

[82] Hanshaw, B. B., verbal presentation, University of Arizona, 1978.

[83] Sugisaki, R., Measurement of effective flow velocity of ground water by means of dissolved gases, American J. Sci., 259, 144-153 (1961).

[84] Andrews, J. H., Lee, D. J., Inert gases in groundwater from the Bunter Sandstone of England as indicators of age and paleoclimatic trends, J. Hydrology, 41, 233-252 (1979).

[85] Mazor, E., Paleotemperatures and other hydrological parameters deduced from noble gases dissolved in groundwaters, Jordan Rift Valley, Israel, Geochim. Cosmochim. Acta, 36, 1321- (1972).

[86] Airey, P. L., Calf, G. E., Campbell, B. L., Hartley, P. E., Roman, D., Habermehl, M. A., Aspects of the isotope hydrology of the Great Artesian Basin, Australia, In: Isotope Hydrology 1978, Interna. Atomic Energy Agency, Vienna, Vol. 1, p. 205-217, 1979.

[87] Caswell, W. B., Maine's ground-water situation, Groundwater, 17(3), 235-243 (1979).

RECEIVED July 23, 1981.

IV. NATURAL ARCHIVES: INTERPRETING THE ISOTOPIC AND CHEMICAL RECORD

Sampling and Precise Dating Requirements for Extracting Isotopic Records from Three Rings

CHARLES W. STOCKTON and WILLIAM R. BOGGESS

University of Arizona, Laboratory of Tree-Ring Research, Tucson, AR 85721

The use of tree-ring width series as the time control for extraction of isotope and chemical information from wood cellulose has become commonplace. However, many researchers are unaware of the need to maximize the signal in the tree-ring series by sampling from those populations that are most sensitive to past environmental conditions. The series which are most environmentally sensitive may also be the most difficult to date accurately. Therefore a sufficient number of samples must be collected to provide adequate dating control. Techniques exist that can assure precise dating of individual ring widths but they depend on adequate replication of samples. Proper site selection and dating techniques for adequate tree-ring analysis will be discussed. Examples will be cited in which necessary sampling depth and ring width analysis appear to be insufficient to justify subsequent conclusions made from the derived isotopic series.

For more than 80 years, tree-ring data have been used to make inferences about past climatic variation. In general, the characteristic most often used has been the variations in widths of the annual growth rings. However, during the past decade other properties, such as cell density (measured by x-ray densiometric techniques), relative widths of early and late wood, and isotopic composition of the cellulose have been used to infer past environmental conditions. It is the isotopic composition that is of interest here.

0097-6156/82/0176-0225$05.00/0

Background

There are certain aspects of tree-ring data that make them
attractive for use in isotope analysis of past environmental
(climatic) variations. Among them is the absolute accuracy in
dating of individual ring widths. If established principles of
crossdating are employed, each annual ring width can be accurately
dated as to the year of its formation and the time scale is
precisely defined to the nearest year. (Crossdating involves
matching the patterns of narrow and wide ring widths between core
samples from the same tree and other trees sampled at a given
location. This essential procedure is a means of identifying
possible multiple or locally absent rings.) Another aspect is the
integrative nature of the environmental inputs that are recorded
in each ring width. Research has shown that trees growing under
stress site conditions tend to be integrators of monthly environmental changes reflective of the site. By careful sampling, it is
possible to maximize the information on a particular environmental
(climatic) variable, locked into the ring-width time series. If
primary interest is in a temperature variation signal, then
sampling would include sites that are most stressed by temperature
fluctuations (e.g., upper timberline). Conversely, if the drought
signal is to be maximized, then sites that are stressed by moisture deficits and high temperatures would be sampled. These
include lower forest border sites, sites with thin soil and
outlier sites, as those occurring along the western boundary of
the Great Plains. A third desirable aspect of tree-ring data
as a vehicle for extracting isotopic indicators of past climatic
variability, is their widespread spatial distribution and relative
abundance.

The close relationship that has developed between dendrochronology and ^{14}C dating in the last 25-30 years has primarily
been one of the dendrochronologist supplying absolute dates for
wood that is used to calibrate the ^{14}C time scale. The divergence
of age as determined by tree-ring and ^{14}C methods as time gets
large is well known. Graphs of these relationships and detailed
descriptions of the ^{14}C-dendrochronology ties are detailed by
Damon, Lerman and Long [1][1]. They point out that the method for
calibrating the radiocarbon scale is by comparing the radiocarbon
ages of accurately dated tree rings with their dendrochronological
ages. Reliability of the method is based on at least two assumptions:

1) the radiocarbon activity of a wood sample
   accurately represents the activity of the
   atmosphere when the wood was formed,
2) the age of the ring-width series is accurately
   known.

---

[1]Figures in brackets indicate the literature references at the
end of this paper.

Inherent in the first assumption is the implication that radio-carbon activity of the wood is site independent. However, as Eddy [2] points out, there are limitations to this implication. If one is interested only in dating the wood, the limitation is in the half-life of the isotope. If the concern is in monitoring changes in the concentration of the isotope in time, then other considerations are essential. The atmosphere and oceans act as low pass filters that delay and dilute real variations in, for example, the radiocarbon production. The tree system also acts as a filter as the atmospheric radiocarbon (as carbon dioxide) enters through the leaves and, by the process of photosynthesis, is preserved as cellulose in the new wood. It is well known that net photosynthesis varies between species and individuals of the same species as well as from site to site. Damon, Lerman, and Long [1] point out "one can expect measured radiocarbon in tree-ring sequences to show an 100-fold attenuation of the production ampli-tude in the 11 year solar cycle". Also, the ^{14}C production is reported to vary by about 22 percent through one 11-year solar cycle [1] and so if there is a 100-fold attenuation of this signal as recorded in tree-ring series, then one is really looking for a 0.2 percent change in the signal of the ^{14}C variation to reveal past variations in the 11-year solar cycle. This seems like an exceptionally small change to be reflected in the tree-ring record based on the variations in net photosynthesis that may occur between species and within individuals on a given site. It might be possible to maximize the strength of such a signal by defining a sampling population (e.g., multiple samples from open, xeric sites) such that the rate of carbon fixation is truly reflective of the atmospheric changes in ^{14}C. We envision such sampling populations as being on open, environmentally sensitive sites as opposed to those within forest interiors where carbon fixation may be influenced by competition and consequently the variations in the ^{14}C signal additionally influenced by biological factors. With refined and new techniques such as the tandem accelerator, this should be possible as smaller amounts of cellulose are necessary and greater numbers of samples can be analyzed.

Dendrochronological experience allows us to delineate those sites where maximum longevity of ring-width series might be expected. Using wood from dead trees, as is done in Bristlecone pine chronology development, limited inferences can be made concerning site characteristics that existed for the tree from which the dead wood is derived. (We do not believe that we are yet in a position to use time-series statistics to prove random selected samples of tree-ring series are from the same population as that from a nearby living stand of trees.) Proven sampling repetition strategies allow us to increase our confidence in absolute date assignments for each ring width as sample size increases. Consequently, the assumption of accuracy of dates assigned to individual ring widths can be controlled by sample density, with error decreasing as sample size increases.

Since Dansgaard [3] demonstrated that stable isotope information can be utilized to infer past climatic conditions, stable isotopes from ice cores and deep sea sediments (extracted from ocean bottom cores) have been used to interpret past climatic variation. However, because of limitations in spatial occurrence of ice cores and deep sea sediments, tree-ring data have become attractive sources for additional stable isotope information. Stable isotope dendrochronology is now receiving considerable attention by numerous researchers [1,4,5,etc.]. The extent to which application of the spatial distribution of tree-ring isotope data can be applied is perhaps unknown at this time, but obviously there is a great potential for a widespread spatial analysis and reconstruction of past environmental events. The use of decay-resistant wood from dead trees may even extend our knowledge into the more distant past. However, before this is possible, the investigator must have some idea of what climatic factors the stable isotopes are recording in the wood cellulose. This requires some knowledge of the climatic factors that are most influential in tree-ring growth on a given site. For example, consider the deuterium/hydrogen ratio expressed in the usual way $\delta(D)$ ^{o}/oo where

$$\delta(D) \ ^o/oo = \left(\frac{(D/H) \ sample}{(D/H) \ standard} - 1 \right) \times 1000.$$

Is the source of variation in $\delta(D)$ as recorded in wood cellulose from meteoric water or is it primarily from ground water sources? Perhaps it is related to growing season temperatures as some have suggested [6]. It seems essential that we demonstrate that the variations in the D/H ratio recorded in the tree system are the same as in precipitation. The ideal experiment would be to compare the D/H ratio in precipitation to the ratio in wood cellulose from trees growing on sites where the maximum growth stress is precipitation related. We envision such a site as one where the trees are growing on an upper topographic position with little soil development. Consequently the only water available to the tree is that falling during or immediately prior to the growing season when the ring is formed. This appears to be similar to the approach taken by White et al., [5] wherein they relate $\delta(D)$ determined from precipitation to that extracted from wood cellulose formed during the same year.

In their experiment, White et al., [5] compared the $\delta(D)$ variation in rain water, sap, and wood cellulose from trees growing on a dry site to that from ground water, sap, and wood cellulose from trees growing in a swampy area. Although some physical explanations of their results remain inconclusive, their results are interesting because they demonstrate 1) $\delta(D)$ variations in the cellulose of tree rings appear to be related to the water available to the tree during the year of formation of the ring; 2) the water available may be from rainfall or from ground water with $\delta(D)$ variations in the wood cellulose comparable to variation of the source. This at least implies that variation of $\delta(D)$ in the wood cellulose is site dependent and it may be possible to maximize environmental information in a ring-width series by site selectivity.

Sampling Sites Reflect Environmental Signal

It appears based on limited research, that the signal inherent in the isotope variability record extracted from wood cellulose may be reflective of the site from which it was collected. Once the relationships between isotopic variation and climate are well understood, isotopes may be usable to decipher past environmental conditions from tree-ring width data that are not ideal for classic dendroclimatology (e.g., complacent tree-ring series). However, it appears that until such is the case, experiments should be designed to test for certain environmental conditions by maximizing the environmental signal of interest and performing the isotopic analysis on resulting data.

Much work has been accomplished in the last ten years demonstrating the association between the growing site of trees and the climatic signal inherent in the ring widths [7]. Drought sensitivity in ring-width series is maximized by sampling trees that are stressed by precipitation and temperature variations. If ring-width series are to be used to obtain drought information, it seems reasonable that the data should be obtained from trees that maximize drought information content. Consequently, one selects the population as being those trees occupying sites where the drought variation signal is maximized. Statistical inferences are then based on the null hypothesis that there is no difference in the sample statistics obtained from a given site and those of the population.

The same applies for temperature: there is strong documentation that trees from certain sites (for example, upper timberline Bristlecone pine and upper northern timberline white spruce) contain strong temperature signals within their ring width series [8]. The limiting factor to growth as expressed in the ring-width series is average temperature and not total precipitation. Consequently, if one would like to test the hypothesis that $\delta(D)$ is primarily a function of temperature, it seems reasonable that the sampling population should be defined as those trees on sites where temperature is the environmental parameter limiting to growth. Again, the null hypothesis for testing a sample statistic against the population parameter is that there is no difference in what the sample shows and what the population reflects. If the hypothesis to be tested is that the D/H ratio extracted from the cellulose and dated by standard dendrochronologic techniques is related to ground water, one would sample accordingly, from sites where the moisture availability to the tree system is from the capillary zone of the ground water reservoirs.

Epstein and Yapp [4] state "it is obviously necessary to calibrate more specifically the relationship between δ(D) records in cellulose nitrate from tree-ring records and known climatic records. This can probably be done best by the analysis of tree rings from widely different, well-documented environments. Such data will allow the comparison of a large variety of trees and determine the versatility of using the isotopic method for climatic temperature determination". We concur with this statement as long as "from the same population" is inserted.

Sample Replication

In dendrochronology, there are well established sample replication limits for a) proper dating control and b) inferences of past environmental conditions based on ring-width variability (for example, see Chapters 5 and 6 of Fritts [7]). In many instances, if a ring-width chronology has been developed for a given region, we can date a single random time series sample of ring widths using a master chronology [9] time series to assign absolute dates to each ring width with a very small chance of error. This provides the needed time control for the sample but it may not provide an adequate sample of the population for extracting the climatic signal. Consequently, analyzing a single series or perhaps two, for isotopic variation and making inferences from this single series can be misleading. One must make certain that the signal being detected in a single tree-ring data series is common to the population (closely defined by microsite conditions) and not necessarily on the basis of cross-datability. In other words, in a statistical sense, cross-datability is a necessary although not always sufficient condition for inferences of past environmental conditions from an isotope series derived from tree rings. One example of such an occurrence follows: some time ago, the Laboratory of Tree-Ring Research dated an apparent drought-sensitive Bristlecone pine sample consisting of two segments in time derived from two different trees. So in effect, one has a single tree-ring series, from two trees. This series was dated, that is, absolute dates were assigned to each ring width using crossdating with a known regional chronology. The wood was analyzed for D/H isotopes and an isotopic series was derived and published by Epstein and Yapp [4]. It turns out that the δ(D) series shows a pronounced 22-year periodicity which is also that of the well known Hale Solar magnetic cycle. Some two or three years after publication of the details of the analysis, Dicke [10-12] studied these results and concluded they provided terrestrial evidence that solar variability is indeed influencing climate. Furthermore this record was used in developing a model for explaining solar variation and its physical relationship to climate. The results may be perfectly valid but, in our view, until they are supported by similar D/H ratio series on other trees showing comparable 22-year periodicity, such data and resulting interpretations contain a degree of skepticism. The argument used by Dicke [11] in analysis of the D/H series for this inference is "the advantage of the D/H ratio as a climate indicator may be the non-local measure of the climate which it provides. The D/H ratio in the precipitated water and incorporated into the plant cellulose depends on the mean surface temperature of the Pacific Ocean where the water evaporates and on the air temperatures where the various precipitations occur". This may or may not be true because 1) the tree-ring data analyzed are complacent indicating a rather constant source of moisture availability (ground water?) and 2) the extremely limited sample size (actually only one tree-ring time series) would lead one to suspect that the resulting 22-year periodicity might and (we stress might) be a statistical artifact and a result of chance alone. We would be gratified to see the results substantiated by additional evidence because they support some of our own research results, in that there appears to be a 22-year rhythm in drought recurrence which may be associated with solar variation of some sort [13]. However, the history of dendrochronology is replete with hopes that have been dashed when the results of large samples have been analyzed. So our advice is 1) analyze samples from well defined homogeneous populations and 2) make sure the sample replication •is sufficient to rule out chance occurrence of the results.

Conclusions

Tree-ring data are used as either time control from which isotope series are constructed, e.g., for calibration of the ^{14}C time scale or to retrieve paleoenvironmental information. In the latter case, it is important to use isotope series from tree-ring populations maximizing the desired signal as existing evidence indicates the concentrations in at least some isotopes is site dependent. Sampling replication is important not only for dating control but also for eliminating chance occurrences in paleoclimatic series.

References

[1] Damon, P. E., Lerman, J. C., Long, A., Temporal Fluctuations of Atmospheric ^{14}C: Causal Factors and Implications, Ann. Rev. Earth Planet. Sci., 1978, 6, 457-494.
[2] Eddy, J. A., The New Solar Physics, Westview Press, Boulder, Colorado, 1978.
[3] Dansgaard, W., Stable Isotopes in Precipitation, Tellus, 1964, 16, 436.
[4] Epstein, S., Yapp, J. C., Climatic Implications of the D/H Ratio of Hydrogen in C-H Groups in Tree Cellulose, Earth and Planet. Sci. Lett., 1976, 38, 252-261.

[5] White, J. W. C., Jacoby, G. C., Jr., Lawrence, J. R.,
Broecker, W. S., An Analysis of the Relationships Between
the D/H Ratios of Rain, Groundwater, Tree Sap and the C-H
Hydrogens of Cellulose and Implications for Tree Rings as
D/H Paleoclimatic Indicators, in preparation, 1980.

[6] Wilson, A. T., Grinsted, M. J., The D/H Ratio of Cellulose
as a Biochemical Thermometer, Nature, 1975, 257, 387.

[7] Fritts, H. C., Tree Rings and Climate, Academic Press,
London, 1976.

[8] LaMarche, V. C., Jr., Paleoclimatic Inferences from Long
Tree-Ring Records, Science, 1974, 153, 1043-1048.

[9] Stokes, M. A., Smiley, T., An Introduction to Tree-Ring
Dating, The University of Chicago Press, Chicago, 1968.

[10] Dicke, R. H., Is There a Chronometer Hidden Deep in the Sun,
Nature, 1978, 276, 676-680.

[11] Dicke, R. H., Solar Luminosity and the Sunpot Cycle, Nature,
1979, 280, 24-27.

[12] Dicke, R. H., The Clock Inside the Sun, New Scientist,
1979, 83, 12-14.

[13] Mitchell, J. M., Jr., Stockton, C. W., Meko, D. M., Evidence
of a 22-Year Rhythm of Drought in the Western United States
Related to the Hale Solar Cycle Since the 17th Century, In:
B. M. McCormac and T. S. Seliga, eds., Solar-Terrestrial
Inferences on Weather and Climate, D. Reidel Pub. Co.,
Holland, 1979, 125-143.

RECEIVED March 27, 1981.

Fluctuation of Atmospheric Radiocarbon and the Radiocarbon Time Scale

PAUL E. DAMON

University of Arizona, Laboratory of Isotope Geochemistry,
Department of Geosciences, Tucson, AR 85721

The basic assumption of constant atmospheric ^{14}C activity in radiocarbon dating is not strictly valid. We now have a record of the fluctuation of atmospheric ^{14}C variations for the last 8,400 years B.P. obtained by measurement of the isotopes of carbon in dendrochronologically dated wood. Prior to contamination of atmospheric ^{14}C activity by fossil fuel combustion and nuclear technology in the 20th century, the first-order secular variation can be closely approximated by a sine curve with a period of 10,600 years and an amplitude of ± 48 per mil. This trend curve is in turn modulated by variations on a time scale of one decade to a few centuries with an amplitude of ± 20 per mil ("deVries").

It is necessary to calibrate the ^{14}C time scale for greater dating accuracy. However, the second-order variations are at least as important as the first-order constancy of atmospheric ^{14}C. For example, they provide a record of prehistoric solar variations, changes in the Earth's dipole moment and an insight into the fate of CO_2 from fossil fuel combustion. Improved techniques are needed that will enable the precise measurement of small cellulose samples from single tree rings. The tandem accelerator mass spectrometer (TAMS) may fill this need.

Radiocarbon (^{14}C) is produced in the atmosphere by the cosmic ray neutron flux interacting with $^{14}N[^{14}N(n,p)^{14}C]$. The ^{14}C 'hot' atom then equilibrates with atmospheric CO_2 which participates in the C-O cycle and passes into the food chain (biosphere). Most of the radiocarbon is taken up by the oceans which constitute the largest reservoir of CO_2 within the secondary geochemical cycle.

Since the work of de Vries [1,2][1], Willard Libby's [3] basic assumption of a constant atmospheric $^{14}C/^{12}C$ ratio has been known

---

[1]Figures in brackets indicate the literature references at the end of this paper.

0097-6156/82/0176-0233$05.00/0

to be not strictly correct. The $^{14}C/^{12}C$ ratio has changed by ± 5 percent during the last nine millennia and corrections must be made to radiocarbon years to reduce them to solar years. For the 8th millennium B.P. (Before Present, i.e., before A.D. 1950), conventional radiocarbon dates are about 800 years too young (figure 1).

Radiocarbon years are calibrated from determinations of the ^{14}C activity and stable isotopic carbon ratios of dendrochronologically dated tree rings [4]. The stable isotope data are required to normalize the dates to average wood with $\delta^{13}C$ value of -25 per mil ($^{13}C/^{12}C$ fractionation relative to PDB reference standard). Photosynthetic and other plant physiological processes may produce differential isotopic fractionation between species, within the same species in different localities and even within the same tree under changing environmental conditions.

The reader may refer to a recent review [4] for a more detailed description of the state of knowledge concerning the causes and implications of temporal fluctuations of atmospheric ^{14}C. Briefly, the two major causes of atmospheric ^{14}C fluctuation are changes in the Earth's dipole field intensity and modulation by solar activity. The effect of these two changing environmental factors can be observed more readily by expressing the fluctuations as Delta values (Δ) which are a measure of the per mil age-corrected deviations from standard pre-industrial mid 19th century wood with $\delta^{13}C$ = -25 per mil. Figure 2 is a plot of measured Δ in per mil from a composite data set [5] for the last 8,000 years. The long period trend curve was generated by a 6th order polynomial regressed on the log of dendrodates vs. the log of conventional radiocarbon ages [6]. This trend can also be approximated closely by a sine-curve with a period of 10,600 years and an amplitude of ±48 per mil [4].

The long-term trend for pre-industrial time expressed by this curve is believed to have been produced by changes in the Earth's dipole magnetic field intensity. When this long-term variation is removed, medium-term fluctuations are observable. The heavy line in figure 3 was produced by Fourier analysis of the residuals around the 6th order logarithmic function [6]. It clearly demonstrates the de Vries effect secular variations [1,2], that have a time scale of a few decades to a few centuries and an amplitude of a few percent. The significance of these secular variations, also known as "wiggles" or "wriggles", has been emphasized, in particular, by Suess [7,8].

The de Vries effect wiggles during the last millennia are shown in expanded form in figure 4 and compared with the high precision data of Stuiver and Quay [9]. The most intense peaks occur during the Maunder minimum (A.D. 1640 to A.D. 1715) and the Spörer minimum (A.D. 1420 to A.D. 1540) when sunspot activity was at a minimum or virtually absent [10]. Lesser, but significant Δ maxima occur at the beginning of the 19th century and between A.D.

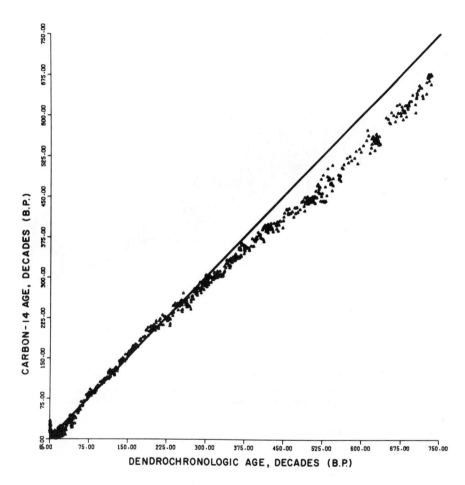

Figure 1.   Carbon-14 age (T1/2 = 5,730 years) vs. dendrochronologic age (dendrodate) for composite data set (5).

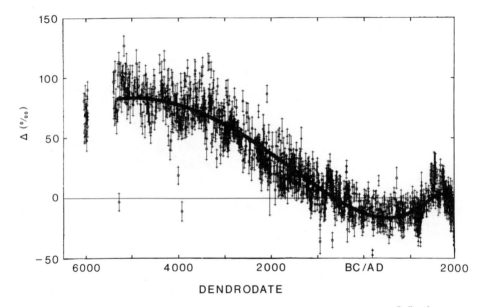

Radiocarbon

Figure 2.   Plot of Δ (per mil) vs. dendrodate for composite data set (6).

The long-term trend curve was generated by a sixth-order polynomial regressed on the log of dendrodates vs. the log of conventional radiocarbon ages. The preindustrial trend is thought to be produced almost entirely by changes in the earth's dipole field intensity. The decrease in the 20th century is due to the combustion of fossil fuels. The medium-term secular variation (wiggles) about the trend are thought to be due to heliomagnetic modulation.

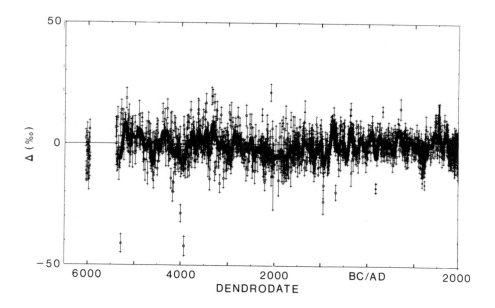

Radiocarbon

Figure 3. Wiggle curve in Δ (per mil) vs. dendrodate with the trend in Figure 2 removed (6).

The heavy line is produced by Fourier analysis of the residuals around the sixth-order logarithmic function. There are about 35 pronounced wiggles in 7,000 years on an average of one every 200 years. Note that some of the wiggles appear to have a greater amplitude than the Spörer and Maunder minima which occurred between A.D. 1450 and 1715 (10).

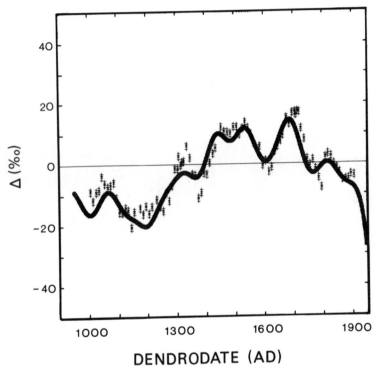

Radiocarbon

*Figure 4.   Segment of Fourier analysis wiggle curve (Figure 2) for the last millen-
nium (6).*

Data points are the high precision (±2‰) measurements of Stuiver and Quay (9). Note
the excellent agreement between the high precision data of Stuiver and the trend line
generated from the lower precision (ca. ±5‰) data of the composite data set. Maxima
occur between A.D. 1020–1080, 1290–1320, 1420–1530, 1660–1710, 1790–1830 or on
the average, every ca. 190 years. The pronounced minimum between A.D. 1100 and
1240 corresponds to the Medieval Warm epoch.

1280 and A.D. 1340 [Wolf minimum, 9]. The ^{14}C minimum between A.D. 1100 and A.D. 1250 corresponds to the Medieval Warm Epoch. A quasi-cyclic relation with the period of ~200 years reported by Suess [11] can be observed in figure 4. There are also approximately 35 maxima observable in the 7,000 year record shown in figure 3, i.e., on the average, one every 200 years.

If the relationship between sunspots and ^{14}C production established for the last three solar cycles [12] is inputed to a simple one-box ^{14}C model [13,14], the de Vries effect "wiggles" can be successfully modeled for the time since the Maunder minimum during which sunspot data are available. However, Stuiver and Quay [9] have questioned the validity of extrapolating the sunspot ^{14}C relationship for past centuries. They have presented model-dependent evidence to support their conclusion that ^{14}C production was three-fold greater during the Maunder minimum than would be predicted by extrapolating the sunspot vs. [14]C production-function back to the Maunder minimum. Lazear et al., [15], on the other hand, have examined the box-diffusion model [16] used by Stuiver and Quay to approximate the carbon cycle and suggest that, as parameterized, the model is over-attenuating. Further work is being done in this laboratory to resolve this problem.

We have calculated the predicted 11-year cycle attenuation for the one box [13,14], three box [17,18], five and six box [19] and box-diffusion [16] models. The predicted attenuations vary from a factor of 58 for the one box model to a factor of 100 for the five box model. The box-diffusion model yields a calculated attenuation factor of 74. The neutron flux and consequent ^{14}C production varies by about 250 per mil from the minimum to the maximum of the sunspot cycle. Consequently, from model predictions we would anticipate a maximum peak to trough amplitude of from about 2 per mil to 5 per mil for the 11-year radiocarbon cycle. The University of Glasgow ^{14}C research group found a much larger maximum variation of up to 30 per mil peak to trough amplitude [20,21]. However, the University of Arizona observed an amplitude that was a factor of 10 lower for the time between A.D. 1940 and A.D. 1954 [22] and since then, results of Stuiver [23], Cain and Suess [24], and Tans et al., [25], have failed to confirm the large variations in atmospheric radiocarbon during the 11-year solar cycle reported by the University of Glasgow research group. In fact, there is no definitive evidence for an 11-year cycle of radiocarbon in the precise (± 2 per mil) data of Lerman [26], or Stuiver [23].

Recently, a U.S.S.R.-Czechoslovokian research group have reported ^{14}C data for dated wine samples from the Caucasus Mountains [27]. Their results are in fairly close agreement with our results for the time of overlapping data (figure 5). If the anomalous data for A.D. 1943 are omitted, the fifth order polynomial fit to the data yields a 5 per mil peak to trough amplitude with a phase lag of 4 years behind sunspot numbers. The amplitude

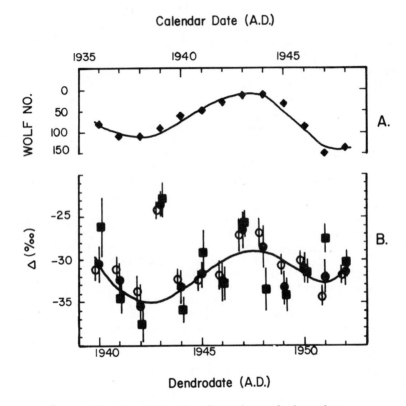

Calendar Date (A.D.)

Figure 5.   Radiocarbon fluctuations and solar cycles.

(A) Annual Wolf Sunspot numbers for the sunspot cycle between A.D. 1936 and 1948. The line is a fifth-order polynomial least squares fit to the data. (B) Radiocarbon content of annual organic samples for the years A.D. 1940 to 1952. The data are weighted averages of analyses of tree rings (■) (22) and wine (○) (27). The line is a least-squares fit of the weighted averages (●) with a fifth-order polynomial. The order of the polynomial was selected according to Damon et al. (22). The time scales of A and B are phase shifted 4 years to conform with the theoretical phase shift predicted by Lerman (26). The anomalous points for A.D. 1943 were not included in the polynomial fit. These data cannot be explained by a process that involved the entire atmosphere because the following data points (A.D. 1944) have returned immediately to normal without decaying.

is somewhat greater than expected but the phase lag is in excellent agreement with theory [26]. The presence of a 5 per mil peak to trough 11-year radiocarbon signal in the composite set of data in figure 5 and its absence in the data of Stuiver [23] appears to provide a contradiction. With the ± 2 per mil precision attained by Stuiver, an 11-year radiocarbon cycle, theoretically, should be observed. In order to explain the difference in their results for the 11-year radiocarbon and results from other laboratories, Baxter et al., [28] suggested that differences in sample location might result in an enhanced or dampened signal. Damon et al., [29] agreed "that such variations exist, but, the available evidence suggests that the magnitude is much less than required to explain their pre-nuclear bomb data." Perhaps, the smaller discrepancies existing between sets of data other than the Glasgow data may be explained by location. In particular, it is interesting that the composite data in figure 5 is for wine from grapes grown in the Caucasus Mountains and Douglas fir tree rings from the Santa Catalina Mountains in Arizona. Both are inland locations and well above sea level. On the other hand, the Douglas fir tree rings used by Stuiver were from a tree cut on the marine west coast Olympic Peninsula from the State of Washington. A possible locational effect might be upwelling of CO_2 (with lower ^{14}C activity) from the Pacific Ocean. The effect of CO_2 upwelling would be to dampen high frequency variations such as the 11-year radiocarbon cycle by mixing ^{14}C depleted CO_2 from deep ocean water with atmospheric CO_2.

With reference again to figure 5, the anomalous ^{14}C activity for the year A.D. 1943 has been reported by both research groups. This anomaly cannot be the result of a global event such as a solar flare because the A.D. 1944 ^{14}C concentration returned to normal. A global event would raise the ^{14}C content of the entire atmosphere and then decay, initially, with a ca. 5-year mean residence time. The Georgia, U.S.S.R., wines were made from grapes that grew at an altitude of ca. 300 meters and a latitude of 30°N compared to an altitude of 2,740 meters and latitude of 32°26' for the Radio Ridge tree from the Santa Catalina Mountains. Thus the postulated event affected latitudes between 32°N and 38°N but, surprisingly, separated by 156° of longitude in opposite hemispheres with an ocean intervening. It is difficult to explain this anomaly. It occurred during the year following the first sustained nuclear reaction (December, 1942) and before the first known atomic explosion and, so, a ^{14}C tagged air mass from a nuclear explosion seems to be ruled out. Another possibility would be an anti-matter meteorite shower with a trajectory between 30°N to 40°N. A third possibility might be variable injection of ^{14}C from the stratosphere into the troposphere within the latitudinal belt from which the samples were collected as proposed by Baxter and Walton [21] to explain the larger variation measured by the Glasgow research group. We are searching for samples from different locations to investigate further this regional effect.

Large samples (e.g., 40 g. wood) are required for precise measurement (±2 per mil) of cellulose from single tree rings using the conventional gas proportional counter technique. Such large samples from individual tree rings are frequently hard to obtain. It will be possible to measure samples a thousand-fold smaller (40 mg.) by ultrasensitive mass spectrometry with accelerators [30,31,32]. A tandem accelerator mass spectrometer (TAMS) is being constructed under contract by the General Ionex Corporation of Massachusetts for The University of Arizona-National Science Foundation Regional Accelerator Facility [33]. However, much further work will be required before ion counting with the TAMS will achieve the high precision obtained on large samples by the best low level beta counting techniques [34].

I am grateful to Drs. Austin Long and Juan Carlos Lerman for helpful discussions and for critically editing the original manuscript, and to Mr. Jeffrey Klein and my other co-authors [6] for permission to republish figures 2, 3, 4. This work was supported by N.S.F. Grant EAR7821813 and the State of Arizona.

References

[1] de Vries, Hl., K. Ned. Akad. Wet., Proc. Ser. B. 61., Variation in concentration of radiocarbon with time and location on earth, p. 94-102, 1958.
[2] de Vries, Hl., Measurement and use of natural radiocarbon, Researches in Geochemistry, P. H. Abelson, ed., Wiley, New York, p. 169-189, 1959.
[3] Libby, W. F., Radiocarbon Dating, University of Chicago Press, Chicago, 2nd Edition, 175 p., 1955.
[4] Damon, P. E., Lerman, L. C., Long, A., Temporal Fluctuations of Atmospheric ^{14}C: Causal Factors and Implications, Ann. Rev. Earth Planet Sci., 6, 457-494 (1978).
[5] Damon, P. E., Lerman, L. C., Long, A., Report on The Calibration of the Radiocarbon Dating Time Scale, Radiocarbon, 22(3), 947-949 (1980).
[6] Klein, J., Lerman, L. C., Damon, P. E., Linick, T., Radiocarbon Concentration in the Atmosphere: 8000-Year Record of Variations in Tree Rings, Radiocarbon, 22(3), 950-961 (1980).
[7] Suess, H. E., Secular variations of the cosmic-ray-produced carbon-14 in the atmosphere and their interpretations, J. Geophys. Res., 70, 5937-5952 (1965).
[8] Suess, H. E., Bristlecone pine calibration of the radiocarbon timescale 5200 B.C. to the present, See Olsson 197a, p. 303-311, 1970a.

[9] Stuiver, M., Quay, P. D., Changes in Atmospheric Carbon-14 Attributed to a Variable Sun, Science, 207(4426), 11-19 (1980).
[10] Eddy, J. A., Maunder minimum, Science, 192, 1189-1202 (1976).
[11] Suess, H. E., The Radiocarbon Record in Tree Rings of the Last 8000 Years, Radiocarbon, 22(2), 200-209 (1980).
[12] Lingenfelter, R. E., Ramaty, R., Astrophysical and geophysical variations in C14 production, In: Radiocarbon Variations and Absolute Chronology, Proc. XII Nobel Symp., New York, I. U., Olsson, ed., Wiley, 513-537, 1970.
[13] Grey, D. C., Geophysical mechanisms for ^{14}C variations, J. Geophys. Res., 74, 6333-6340 (1969).
[14] Grey, D. C., Damon, P. E., Scientific Methods in Medieval Archaeology, Sunspots and radiocarbon dating in Middle Ages, R. Berger, ed., University of California Press, p. 167-182.
[15] Lazear, G., Damon, P. E., Sternberg, R., The Concept of DC Gain in Modeling Secular Variations in Atmospheric ^{14}C, Radiocarbon, 22(2), 318-327 (1980).
[16] Oeschger, H., Siegenthaler, U., Schotterer, U., Gugelmann, A., A box diffusion model to study the carbon dioxide exchange in nature, Ann. Rev. Earth Planet Sci., Tellus 27, 168-192 (1975).
[17] Houtermans, J. C., Suess, H. E., Oeschger, H., Reservoir models and production rate variations of natural radiocarbon, J. Geophys. Res., 78, 1897-1908 (1973).
[18] Sternberg, R. S., Damon, P. E., Radiocarbon Dating, Sensitivity of Radiocarbon Fluctuations and Inventory to Geomagnetic and Reservoir Parameters, p. 691-717, 1979.
[19] Keeling, D. C., Chemistry of the Lower Atmosphere, The carbon dioxide cycle: reservoir models to depict the exchange of atmospheric carbon dioxide with oceans and land plants, S. I. Rasool, ed., Plenum, New York, p. 251-329, 1973.
[20] Baxter, M. S., Farmer, J. G., Radiocarbon: short-term variations, Earth Planet Sci. Lett., 295-299 (1973).
[21] Baxter, M. S., Walton, A., Fluctuations of atmospheric carbon-14 concentrations during the past century, Proc. R. Soc. London Ser. A., 321, 105-127 (1971).
[22] Damon, P. E., Long, A., Wallick, E. I., On the magnitude of the 11-year radiocarbon cycle, Earth Planet Sci. Lett., 20, 300-306 (1973).
[23] Stuiver, M., Radiocarbon timescale tested against magnetic and other dating methods, Nature, 273, 271-274 (1978).
[24] Cain, W. F., Suess, H. E., Carbon 14 in tree rings, J. Geophys. Res., 81, 3688-3694 (1976).
[25] Tans, P. P., Natural atmospheric ^{14}C variation and the Suess effect, Nature, 280(5725), 826-828 (1979).
[26] Lerman, J. C., Radiocarbon Variations and Absolute Chronology, Discussion of causes of secular variations, Proc. XII Nobel Symp., New York, p. 609-610, 1970.

[27] Burchuladze, A. A., Pagava, S. V., Povinect, P., Togonidze, G. I., Usacevt, S., Radiocarbon variations with the 11-year solar cycle during the last century, Nature, 287, 320-322 (1980).

[28] Baxter, M. S., Farmer, J. G., Walton, A., Comments on "On the Magnitude of the 11-Year Radiocarbon Cycle", Earth and Planet. Sci. Lett., 20(3), 307-310 (1973).

[29] Baxter, M. S., Farmer, J. G., Damon, P. E., Long, A, Wallick, E. I., Comments on "Radiocarbon: Short-Term Variations", Earth and Planet. Sci. Lett., 20(3), 311-314 (1973).

[30] Hall, E. T., Advances in carbon dating using high energy mass spectrometers, Contemp. Phys., 21(4), 345-358 (1980).

[31] Litherland, A. E., Ultrasensitive mass spectrometry with accelerators, Ann. Rev. Nucl. Part. Sci., 30, 437-473 (1980).

[32] Hedges, R. E. M., Radiocarbon dating with an accelerator: review and preview, Archaeometry, 23(1), 3-18 (1981).

[33] Purser, K. H., Hanley, P. R., A carbon-14 dating system, In: Proc. 1st Conference on Radiocarbon Dating with Accelerators, University Rochester, 165-186, 1978.

[34] Stuiver, M., Carbon-14 dating: a comparison of beta and ion counting, Science, 202(24), 881-883 (1978).

RECEIVED May 14, 1981.

Tree Thermometers and Commodities: Historic Climate Indicators

L. M. LIBBY

University of California—Los Angeles, Environmental Science and Engineering, Los Angeles, CA 90024

L. J. PANDOLFI

Global Geochemistry Corporation, Canoga Park, CA 91303

In four modern trees, we find that the carbon, hydrogen, and oxygen isotope ratios track the modern temperature records; namely we find that trees are recording thermometers.

In a 200 year sequence of a Japanese cedar, we find that there are the same periodicities of variation of D/H and $^{18}O/^{16}O$ as have been found in $^{18}O/^{16}O$ in a Greenland ice well.

We find the same periodicities in uranium and organic carbon concentrations versus depth in a sea core from the Santa Barbara Channel, and in carbon-14 variations in a sequence of Bristlecone pine from southern California.

We find in a 2000 year sequence of Japanese cedar and in a 1000 year sequence of European oak that D/H and $^{18}O/^{16}O$ are related to each other by a slope of 8, just as they are in world-wide precipitation.

In a 72 year sequence of Sequoia gigantea, measured year by year for its oxygen isotope ratios, we find the 10.5 year cycle of sunspot numbers, but we do not find the 21 year cycle of sunspot magnetism. This we believe indicates that the sun is affecting the earth's climate with non-magnetic particles, probably photons.

All these phenomena, we believe, are related to periodic changes in sea surface temperature caused by periodic changes in the sun, as are the variations in commercial commodities, and consequent variations in prices and wages.

Furthermore we find that the catch of blue crab in the Chesapeake Bay shows a periodic variation of 10.7 years in agreement with the solar photocycle of 10.5 years, but does not show a variation periodic with the 21 year solar magnetic cycle.

0097-6156/82/0176-0245$13.55/0

We further find that the price of wheat and the laborer's wage vary in agreement with the temperature record in Europe since 1250 A.D.

Principles

The onset of testing of hydrogen bombs in the atmosphere led to an understanding of isotope fractionation in the water vapor that distills from the ocean surfaces throughout the world.

This information derived from the establishment of a global network of 155 collecting stations in 65 countries in the period beginning in 1953 and continuing to the present by the International Atomic Energy Agency and the World Meteorological Organization. Monthly meteorological data (amount of precipitation and temperature) were reported, and monthly samples of precipitation were measured for tritium, deuterium to hydrogen ratio, and oxygen-18 to oxygen-16 ratio [1][1].

The measured ratios are expressed as:

$$\text{delta } D = (((D/H)_s - (D/H)_{std})/(D/H)_{std})) \times 10^3 \text{ ppt}$$

$$\text{delta } ^{18}O = (((^{18}O/^{16}O)_s - (^{18}O/^{16}O)_{std}))/(^{18}O/^{16}O)_{std}) \times 10^3 \text{ ppt}$$

where subscript s refers to the sample, subscript std refers to standard mean ocean water (SMOW), and ppt means parts per thousand.

The error of measurement of delta D is about ±2 ppt and of delta ^{18}O is about 0.2 ppt.

The plot of world data, of delta D versus delta ^{18}O (figure 1), shows that all the measurements for terrestrial surface waters lie on a line with a slope of eight characterizing Rayleigh distillation of water vapor from the sea surface to form atmospheric precipitation. This plot with its slope of eight was originally demonstrated by Harmon Craig [2,3] and by W. Dansgaard [4]. The line is expressed by the relation

$$\text{delta } D = 8 \text{ delta } ^{18}O + \text{constant; constant} \simeq 0$$

where the slope of eight can readily be computed from the measured temperature coefficients for $((D/H)_{liquid}/(D/H)_{vapor})$ and for $((^{18}O/^{16}O)_{liquid}/(^{18}O/^{16}O)_{vapor})$ [5].

Rayleigh distillation is a process in which the condensate is immediately removed from the vapor after formation (by fallout of rain and snow in the meteorological case) and leads to a higher

---

[1]Figures in brackets indicate the literature references at the end of this paper.

fractionation than processes which occur at equilibrium, due to kinetic effects which are not theoretically understood.

To each point on the line of figure 1 there corresponds a temperature of distillation (see figure 2).

In figure 1, the points at very large isotope depletions (delta D ∿ -300 ppt and delta ^{18}O ∿ -40 ppt) have been measured in very cold ice from the bottom of the Antarctic ice cap, laid down in ice ages. Points at small depletions (delta D ∿ 0 ppt and delta ^{18}O ∿ 0 ppt) have been measured in tropical precipitation distilled from warm oceans. Points between have been measured in middle latitudes.

The IAEA monthly measurements show seasonal variations in that the heavy isotopes are depleted in precipitation when water vapor distills off cold oceans in the winters and enriched in precipitation when water vapor distills off warm oceans in the summers. See figure 3 for monthly isotope variations in precipitation, for example in Stuttgart.

This effect was found in the successive seasonal layers of ice of both the Greenland ice cap and the Antarctic ice cap, showing variations like those in precipitation in temperate regions, but on the average more depleted in the heavier isotopes.

Moreover, in the large scale, in the great depths of the ice cap containing ice laid down 10,000 years ago and more in the last ice age, the ice is more depleted in the heavy isotopes than can be found in any modern day precipitation.

Thus it becomes evident that in the polar ice caps there is stored the history of the surface temperatures of the far northern and far southern oceans, from which distilled, for the most part, the historic precipitation laid down in the ice caps.

For temperate regions, the history of the surface temperature of the oceans is stored in the glaciers of those regions, but glaciers have random advances and retreats which spoil the orderly sequence of the historic yearly ice layers.

However, the history of the surface temperatures of the temperate oceans should be stored in the rings of trees which grew in the temperate regions of the world and which subsisted on precipitation which distilled from those oceans. Each tree ring should contain some kind of average annual value of the isotope ratios in the precipitation of the year corresponding to the ring.

The wood in each ring is formed according to the reaction,

$$CO_2 + H_2O \rightarrow wood + oxygen\ gas$$

As for the carbon isotope ratios in tree rings, these derive from and reflect carbon isotope ratios in atmospheric carbon dioxide. There is some evidence suggesting that the ratio ^{13}C/^{12}C in atmospheric carbon dioxide varies seasonally in such a way that the isotope ratio is large in the summer. For example, figure 4 shows monthly variations in the stable carbon isotope ratio in atmospheric carbon dioxide at Spitsbergen, on the Pacific

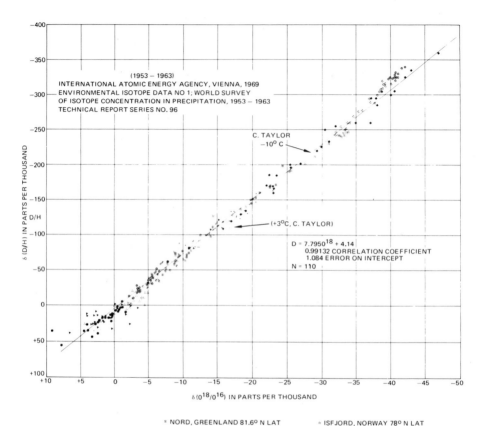

Proceedings of the National Academy of Sciences

Figure 1.   Deuterium isotope ratio vs. oxygen isotope ratio for world-wide precipitation (IAEA data), showing the slope of eight (31).

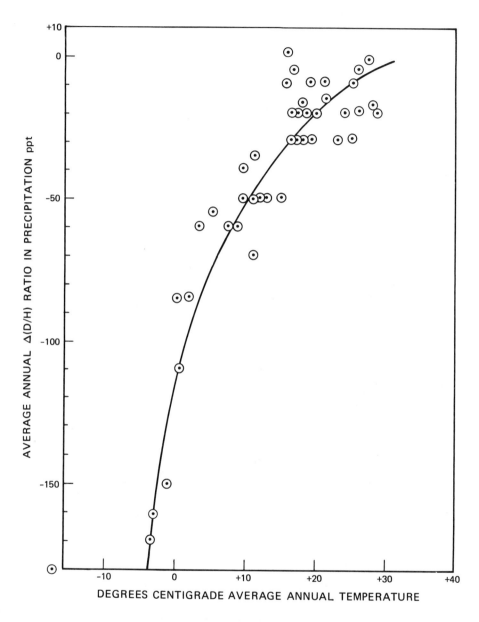

Figure 2.   Deuterium isotope ratios in world-wide precipitation vs. monthly aver-age air temperatures showing that for every point on the line in Figure 1, there is a corresponding average air temperature.

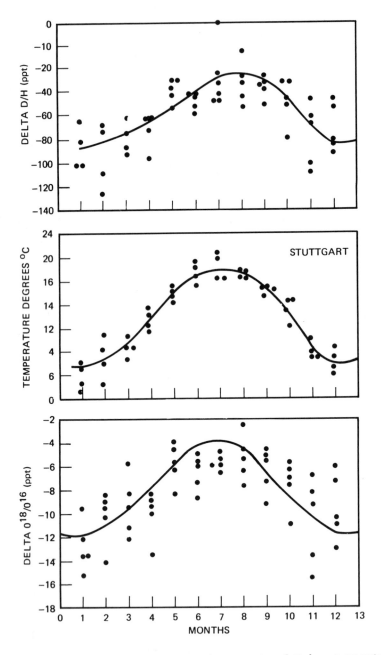

*Figure 3. Monthly oxygen and deuterium isotope ratios plotted vs. temperature
in Stuttgart precipitation.*

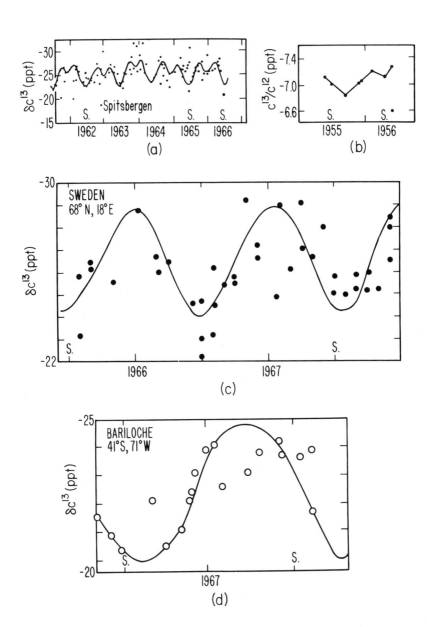

Figure 4.   Monthly variations in the stable carbon isotope ratio in atmospheric CO_2 at Spitsbergen, on the Pacific Coast of the United States, in Sweden, and at Bariloche, Argentina.

Coast of the United States, in Sweden, and at Bariloche, Argentina [6-10], but such variations are not primarily related to ocean surface temperatures.

Keeling interprets these seasonal isotopic variations as caused by trees preferentially removing $^{12}CO_2$ from the atmosphere in summer when they are growing but not in winter when they are dormant.

Wood is composed approximately of cellulose and lignin. Cellulose is a multiple alcohol of schematic formula $(H-C-O-H)_n$ so that the reaction for formation of cellulose may be written,

$$CO_2 + H_2O \rightarrow (H-C-O-H)_n + O_2$$

Lignin contains interconnected aromatic and aliphatic rings [11] (see figure 5) and aliphatic chains containing about 30 percent oxygen by weight in the form of ether, carbonyl, and hydroxyl bonds. Wood is approximately 25 percent lignin [11], its percentage varying somewhat from spring wood to summer wood, and it is possible that its percentage may vary somewhat from ring to ring, so that in principle, its variation might affect the temperature coefficient of wood formation.

Assuming the principle of thermodynamic equilibrium to hold in the formation of wood, a very slow process, we have estimated what effect as much as 10 percent variation in the percentage of lignin may have on the temperature coefficient for the formation of wood. We find it to be about 1.5 percent.

We, Libby and Pandolfi, have felt that a 1.5 percent uncertainty is tolerable within the limits of other errors inherent in the method, and therefore we have always analyzed whole wood in our study of isotope variations in tree-ring sequences.

In analyzing whole wood, one is confronted by the question of whether to use wet or dry chemistries. Of course if one decides to separate cellulose from lignin, then one is forced to use wet chemistries. It is only in whole wood analysis that dry chemistry becomes possible. With wet chemistries, performed necessarily with hydrogen- and oxygen-containing solvents, there is always the risk of isotope exchange with the solvent. See for example the review article of H. Taube [12]. He shows that there exists intimate exchange of hydroxyl oxygen (-O-H) with carbonyl oxygen (-CO-OH) under all conditions of acidity and alkalinity in liquids such as water, ketones, aldehydes, and alcohols. Sepall and Mason [13] describe exchange of cellulose and whole wood hydrogens with hydrogen in water as being rapid and effective, leading one to expect similar exchanges with other solvents containing OH groups. The exchange of hydrogen in cellulose with water was found to be 50 percent in one hour at 25 °C. With dry chemistry there is no possibility for isotope exchange with the reagents, and for this reason we have always used dry chemistry.

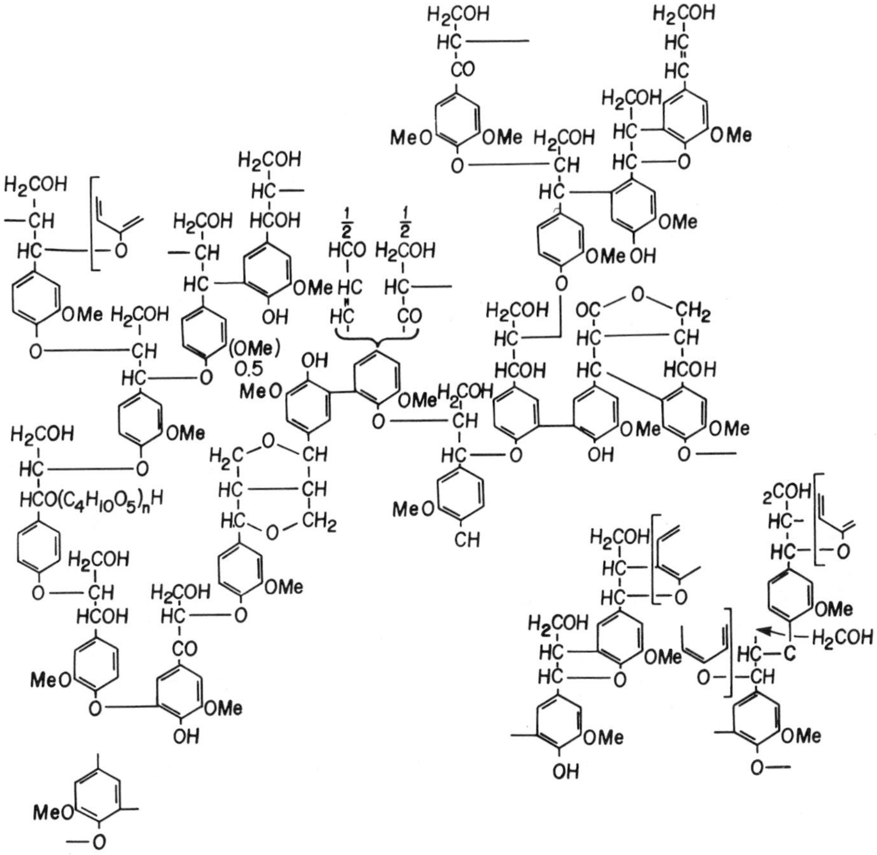

Figure 5.   Schematic molecular structure of lignin.

Thus in our study of isotope variations in lengthy chronological sequences of tree rings, we are evaluating fluctuations in
the sea surface temperatures, from which distilled the precipitation which nourished the trees; and the sea surface temperatures,
in turn, are affected by variations in the ultraviolet spectrum of
the sun.

The variations in the organic matter content of sea cores
are caused by variations in the amount of living matter growing
in the sea surface and becoming more abundant as the sea surface
grows warmer, but dropping to the bottom to form sediments when
death occurs. Uranium ions, always present in sea water, attach
themselves readily to organic matter in the sea water and fall to
the bottom to form sediments just as does organic matter, so that
when organic matter is more abundant in sediments, uranium is
also.

Therefore, we are studying a new technology which may offer
the first advance in evaluation of solar physics since the invention of the solar coronograph.

Very little is known about fluctuations of the solar constant
and even less about fluctuations of the ultraviolet part of the
solar constant. Until now, periodicities of the sun have been
evaluated solely from the sun spot cycles, observed over some four
hundred years (except for a few desultory observations in ancient
China), but the sun has many ways to vary besides sunspot variations.

As was concluded by the Department of Transportation's CIAP
study, "All that is really known about the solar constant is that
it has not changed by as much as a factor of two throughout the
ecological history of the earth, that over the past 50 years it
has not changed by as much as 10 percent, and that over the last
10 years, it has not changed by as much as 1 percent. These known
limitations do not rule out long-period (hundreds of years or
longer) changes of, say, less than 30 percent over the entire solar
spectrum. Neither do they rule out short period changes in the
near UV nor UV flux (less than 2500 Å) of the order of a factor of
two." [14]

By the method of measuring stable isotope ratio variations in
tree ring sequences, we hope to set limits on such solar changes
and evaluate their periodicities back as far in time as tree ring
sequences exist, for at least 8000 years, in a sequence tied to
the present by overlapping ring patterns onto present patterns.
In "floating" sequences in pre-glacial trees, sequences not fitted
to present-day sequences, but instead dated by radio carbon, we
have another data base where again evaluation of solar variations
becomes possible for the more distant past.

Varves suggest themselves as another data base from which
to evaluate the history of the climate. These are sequences of
sediment layers deriving from fresh water streams, the summer
sediment having a different particle size and color than does the
spring sediment, so that the deposit for each year becomes visible

as a distinct bicolored layer, and the years of sediment can be counted back from the present. Varve sequences are known which extend over ten thousand years in a single sequence. In their yearly layers there are willow leaves and twigs, and there are insect wings of beetles, each wing being large enough for a sample for measurement of the hydrogen and oxygen isotope ratios. If the insects have subsisted on rain water, study of the isotope ratios in their wings should allow one to deduce the history of climate for the entire varve sequence, as would also measurement of willow leaves.

Sea cores offer a data base which should in principle allow deduction of the history of the local sea surface temperature immediately above the deposition site of the core, for there is enough organic material in sea cores to provide the necessary samples for isotope measurement at frequent intervals versus depth in the core, but the time resolution is far less accurate than in varves and tree rings because burrowing sea bottom animals smear the record of the layers.

Any of these data banks, those parts from the ice ages, can have their stable isotope ratios perturbed by the huge ice reserves which were removed from the sea and piled up on land, because the ice depletes the oceans in the light isotopes, and therefore significantly enriches the sea in the heavy isotopes, so that sea sediments and continental precipitation, rain and snow, reflect this perturbation as well as perturbations caused by temperature changes alone.

Tree ring sequences extending back from the present to some 8,000 to 10,000 years ago are being prepared by Bernd Becker of the Universität of Hamburg, from huge trees discovered in the beach sediments of the great rivers of Europe, the Rhine, Rhone, and Danube, which grew 10,000 years and more before the present. These scientists fit the tree ring patterns together in overlapping records, and have quantities of the authenticated wood dated by pattern recognition to give to analytic laboratories for analysis of climate variations. They have given us some samples.

We obtained pieces of counted and dated Sequoia gigantea from Paul Zinke of the Forestry Department at the University of California at Berkeley, from Henry Michael of the University Museum of the University of Pennsylvania and from the United States Forestry Service in Kings Canyon National Park.

Sequences of bog oak have been prepared from the ancient oaks dug out of the bogs of England and Ireland by laboratories in those countries and are being analyzed in Scotland and North Ireland.

We hope that tree ring sequences will be prepared from trees of the southern hemisphere from which one could learn whether climate changes have been simultaneous in both hemispheres. There are large tree stumps in New Zealand, and perhaps similar material could be found in Australia and other southern lands, and there may be varve sequences as well in the southern hemisphere.

History and Technology

Long term changes in precipitation, caused by changes in climatic temperature, are well documented in polar ice caps; the heavier of the stable isotopes is depleted in ice laid down in the ice age by comparison with present day ice. In 1970 we extended this concept to trees, suggesting that they, also, are thermometers. Trees grow from water and atmospheric CO_2. In trees which grow on rain water, isotope variations in their rings should be climate indicators because the isotope composition in rain and CO_2 varies with temperature.

On May 17, 1971, the Defense Advanced Research Projects Agency [15] funded our proposal that "temperature variations may be evaluated by measuring stable isotope ratios in natural data banks such as tree rings and varves". L. M. Libby had previously calculated [16] the theoretical temperature coefficients of the stable isotope fractionations in manufacture of wood from CO_2 and H_2O, finding that the coefficients are small compared with those measured in rain and snow [17].

We considered whether to measure whole wood, lignin, or cellulose, wood being about 25 percent lignin on the average [18].

But isotope fractionation at climatic temperatures is a function of the frequencies of the chemical bonds [16]. We quote from Herzberg [19] as follows: "One would expect the -C-H bond to have essentially the same electronic structure and therefore the same force constant in different molecules, and similarly for other bonds. This is indeed observed". For the -C-H bonds the vibrational frequencies in lignin and in cellulose are almost equal, but in fact differ by 6 percent [19] because cellulose is a multiple alcohol $(H-C-O-H)_n$ and lignin is a polymer containing both aromatic and aliphatic carbons connected to hydrogen. Therefore, assuming thermodynamic equilibrium, variations of the lignin concentration in tree rings might affect the hydrogen isotope ratio by as much as 25 percent of 6 percent, namely by as much as 1.5 percent.

Likewise for the -C-O-H bonds, the vibrational frequencies are equal in lignin and in cellulose within a few percent [19]. But lignin, different from cellulose, also contains ether linkages (-C-O-C-). The -C-O-H linkage has a C-O bond distance of 1.427 Å, and the ether linkage has a C-O bond distance of 1.43 Å [20]. Therefore the presence of lignin, containing 14 percent oxygen of which 16 percent is ether linked [18] (see our figure 5) might affect the isotope ratio of oxygen in whole wood by 0.3% x 25% x 14% x 60%, equal to 6×10^{-3}%, by variation from its average concentration of 25 percent.

The carbon-carbon bond linkage in cellulose is 1.541 A and in the aromatic groups of lignin is 1.395 A, a difference of 10 percent [20]. Therefore the presence of lignin, containing 75 percent carbon [18] of which 60 percent is aromatic, may affect the carbon isotope ratio by 10% x 75% x 60% x 25%, or about 1 percent, by variation from its average concentration of 25 percent.

On these numerical arguments and on the necessity to avoid isotope exchange with liquids, we based our decision to measure stable isotope ratios in whole wood.

The next problem was which trees to measure. Many tree ring sequences can be counted with an accuracy of about one year. Those which are not yet tied to modern sequences by overlapping ring patterns (said to be "floating"), can be dated in favorable cases with an accuracy of about 30 years by radiocarbon dating, depending on the age and the number of radiocarbon measurements which are made.

But to prove our hypothesis that trees are thermometers, we needed to compare our measurements of stable isotope ratios in the tree rings with mercury thermometer records near where the trees grew. Thus we could not use bristlecone pines because there is no lengthy temperature record for hundreds of miles near their home in the White Mountains of California[2]. Because the longest temperature records are in Europe, we obtained a German oak from the only tree ring laboratory then existing in Europe, the laboratory of Bruno Huber in Munich [21,22], where the oak rings were counted and labeled with the numbers of the years in which they grew. More recently his students, Bernd Becker and Dieter Eckstein, who have established tree ring laboratories in Stuttgart and Hamburg, respectively, have sent us additional sequences of German oaks in which they have counted and labeled the rings.

From the laboratory of K. Y. Kigoshi in Tokyo, we obtained a 2000 year ring sequence of a cedar from the southern tip of Japan, in which Kigoshi had counted and labeled the rings; in addition he verified his dates by making 50 radiocarbon measurements in its wood. Although radiocarbon dating's accuracy is only about 40 years, from 50 measurements the verification achieves an accuracy of $40/(50)^{1/2}$ or about 6 years.

The temperature records needed for comparison with the German oaks exist at nearby Basel and Geneva, extending back more than two centuries, and in middle England, extending back three centuries [23-25]. Temperature records for the cedar exist at nearby Miyazaki, Japan, since 1890. In addition there are surrogate climate records for the far east in the form of records of dates of cherry trees blooming, of first freezing dates of lakes, and of number of snowy days per year [26].

---

[2]Nevertheless some measurements of D/H in bristlecones have been published [27].

We had an idea of the magnitude of the oxygen and hydrogen isotope variations we could expect to find in these trees because, for 15 years the International Atomic Energy Agency in Vienna (IAEA) has measured and published the stable isotope variations in rain and snow, month by month, for some 100 world-wide weather stations, including those in Germany, Austria, and Japan, see figure 1.

We considered whether old heart wood could exchange isotopes with modern sapwood. On the contrary there is compelling evidence that when sapwood passes into heart wood it becomes sealed against sap and therefore against isotope exchange with sap, at least in species having tight rings. For example Huber has shown, using biological dyes in many tree species, that water conduction remains limited to the outermost annual ring.

Additional evidence is that radiocarbon sugar injected into or ingested by a living tree does not move into neighboring rings but remains in the ring in which it was injected or ingested. Furthermore we have observed that when an intact block of Sequoia gigantea was soaked for a month in an air atmosphere saturated with water previously labeled with delta D_{SMOW} = 1,170 ppt, no deuterium exchange with wood was observed. This is reasonable to expect because wood is a remarkable polymer, containing very large molecules, cross linked internally and to each other with hydrogen bonds.

Hence we concluded that the climate record in heartwood cannot be modified nor perturbed by the sap in the outermost ring, of the current active year.

For our first tree sequence [28-32] we measured D/H by reacting sawdust with uranium to produce H_2, 99 percent quantitatively. For measurement of $^{18}O/^{16}O$, we modified the method of Rittenberg and Pontecorvo [33] by carrying it out at very high temperatures, 99 percent quantitatively. The temperature must be 525 °C; if it is lower, the reaction is not quantitative; see the section on our chemistry later in this paper. To measure the stable isotope ratio in carbon, we burned sawdust to completion in oxygen.

Whether the oxygen in tree rings comes from water or from CO_2 is a non-question, because Cohn and Urey [34] showed that isotopic equilibrium between the two substances is obtained in a damp atmosphere within a few hours at room temperature.

For measurement of the oaks we used, per force, a mass spectrometer of somewhat low accuracy, and achieved the accuracy to demonstrate that trees are thermometers by making many measurements on each sample. On the tree sequences which we measured later, we used high precision spectrometers with accuracies of ± 0.1 parts per thousand (ppt) for $^{18}O/^{16}O$ and $^{13}C/^{12}C$, and ± 2 ppt for D/H. The measurements are expressed in terms of δD and δ_{18}.

Sample Preparation

We mill a groove perpendicular to the tree rings, that is, along the radius of the tree; sawdust from each few rings is collected into an individual vial with a camel's hair brush. The vials are dried at 50 °C and capped off to protect the dried sawdust from damp air. Our chemistry is now as follows:

To evolve CO_2 for measurement of $^{18}O/^{16}O$: pump for 4 hours on 3 mg of sawdust mixed with 120 mg $HgCl_2$ in vacuo. Seal container. Heat at 525 °C for 4 hours; if T is less than 525 °C, production of CO_2 does not quantitatively remove oxygen. React with triple-distilled quinoline at boiling temperature until quinoline turns yellow. Freeze in ethanol-dry ice-slurry at -120 °C. Pass gas through two dry ice-acetone traps.

To evolve H_2 for measurement of D/H: burn 5 mg dry sawdust in 1 atm. O_2 in a cupric oxide furnace at 750 °C. Use oxygen purified over silica gel and cupric oxide to ensure that the O_2 is hydrogen free. Freeze out H_2O and CO_2 in a liquid oxygen trap. Release CO_2 at dry ice temperature. React H_2O vapor on clean uranium shavings at 950 °C, thus producing H_2 quantitatively.

To evolve CO_2 for measurement of $^{13}C/^{12}C$: burn 3 mg sawdust in dry, clean oxygen gas.

We thank Willard F. Libby for advising how to make this dry chemistry quantitative.

Recent Trees and Thermometer Records

To prove that trees are thermometers, we needed to measure modern trees whose rings have been correctly counted, and to obtain lengthy mercury thermometer records from places near where the trees grew, preferably from the same altitude. The oldest temperature records in the World Weather Records [35] are for Basel and Geneva, each at about 300 meters altitude. We fortunately obtained from the laboratory of Bruno Huber in Munich a German oak grown in middle Germany at an altitude of about 300 meters, which had been correlated with the fiducial oak tree ring sequential pattern developed by Huber and his colleagues [22]. Later we obtained samples of similarly calibrated German oaks from his students, B. Becker and D. Eckstein [31,32].

So far, we have measured isotope ratios in four modern trees of four different species at four different altitudes, latitudes, and longitudes, and compared them with local temperature records from mercury thermometers: German oaks Quercus petraea, Bavarian fir Abies alba, Japanese cedar Cryptomeria japonica, and California redwood Sequoia gigantea. For the parts of the oaks and the cedar extending beyond the beginning of mercury thermometer records, we compare the measured isotope ratios with surrogate evidence of climate change such as lateness of cherry tree bloomings, number of snowy days per year, and number of days per year when lakes were frozen. We find significant correlations.

In figure 6 we compare carbon, hydrogen, and oxygen isotope measurements for the German oaks with English temperature records [23-25] back to the time of the invention of the mercury thermometer in the late seventeenth century. Temperature records of Basel and Geneva from 1850 resemble those of middle England and thus fit the isotope ratios well, but we show the English thermometer records here because they extend further into the past. The carbon isotope ratios for the years 1890-1950 have been corrected for the effect of fossil CO_2 production (Suess effect); the maximum correction, that for the year 1950 A.D., was taken as 8.4 percent or a maximum increase of 2.1 ppt in $^{13}C/^{12}C$, because in wood from rings of 1920 A.D., two radiocarbon dates were measured by Dr. Rainer Berger, U.C.L.A., as 375 ± 35 years old with respect to 1950 A.D., whereas the actual age is 1950 - 1920 = 30 years, corresponding to 4.2 percent dilution of atmospheric $^{14}CO_2$ by inert CO_2 produced by man's burning of coal and oil up to 1920. In 1950 the correction for C^{14} in this particular tree should be 8.4 percent and the correction for ^{13}C dilution should be 8.4 percent of 25 ppt or 2.1 ppt.

Direct measurements of $^{18}O/^{16}O$ in rain and snow have been made and are available in publications of the International Atomic Energy Agency [17]; figure 7 shows the ^{18}O and D correlations with temperature for Austrian stations for the past 15 years. The isotope concentration of precipitation varies similarly with temperature in many other places as shown by plots of the IAEA isotope measurements against air temperature.

Figure 8 shows the $^{18}O/^{16}O$ ratio for a Bavarian fir, Abies alba, the rings of which were counted by Becker and Siebenlist [36], compared with temperature records made near where the tree grew, both coming from 1000 meters altitude on the north slopes of the Alps. This is mountainous country so that local differences in temperature may be expected.

Figure 9 shows $^{18}O/^{16}O$ in a Cryptomeria japonica compared with a local temperature record, in southern Japan, from the weather station at Miyazaki. The tree grew at 1,350 meters altitude whereas Miyazaki is at sea level.

Figure 10 shows $^{18}O/^{16}O$ in a modern part of a Sequoia gigantea which grew in the Giant Forest of Sequoia in King's Creek Canyon National Park, Three Rivers, California, at 1940 meters altitude [37,38] compared with air temperatures for the same years measured in Yosemite National Park, about 100 miles south, at an altitude of 1200 meters. We compare with summer air temperatures because the Sequoias in King's Canyon grow for the most part in summer when many inches of rain fall from clouds coming from the Gulf of Mexico.

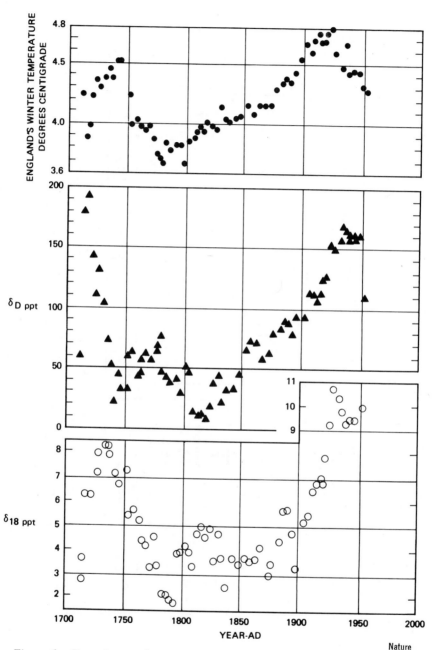

Nature

Figure 6.   Deuterium, carbon, and oxygen isotope ratios vs. temperature in a German oak, A.D. 1700–1950 (26).

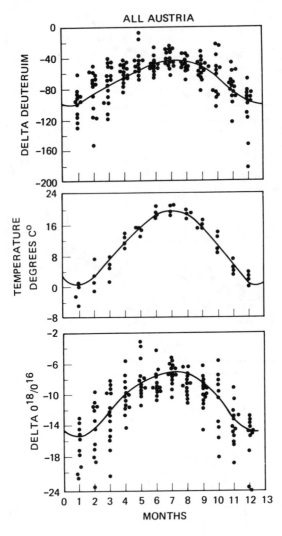

Nature

Figure 7. Oxygen and deuterium isotope ratios vs. temperature in Austrian precipitation (26).

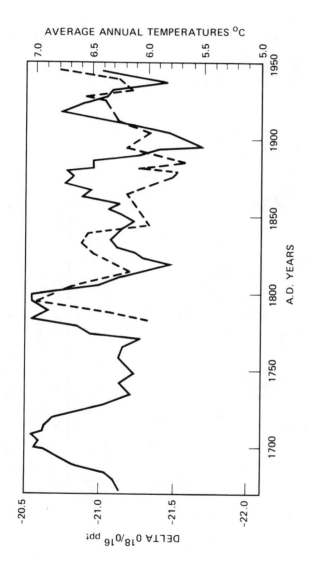

Figure 8.   Oxygen isotope ratios vs. temperature in a Bavarian fir. Key: – – –, average annual temperatures at Hohenpeissenberg (°C); and ———, $^{18}O/^{16}O$ ratios in Bavarian fir.

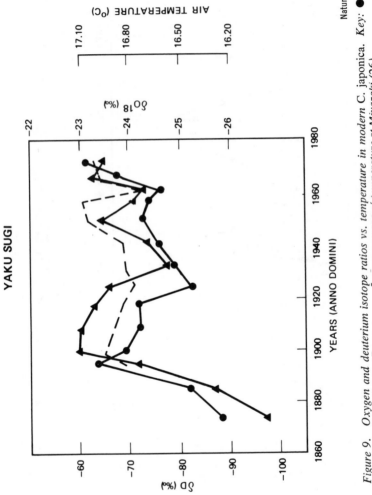

Figure 9. Oxygen and deuterium isotope ratios vs. temperature in modern C. japonica. Key: ●, $\delta^{18}Opdb$; ▲, δD_{SMOW}; – – –, *5 years average of temperature at Miyazaki (26).*

Nature

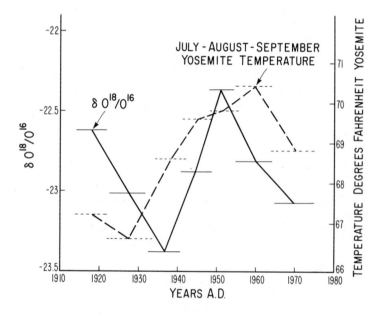

Figure 10.   Oxygen isotope ratio vs. temperature in S. gigantea.

Old Trees and Surrogate Evidence

In figure 11 we show oxygen measurements for oaks which grew before thermometers were invented, indicating warm intervals at A.D. 1530, 1580 and 1650, and cold periods at about A.D. 1700 and 1800 in agreement with Bergthorsson's deductions of climatic variations in Iceland [39] and in agreement with the historical evidence of severe climate deteriorations in the First and Second Little Ice Ages in Europe.

In figure 12 we show oxygen and hydrogen isotope ratios in a Japanese Cryptomeria japonica, also from the island of Yaku at the southern tip of Japan as is the modern cedar (figure 9), for the years 137-1900 A.D., compared with temperatures deduced from old diaries for Japan and China.

In figure 13 we show the oxygen isotope ratio for the same Sequoia gigantea (figure 10) for the years 1750-1975 A.D. It shows the same negative slope indicating deteriorating climate as does the Japanese cedar in figure 12 amounting to a long term decline of about 1.6 °C in Miyazaki in the last 2000 years. This is significant especially considering that there was a decline of about 10 °C in the last ice age.

Climate Periods

We have made Fourier transforms [40] of the data in figures 12 and 13, to deduce the power spectra of periodicities. The results are shown in figures 14 and 15. The same periods are found in deuterium and oxygen in the Japanese cedar, within experimental error, as indicated from the widths of the peaks of the power spectra, figure 14; they are listed in Table 1.

Table 1.   Periods yielded from several sets of measurements
           by fourier transform (years).

Cryptomeria japonica		Santa Barbara sea core		Bristlecone Pine	Greenland Ice Well
D/H	$^{18}O/^{16}O$	Organic carbon	Uranium	Carbon-14	$^{18}O/^{16}O$
156	156	161	156	162	179
110	124	121	118	108	–
97	97	95	95	–	100
86	88	82	81	–	78
65	70	71	70	–	68
58	55	55	53	–	55

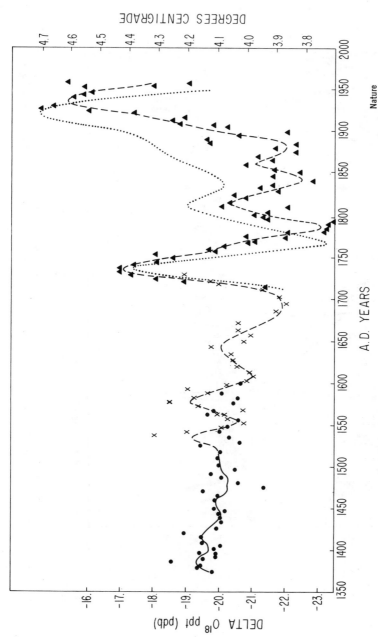

Figure 11.   Oxygen isotope ratios in a sequence of German oaks, A.D. 1350–1950 (26).

Key: *Marburg, Germany oak:* ———, 5 samples (~ 30 years) running average; ●, 4–7 year samples;
· · ·, average of Jan., Feb., and Mar. temperatures in England. *Spessart Oak No. 2:* ×, 5 samples (30
years) running average; *Spessart Oak No. 1:* ▲, 7 samples (30 years) running average; – – –, smoothed
                                                     by eye.

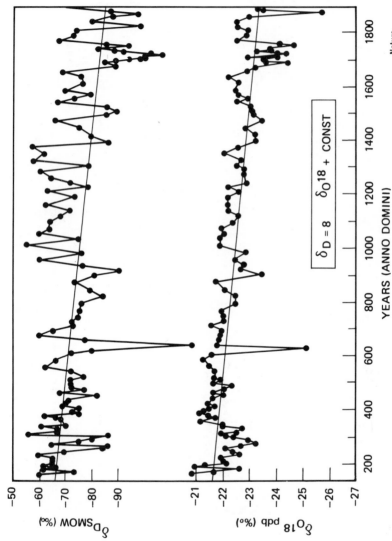

Figure 12.A. Deuterium and oxygen isotope ratios in a Japanese C. japonica (Yaku Sugi) from Yaku Island, A.D 150–1900(26). Key: top, $\delta D_{(SMOW)} = -0.01053$ (years) -64.9; bottom, $\delta^{18}O(pdb) = -0.00129$ (years) -21.1.

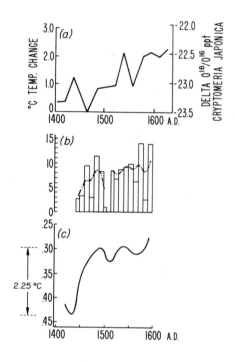

Nature

Figure 12B. Oxygen isotope ratios in C. japonica calibrated for temperature change using temperatures obtained from (a) time of cherry tree blooms, (b) first freezing of Lake Biwa, and (c) number of snowy days per year (26).

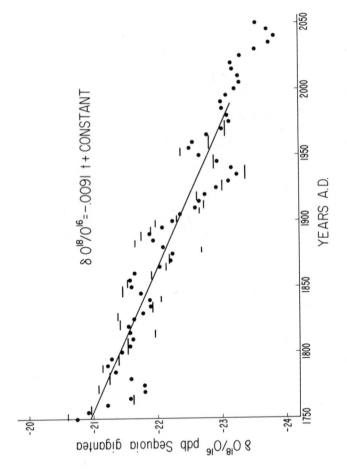

Figure 13. Oxygen isotope ratios in 250 years of S. gigantea ring sequence from California.

The horizontal lines represent measured values. The black circles represent values predicted from the reconstruction using periods, phases, and amplitudes obtained from the Fourier transform. The prediction has been projected into the future.

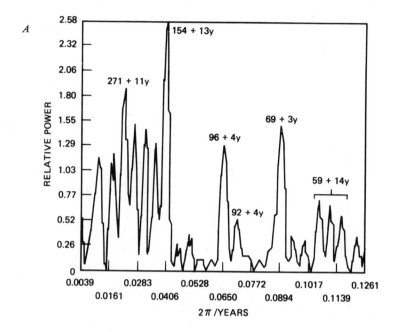

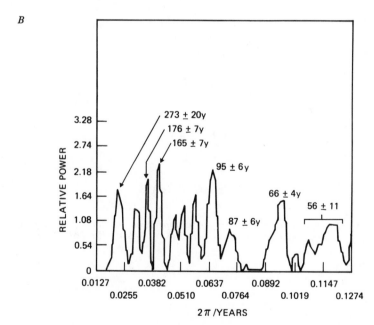

Nature

Figure 14.   Fourier transform of (A) the oxygen isotope ratio vs. time in C. japonica *and (B) deuterium isotope ratio vs. time in* C. japonica, *transformed into power vs. reciprocal period (average slope subtracted) (26).*

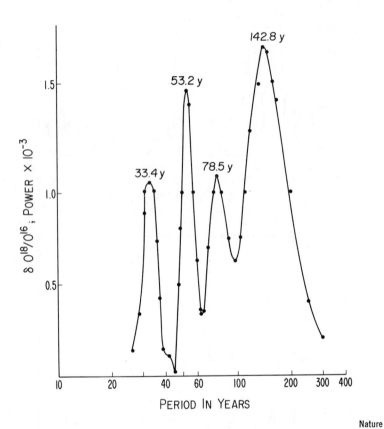

Figure 15. Fourier transform of oxygen isotope ratio vs. time in S. gigantea, transformed into power vs. period (A.D. 1749–1975) (26).

Our samples consist of wood from about 5 years each, so we cannot expect to find meaningful evidence for periods of less than about 40 years in these data. The Japanese cedar spans only about 1800 years, so we cannot ask for meaningful evidence for periods of more than about 250 years. In making the Fourier transforms, we have used the deviations of the isotope ratios from the long term slopes--that is we have corrected for the slopes. We have tested the meaningfulness of the periods in Table 1 by manufacturing data sets consisting of a number taken from the least squares fit to the data, namely the straight line fit, plus a random number varying over the numerical range of the deviations from the line. Each set of manufactured data was subjected to the Fourier transform. In each case, in the transforms for 30 such manufactured data sets, no peaks were generated of significance, indicating a confidence level for the periods in Table 1 of better than 96 percent.

The Fourier transforms were performed in the standard way. No smoothing nor filtering was employed. Subtraction of the data from the least squares fit removes the constant or linear term characterizing a Markovian process. Fourier transform of the differences from the linear fit suppresses the enhancement of both the power and amplitude spectra at low frequencies.

The periods derived from the transform of the oxygen isotope ratio in the S. Gigantea appear to agree with those in the C. japonica as shown in Table 2. In the Sequoia, the samples were taken for as few as two years at a time, so that it is possible to have confidence that the 33 year period found in it is meaningful. But because the tree has been analyzed only for about a 225 year span, we can not ask for meaningful evidence of periods of more than 143 years, and although it appears to agree with the period of 156 years found in the C. japonica, and although in our experiments with artificially generated isotope ratios in the Fourier transform no such large amplitudes were produced, still one must take caution.

In the Fourier transform, amplitudes $P^{1\,2}$ and phases ϕ are generated for each period. We have put the periods, amplitudes and phases back into the summation of transcendental functions which represents the function of the data, and have used this expression to generate the original data, as shown by the black spots in figure 13. By running this function into the future we have made a prediction of the climate to be expected in King's Canyon; the prediction is that the climate will continue to deteriorate on the average, but that after our present cooling-off of more than the average decay in climate, there will be a temporary warming-up followed by a greater rate of cooling-off.

Naturally, a complete analysis of the 3000 year span of the S. gigantea which we have in hand will yield a more reliable prediction of the future climate for King's Canyon.

Table 2.

Cryptomeria Japonica 168 A.D. - 1885 A.D. Yaku Island, Japan	Sequoia Gigantea 1749 A.D. - 1975 A.D. Three Rivers, California
τ 156 y	143 y
P 2.72	1.59
ϕ 1.76 rad	1.73 rad
τ 70.0 y	78 y
P 1.47	1.09
ϕ 2.93 rad	5.86 rad
τ 58 y	53 y
P 0.79	1.09
ϕ 3.09 rad	3.23 rad
τ --	33 y
P --	1.59
ϕ --	3.06 rad

Eight of the periods found in the oxygen and hydrogen isotope ratios of the Japanese cedar are also evidenced in variations of the oxygen isotope ratio versus depth in the Greenland ice [41] as listed in Table 1 [42]. The remarkable agreement between our tree records of oxygen and hydrogen isotopes and the ice record of oxygen isotopes shows itself in yet another way: we have found the D/H and $^{18}O/^{16}O$ ratios for the Japanese cedar to be significantly correlated in opposite phase to the ^{14}C variations in bristlecone pines of southern California [26] as measured by Suess [43] (See Fig. 27). Similarly Dansgaard et al., [41] found the oxygen isotope record in Greenland ice to be significantly correlated in opposite phase with bristlecone carbon-14. Furthermore the Fourier transform of the carbon-14 variations shows four of the same periods as in our tree and in the Greenland ice; see Table 1.

We conclude from the correlations of these four sets of data that the calculation of Dansgaard et al., [41] of the age of ice versus depth in the Greenland ice cap seems to be correct with an error of not more than a couple of years, at least over the last 800 years. We conclude that the climate variations in Greenland, southern Japan, and southern California have had the same periodicities for the last 800 years or more.

A logical explanation for the global nature of these correlations is that they are all related to variations of the sun, which cause variations in the temperature of the sea surface, thus causing variations in the isotopic composition of water vapor which distills off the sea and is stored as wood in trees and also forms the annual layers of the ice cap. The variations of the sun are furthermore related to the flux of solar neutrons in the earth's atmosphere and so cause small variations in the carbon-14 content of the bristle cones. During times of a quiet sun the average carbon-14 production is about 25 percent larger than when solar activity is high [43].

We now report two additional data sequences [44,45] versus age, in which the same periodicities are revealed, namely the organic carbon and uranium concentrations in a sea core from the Santa Barbara Basin off California. Preservation of annual varves in this anoxic sediment provides a record of the age of the sediments, there being one varve or distinct layer for each year. The concentrations, versus depth, of organic carbon and uranium were measured in a continuous sequence of samples, each containing seven years of sediment, in sea core PT-8G, spanning the years 1264-1970 A.D. The age of sediment versus depth in the core was determined by comparison of its varves with those in core 214 in which the varves had been counted [46,47].

The age determination allowed Fourier transforms to be made, transforming concentration versus depth into signal power and amplitude versus the period expressed in years. The periods found are listed in Table 1. The sea core spans 700 years, with each sample containing seven years of sediment, so that evidence for periods between 40 years and 200 years should be meaningful [48].

Table 1 shows that eight periods found in uranium and organic carbon variations in the sea core are also found in stable isotopes of hydrogen and oxygen in the tree and in oxygen isotope variations in the ice, enriching the interpretation of climate variations on a global scale, and enriching the attribution of these periodicities to variations of the sun causing changes in sea surface temperature. The temperature of the sea surface determines the abundance of life and therefore the abundance of organic matter which falls to the ocean bottom and binds uranium ions in the sea water to it as it falls. (It is well known that in uranium ores the amount of organic carbon is proportional to the amount of uranium.)

The recently completed CIAP study of the stratosphere [49] indicated mechanisms by which the sun's variation may influence temperature on the earth's surface. In particular, solar energy absorbed in the stratosphere is rapidly converted by chemical reactions, producing a variety of chemical species which reradiate both to space and to the ground. Thus climate on the ground is sensitive to variations in stratospheric chemical components. The solar light is known to vary by factors of 2 and 3 in intensity at short wave lengths, affecting the concentrations of those species. Re-radiation to the ground is especially sensitive to ozone and water concentrations in the stratosphere and their variations are only just beginning to be assessed. Study of stable isotopes in trees and of uranium and organic carbon in sea sediments as a function of time in the past may be a powerful way of studying past misbehaviors of the sun.

The most well known misbehavior, i.e., variation of the sun, is by its varying the number of its sun spots, with a change in field between the north and south hemispheres of the sun every 10.5 years, and a completed magnetic sun spot cycle every 21 years on the average. Supposing that the sun's variations have caused the variations of the stable isotopes in our trees, the stable oxygen isotopes in the ice cap, the organic carbon and uranium in the ice cap, and the carbon-14 in the bristlecone pines all to vary with the same periodicities, then it seemed reasonable to us to look for evidences of the 10.5 and 21 year sunspot cycles in our trees. In order to find meaningful evidence for such short periods, it is necessary to measure the stable isotope ratios in every single year of a tree ring sequence.

Accordingly, our student, Paul Hurt [50] measured the oxygen isotope ratio in a ring sequence of 72 years in a B.C. sample of Sequoia gigantea. He chose this sequence because the rings are unusually wide and consequently easy to separate. He analyzed it year by year; his results, shown in figure 17, have significant power signals at 6.69, 10.4, and 37.4 years. To determine the significance of these signals, we generated 30 random number sequences of 72 numbers each having δ_{18} values between -18 and -21 ppt, the range of the values of the measurements, figure 16, and made their Fourier transforms, but in none of these computer experiments were produced peaks with more than half the amplitude of those in figure 17 indicating that the periods of 6.69 and 10.4 years are real, with a confidence of 96 percent or better (no peaks in 30 experiments with random number sets).

The reality of the large peak at 37.4 years might be questioned because 37.4 years is a large fraction of 72 years, however it appeared in the Fourier transform of the oxygen isotope ratios of the modern part of the same Sequoia (1749-1975 A.D.) at 33.4 years so we believe it to be real. Furthermore we cut the data sequence from 72 years to 63 years and made its Fourier transform, and found that the 37 year signal remained unaffected, indicating that it is not an artifact of a half harmonic of 72 years.

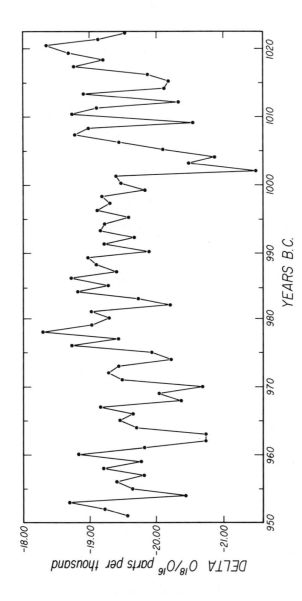

Figure 16.   Oxygen isotope ratio vs. time in S. gigantea, measured for each year.

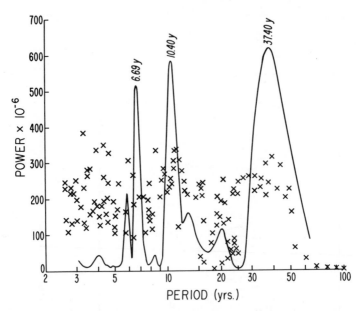

Figure 17.   Fourier transform, power vs. period, of oxygen isotope ratios vs. time, measured year by year. Spectral distribution of $\delta^{18}O$ (———) in tree and random numbers (×).

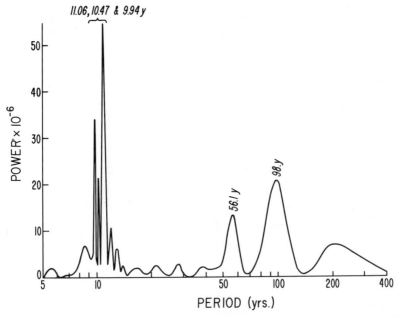

Figure 18.   Fourier transform of sunspot numbers, counted year by year for 250 years, transformed into power vs. period, assuming all sunspot numbers to be identical in magnetic polarity (positive numbers).

For comparison with the periodicities in the sunspot cycle, we have made Fourier transforms of the 350 years of the sunspot counts. By assuming all sunspot counts to be positive numbers, [51], we obtained the power spectrum shown in figure 18 with signals at 9.94, 10.47, and 11.06 years, for a strong signal averaging to 10.6 years. By assuming sunspot counts to have opposite magnetic polarity in the northern than in the southern hemisphere of the sun [52], we obtained the power spectrum shown in figure 19 with a strong signal at the well known period of 21 years [52].

The fact that the oxygen isotope power spectrum in figure 17 does not show a significant signal at 21 years but shows a strong signal at 10.4 years (which could easily by 10.5 years within our errors) we interpret as indicating that the influence of the sunspot cycle on the earth's climate is effected by neutral particles (which are affected by the number of spots but not by their magnetic polarity, because they are neutral), probably photons. Neutral particles are not deflected by the earth's magnetic field, so they are able to come to the earth symmetrically from either hemisphere of the sun.

The signals in the oxygen isotope record from the S. gigantea's rings at 6.69 and 37.4 years may or may not be related to periodicities, of the sun. The sun has many ways to vary, apart from the sun spot cycle, such as fluctuations in frequency of solar flares and plages, and misbehaviors of the overall magnetic field of the sun and of the solar corona.

The conclusions of Hurt's study of year-by-year oxygen isotope ratios in 72 years of S. gigantea are thus supportive of the conclusions of the CIAP study [49] that solar variations influence the abundances of many kinds of chemical species in the stratosphere, and therefore influence the amount of solar energy they absorb and re-radiate to earth, and therefore influence the surface temperature of the earth and especially the surface temperatures of the oceans. It is the surface temperature of the oceans which produces the phenomena we have discussed: the isotope ratio variations in rain and hence in tree rings, the isotope ratio variations in the Greenland ice cap, in the organic carbon and uranium concentrations in sea cores, and furthermore variations of the sea surface temperature produces variations in the carbon-14 to carbon-12 ratio fractionation at the sea air interface and hence in the carbon-14 content of atmospheric carbon dioxide and hence in the carbon-14 content of tree rings.

The Bio-Organic Reservoir

In the foregoing discussion, we have been concerned with the effects of climate on the biosphere. Now we turn our attention to the effect of the biosphere on climate.

The biosphere is an important reservoir of carbon, derived from atmospheric carbon dioxide, which it both obtains from the

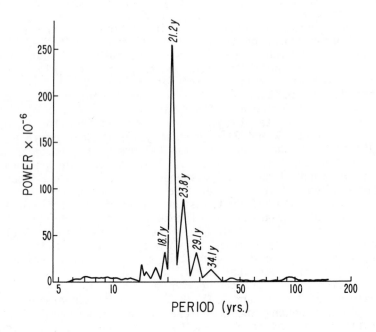

Figure 19.   Fourier transform of sunspot numbers, counted year by year for 250 years, transformed into power vs. period, assuming sunspots to have opposite magnetic polarity in northern and southern hemispheres.

atmosphere and stores and returns to the atmosphere as the stored material decays. Its mass depends on the CO_2 content of the atmosphere, on air temperature, on sea surface temperature, namely on climate, and on concentration of carbonic-anhydrase in the surface sea water. The dependence of rate of absorption of CO_2 into surface sea water had been indicated by Berger and Libby [53].

In turn, the concentration of CO_2 in the atmosphere depends on the mass of the biosphere and its rate of decay after death, and on the carbonic-anhydrase concentrations in the sea surface. In future predictions of the rate of increase of CO_2 partial pressure in the atmosphere due to burning fossil fuels, it will be important to include the interaction of the atmospheric CO_2 with the bio-organic reservoir and the catalyzation of its absorption into the sea by means of the action of carbonic-anhydrase dissolved in sea water, considerations which have not been taken into account in past computations.

In Table 3 are listed data which describe the interaction of the system: atmospheric CO_2 with CO_2 dissolved in sea water and with CO_2 equivalent stored in the bio-organic reservoir. From these data we have shown [54] that changes in climate by ± 10 °C can cause changes in the carbon-14 content of bio-organic matter, and therefore in apparent carbon-14 age of as much as ± 100 years. Such changes are well able to account for Suess' "wiggles" of about a hundred years duration each [43] in the carbon-14 concentrations in the 5000 year sequence of bristlecone pine rings constructed by W. Fergusson.

Similarly the $^{13}CO_2$ concentration of the atmosphere is affected by variations of the biosphere mass just as is the carbon-14 content, although the amplitude of the perturbation is only half that for carbon-14 [55].

If the stable isotope ratio of $^{13}C/^{12}C$ is to be further measured in tree rings and interpreted as an indicator of climate variation, (and we have barely begun to initiate its use as a thermometer in the present work, confining our measurements to the stable isotopes in water, because water is so abundant compared to carbon dioxide and because the dependence of its isotope ratios is relatively simple compared with those of carbon dioxide), some more sophisticated considerations must be given to the distribution of carbon dioxide among the reservoirs on the surface of the earth.

In the process of making bio-organic matter, plants store carbon depleted with respect to carbon dioxide by about 2.7 percent in carbon-13 [56-58]. If the total amount of depleted carbon stored in the biosphere in the past has been different from what it is now, interpretation of $^{13}C/^{12}C$ variations in tree rings as caused solely by temperature changes will reflect an error.

Table 3.   Quantities of Importance for the Kinetics of
           Atmospheric Carbon Dioxide.

Total Atmospheric CO_2 = 0.13 g C/cm^2

Mixed Layer of ocean (100 meter thick surface waters) contains
    0.15 g C/cm^2

0.013 g C/cm^2 exchanged from atmosphere to ocean each year

0.001 g C/cm^2 returned to atmosphere per year by plant decay

storage time in biosphere is about 140 years

storage time in oceans is about 1000 years

Total depth of oceans of 2800 meters contains 0.14 g organic
    carbon/cm^3

    The error results in the following way. The amount of
organic carbon dissolved in the oceans is 0.14 gm/cm^2, about equal
to the amount of carbon in atmospheric carbon dioxide of 0.13
g C/cm^2. Since the organic carbon is 2.7 percent depleted in
carbon-13, the atmospheric reservoir is correspondingly enriched
by 2.7 percent.
    In past times the total mass of stored organic carbon may
have been larger or smaller than it is now, depending on the past
climate. Let us define as the norm the amount of carbon presently
stored, M^*, and define a time dependent factor, M/M^*, by which the
organic carbon reservoir may be increased ($M/M^* > 1$) as in a lush,
tropical coal age, or decreased ($M/M^* < 1$) as in an ice age, where
M is the mass of organic carbon at the time in the past when the
material was alive. We assume that the total amount of atmospheric carbon dioxide has always remained the same (which may or
may not be true).
    Then the atmospheric reservoir is enriched by about 2.7
percent multiplied by the ratio of the mass of stored organic
carbon to its mass today. That is the enrichment E is given by,

$$E = 0.027 \ (M/M^*)$$

so that at the present time $M/M^* = 1$ and $E(1979) = E^* = 0.027$.
    Thus the enrichment correction depends on the numerical
value of 0.027 depletion of carbon-13 in plants, which in turn
depends on the temperature at which the bio-organic material
grew. The temperature coefficient for $^{13}C/^{12}C$ in marine plankton
has been measured as 0.35 ppt/°C [59] and independently as
0.5 ppt/°C [60]. In the absence of more measurements we may
assume it to be the larger of the two experimental values, namely

0.5 ppt/°C, and this number enters in the calculation of the enrichment, E(t), depending on the average temperature at time t in the past.

Another effect enters also, caused by the presence of ocean carbonate. The sea contains dissolved carbonate that is fractionated in $^{13}C/^{12}C$ in comparison with that in the atmosphere; it is enriched in ^{13}C and is less so as the temperature increases, causing a small correction to the enrichment E.

This perturbation comes only from the carbonate in the approximately 100 meters thick surface mixed layer; it contains about 0.15 g carbon/cm^2, approximately equal to that in the atmosphere and to the organic carbon dissolved in the total depth of the sea. In the case of organic carbon, the total depth of the sea is involved because bacterial decomposition occurs at all depths, producing methane and carbon monoxide with both of which the sea is saturated, so that these gases are bubbling up from all depths. In comparison the sea is not saturated with CO_2 at any depth.

Carbon isotope fractionation by CO_2 absorption at the air water interface has been measured [61], and the ratio C^{13}/C^{12} has been found to vary from 9.2 to 6.7 ppt over the temperature range 0° - 30 °C. Thus we may write the enrichment at time t as,

$$E(t) = 0.027 \ (M/M^*) \ (1 - 0.5 \times 10^{-3}(T - T^*))$$

$$(1 + 8.5 \times 10^{-3} \ (T - T^*))$$

where the number 8.5×10^{-3} is taken for the average fractionation of carbon-13 at the sea-air interface and where T is the average temperature at time t and T^* is the average temperature today; we assume $T^* = 20$ °C.

The values of E(t) so computed are listed in Table 4. The correction for fractionation of carbon dioxide at the sea surface is a serious one. It makes the interpretation of $^{13}C/^{12}C$ variations in wood difficult and militates against the use of the isotope ratio of carbon as a thermometer. This correction, when applied to variations of carbon-14 in wood, is able to explain the Suess radiocarbon "wiggles" of about 100 years duration each, without the need to invoke changes in the neutron flux from the sun [54].

Table 4.   Values of enrichment of $^{13}C/^{12}C$ computed for atmospheric carbon dioxide versus plant mass M and versus temperature.

M/M*	10 °C	20 °C	30 °C
0.5	0.015	0.014	0.010
1.0	0.030	0.027	0.020
2.0	0.060	0.054	0.040

Introduction to The Study of Commodities

        In order to interpret, verify, and calibrate the temporal
changes in the stable isotope ratios of hydrogen and oxygen in
tree rings, we look to historic records of crops that mankind has
reaped from the environment without cultivating it intensively,
depending instead on the vagaries of climate to produce a good
or bad crop.
        The numbers of kinds of major crops have been few, and the
records of crop yields have become more reliable only in the last
few centuries. We use what we have been able to find in the
libraries of the Department of Agriculture at the Hague, Holland,
in the British Department of Agriculture, London, and in the much
less lengthy records in the United States.
        We analyze here two kinds of records indicative of climate
change, the yearly record of wheat prices in four European
countries since 1200 A.D., and the yield of blue crab in the
Chesapeake Bay since 1920, as examples of mathematical treatments
that could be used on other temporal records of commodities
derivable from archives.
        In Europe, the archivists have pioneered the evaluation of
past climate changes from records of wine harvests and from draw-
ings showing the advance and retreat of glaciers. This great
body of work is excellently reviewed by Ladurie [62].
        In America, researchers at the Laboratory of Tree Ring
Research at the University of Arizona, Tucson, Arizona, have
developed the measurement and analysis of tree ring widths using a
score or so of trees in each stand to provide pluviometric maps
versus time, and find that these maps have to be established
separately for each region and for each set of trees, so that we
can not look to ring width records versus time for information
about the cyclic evolution of climate.
        Unlike records of stable isotope variations, records of ring
width variations depend on many factors besides temperature alone,
for example also on amount of rainfall and for example on amount
of volcanic dust fall-out, forest fire, animal use of the forest
floor, and it is impossible to deduce periodicities from them
[63].

Other Climate Indicators: Commodities, Prices, Wages

        The blue crab (Callinectes sapidus) catches in the Chesapeake
Bay have been measured [64] for the years 1922-1976 A.D., see
figure 20. We have made a Fourier transform of the yield in
millions of pounds per year (1 million pounds = 454 metric tonnes)
versus years into amplitude versus period in years, see figure
21 and find principle cyclic periods of 8.6, 10.7, and 18.0
years.
        The Fourier transform of the record of annual air temperature
versus time shows principle periods at 7.4, 9.8, and 17.5 years;

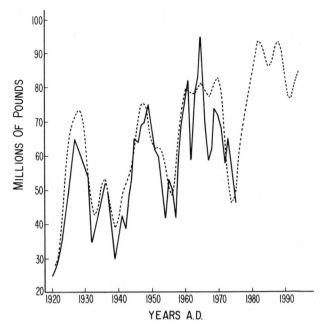

Figure 20.   Annual catch of blue crab measured in millions of pounds per year in the Chesapeake Bay, A.D. *1920–1976. Key: ——, measured; – – –, predicted.*

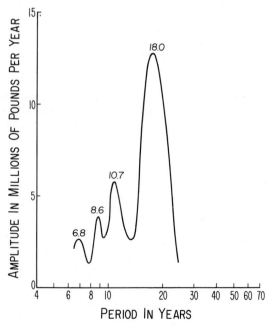

Figure 21.   Fourier transform of annual catch of blue crab vs. time into amplitude vs. period in the Chesapeake Bay.

the Fourier transform of the record of annual rainfall at Philadelphia shows period of low rainfall of 8.3, 9.5, and 18.5 years [65,66].

Tidal forces have periods of 8.8 and 18.6 years [67]. The variations in the tidal forces arise because the orbital planes of the Moon and the Earth are slightly inclined with respect to each other and because the Sun, Moon, and Earth form a mutually rotating gravitational system so that the magnitude of the tidal force depends on their relative positions.

For comparison, the periods of enhanced blue crab yield, of maxima of air temperatures at Philadelphia (which is close to the Chesapeake Bay), of minima of rainfall at Philadelphia, and of enhanced tidal forces (leading to high tides) are listed in Table 5. The explanation we offer for the agreement among the periods so listed is that high tides wash nutrient into the surface waters of the Bay, and higher temperatures warm the surface waters, and minimum rainfall allows the surface waters to become more saline, all of which factors are salubrious for crab growth.

Variation of temperature and rainfall with the sunspot periodicity of 10.5 years is explained by the findings of the CIAP study [49], in that climate on the ground is affected by variations in the chemical species in the stratosphere whose concentration varies as the sun's spectrum and intensity vary.

Why temperatures and rainfall near Chesapeake Bay should be affected by variations of the tidal forces is not so clear. However the atmosphere and stratosphere are pulled away from the earth by tidal forces just as are the waters of the earth. These forces vary by as much as 10 percent during the tidal periods [67] resulting in density variations in the stratosphere with the same periods; the consequent density variations may affect the relative rates of stratospheric chemical reactions, causing disturbances of temperature and rainfall on the ground with the tidal periodicities.

The periods of fluctuation of blue crab and their absolute phases and amplitudes listed in Table 5 have been used to reconstruct the annual yield of blue crab, Y, and project it into the future (shown by the dashed line in figure 20). The prediction formula is:

$$Y = \sum_n A_n \sin (2 \pi t/\tau_n + \phi_n)$$

where ϕ_n is the phase of the nth period τ_n and A_n is its amplitude.

The prediction in figure 20 shows the recovery of the crab crop already observed in 1977 [64], to increase until 1982 and to remain at high yield until 1988, followed by a decrease. We caution that the absolute phases and amplitudes are to be regarded as inaccurate, because their values depend sensitively on the accuracy of the input data. This may be the reason for the disagreements of the relative phases shown in Table 5 [68].

Table 5.

Annual Average of Philadelphia Temperature			Annual Average of Philadelphia Rainfall (minima)			Annual Average of Blue Crab Catch in Chesapeake Bay			Earth-Sun-Moon Tidal Forces
t	φ	A	t	φ	A	t	φ	A	t
17.5	0	1.0	18.5	0	1.0	18.0	0	1.0	18.6
9.8	-1.9	1.2	9.5	-4.1	1.1	10.7	-0.9	0.5	--
7.4	+1.4	0.9	8.3	-2.5	0.9	8.6	+2.1	0.3	8.8

t in years, A relative amplitude, φ phase in radians.

As with all of our Fourier transforms of real data, we test the statistical significance of the periods so revealed by generating appropriate sets of Markovian data, each datum consisting of a constant, a, plus a random number ε_j, where the random number ε_j is small compared with a, and varies in such a way that the artificial data set so generated ranges between the maximum and minimum of the values of the real data. Each set so generated we subject to the Fourier transform, and we look for peaks of amplitude corresponding to those found for the real data. If, in 30 such experiments, we find no comparable peaks, we conclude that the periods found in the true data set have a significance of better than 1/30, namely better than 96 percent confidence level. This has been the case of all our Fourier transforms of sets of real data until now.

Records of delicacies such as crab extend for less than 100 years. The commodity for which quantitative records of price (contrasted to yield) extend over the longest span, to our knowledge, is European wheat, for which we have found historic prices in England, Netherlands, France, and northern Italy since 1250 A.D. in units of guilders per 100 kilograms, figure 22 [69]. In all four countries there were many-fold increases in prices at around 1600 and 1800 A.D., caused by something more general than local political and economic maneuvering.

Economists agree that there exists a definite relation between the market price of a commodity such as wheat and the quantity available [70]. Thus the price of wheat, in historic times of plenty and in times of short supply, should be an indicator of abundance in good climate and of shortage in bad climate, and therefore an indicator of climatic changes.

Here we discuss prices of wheat and wages (ability to pay for wheat) from 1250 A.D. to the present, and compare their variations with variations of average air temperatures from 1650 A.D. to the

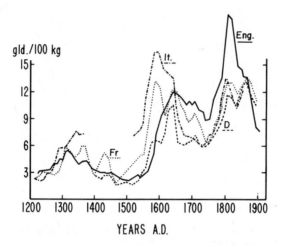

Figure 22.   Price of wheat in Italy, France, Netherlands, and England, expressed in guilders per 100 kg vs. time, A.D. 1200–1900.

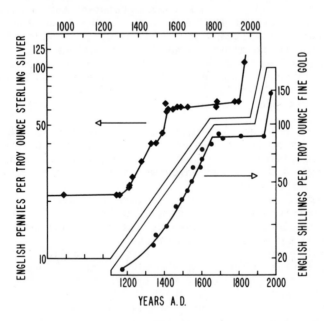

Figure 23.   Cost of sterling silver and of pure gold vs. time, expressed in English pennies per troy ounce, A.D. 1000–1950.

present as a calibration of the modern part of the wage and price record. The population has been increasing during this time span; for this it is necessary to make corrections and also necessary to correct for inflation of the money. In figure 23 are shown the prices per troy ounce for sterling silver (90% pure) and for fine (pure) gold, in English money, since 1250 A.D. Sterling prices are more basic than gold prices because sterling was the only official coin of the realm until 1774.

One way of normalizing grain prices to take into account inflation is to compute the ratio R_s of wheat prices to price of sterling, and this we have done. A second way of normalizing is to compute the amount of grain which a laborer's daily wage can buy, namely R_w, the ratio of labor's daily wage to wheat price. R_s and R_w are shown plotted in figure 24 versus centuries. R_s has been computed from data in figures 22 and 23. R_w has been taken from Meredith [71] and Steffen [72]. The third part of figure 24 shows a 25 year running average of winter air temperatures in England computed from Manley [25].

The warming trend between 1800 and 1930 is a general phenomenon displayed in temperature records of Holland, Edinburgh, Stockholm, Vienna, Berlin, Copenhagen, Greenland, Ukraine, Siberia, Basel, and Geneva; in the United States the trend is evident in New Haven, Philadelphia, St. Paul, St. Louis, and Washington, D.C. The amplitude of the change varies locally from 1 to 2 °C. In particular the warming of the entire northern temperate zone is estimated at 0.64 - 0.70 °C [65] for the entire period.

The major declines of R_w and R_s with a minimum at 1630 A.D. is known as the First Little Ice Age, a time famous for famines killing millions of people, cold summers, bitter winters, failed harvests, wars and civil unrest. The declines of R_w and R_s reaching a minimum at around 1800 A.D. correspond to the Second Little Ice Age, which caused civilian hardships which have been extensively documented, and remain in the political memories of the Irish and Scots today.

Our purpose here was to obtain the temperature scale shown in figure 24, applicable to times as far back as 1250 A.D. when there were no thermometer records, in order to compare the temperature record so obtained with the record of Tree Thermometers. To this end, we measured ΔR_w, ΔR_s, and ΔT (from figure 24), from peaks to valleys, since 1700 A.D. and from these, computed the temperature coefficients $\Delta T/\Delta R_w$ and $\Delta T/\Delta R_s$; using these coefficients, we computed the temperatures for 1250 - 1600 A.D. The resultant temperature scale, shown on the right hand ordinate of figure 24 [73], is in agreement with that obtained from variations of the stable isotope ratios in the European oaks.

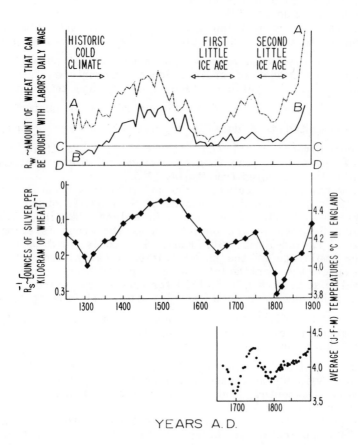

Figure 24.   Variations in amount of wheat purchasable with daily wage of (A) a carpenter and (B) an agricultural laborer, compared with (CD) amount necessary to nourish a family, and compared with cost of wheat and with temperatures, A.D. 1650–1900.

The Slope of Eight

W. F. Libby, L. M. Libby, and L. J. Pandolfi

The linear relationship of the isotope ratios of hydrogen and oxygen with a slope of 8 in rain and snow was discovered by Craig [74] to be

$$\delta D = 8 \; \delta_{18} + C$$

where C is a very small constant,

$$\delta D \equiv ((D/H)/(D/H)_{smow} - 1) \times 10^3$$

$$\delta_{18} \equiv ((^{18}O/^{16}O)/(^{18}O/^{16}O)_{smow} - 1) \times 10^3$$

and where the ratios labeled "smow" are those for standard mean ocean water.

This relation has been abundantly verified for rain and snow collected world-wide, and measured by the IAEA since 1969 [17]. The IAEA data are shown in figure 1 as a plot of δD versus δ_{18} with the characteristic slope of 8.

This slope can readily be computed from laboratory measurements (e.g., Stewart [75], and references quoted therein); for the temperature range $0° - 35 °C$ these measurements can be represented as,

$$(D/H)_{liq}/(D/H)_v = -9 \times 10^{-4} \, T + 1.1035$$

$$(^{18}O/^{16}O)_{liq}/(^{18}O/^{16}O)_v = -9 \times 10^{-5} \, T + 1.01135$$

where the subscripts "liq" and "v" mean liquid and vapor and where T is the temperature in degrees C. We take the ratio for the liquid to be equal to the ratios for ocean water (smow).

So, for rain (made from vapor distilled from the ocean),

$$\delta D = ((D/H)_v/(D/H)_{liq} - 1) \times 10^3 = 10^3 \, [1/(-9 \times 10^{-4} \, T + 1.1035) - 1]$$

$$\delta_{18} = ((^{18}O/^{16}O)_v/(^{18}O/^{16}O)_{liq} - 1) \times 10^3 = 10^3 \, [1/(-9 \times 10^{-5} \, T + 1.01135) - 1].$$

Thus the derivatives of δD and δ_{18} with respect to temperature are,

$$\Delta(\delta D)/(\Delta T) = 0.9 \, / \, (-9 \times 10^{-4} \, T + 1.1035)^2$$

$$\Delta(\delta_{18})/(\Delta T) = 0.09 \, / \, (-9 \times 10^{-5} \, T + 1.01135)^2.$$

The ratio of these two derivatives gives the slope of 8 of the rainwater curve in figure 1. In fact the IAEA data yield

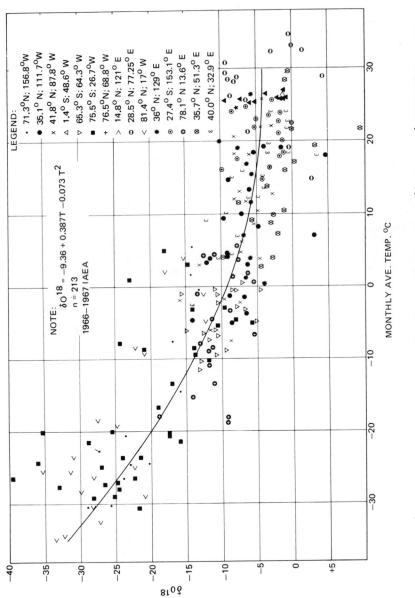

Figure 25. Oxygen isotope ratio in world-wide precipitation vs. monthly average air temperature, showing that for every point on the line in Figure 1, there is a corresponding average air temperature.

a far more accurate determination of the slope than can be made in the laboratory.

Using the above laboratory relations we compute the values of δD and δ_{18} expected for rain made by a single distillation from the ocean and for rain made by a subsequent distillation for sea rain which fell on the land. These values show that when rain and snow exhibit very large depletions of the heavy isotopes, e.g., δD of -200 and -300 ppt, two or three distillations have occurred. It is known from measurements of tritium in rain [76] that two or three distillations occur in the U.S. between evaporation from the Pacific Ocean and precipitation in the eastern U.S. The agreement between the number of distillations deduced from stable isotopes and deduced from tritium is gratifying.

For every point on the curve in figure 1 characterizing rain and snow, there is a corresponding average air temperature, as may be seen in figure 25. Consequently we looked for a similar dependence and the slope of 8 to occur in trees which ingest rain. We found the slope of 8 in a Japanese cypress [25] measured for the years 150 - 1970 A.D. But in the sequence of German oaks which we measured [28-31] we reported that we did not find a slope of 8, and we listed our measurements.

We now state that a slope of 8 obtains in all of the trees which we have measured including the oak. The explanation is as follows. For the oaks, we measured all isotope ratios using wood for the years 1712 - 1714 A.D. as our standard (as described in [31]). In those years Europe was very cold, at the bottom of the First Little Ice Age according to our tree measurements (see figures 6 and 7, and according to European thermometer records. The wood we used as our standard was consequently greatly depleted in heavy isotopes, in deuterium by -180 ppt, because the rain ingested was correspondingly depleted.

When our measured, listed, and published, isotope measurements, referred to wood of 1712 - 1714, are referred instead to ocean water (smow), then δD_{smow} and $\delta_{18(smow)}$ are related to each other by the slope of 8 for the oak sequence just as they are in the cedar sequence, see figure 26.

The relation between the tree δD and smow δD is, in ppt,

$$\delta D_{smow} = 0.82\ \delta D_{tree} - 180$$

which follows from the relations, for deuterium,

$$\delta D_{smow} = 10^3 \times (R_{tree}/R^*_{smow} - 1)$$

$$\delta D_{tree} = 10^3 \times (R_{tree}/R^*_{tree} - 1)$$

$$(10^{-3}\ \delta D_{smow} + 1)\ R^*_{smow} = (10^{-3}\ \delta D_{tree} + 1)\ R^*_{tree}$$

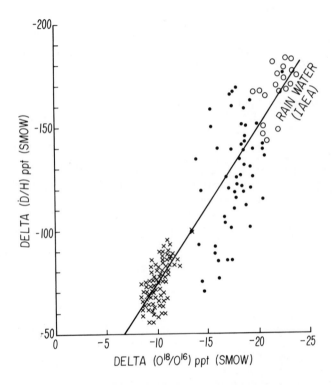

*Figure 26.   Deuterium isotope ratio vs. oxygen isotope ratio measured in our
sequence of German oaks and Japanese C. japonica.*

*All ratios are related to each other by the slope of eight, namely, according to $D/H =
8\ {}^{18}O/{}^{16}O$ + constant. We subtracted the constant in order to make all points lie on the
line for world-wide precipitation (IAEA). Key: ×, Japanese cedar (A.D. 150–1970); ●,
German oak I (A.D. 1712–1950); and ○, German oak II (A.D. 1530–1725).*

so that for 1712-1714, when δD_{tree} = 0 and δD_{smow} = -180,

$$R^*_{tree}/R^*_{smow} = 1 - 0.180 = 0.82.$$

Finally, we now predict that for all trees which subsist on rainwater, the deuterium and oxygen isotope ratios will be related linearly on their rings with a slope of 8 [77], namely,

$$D/H = 8 \; (^{18}O/^{16}O) + constant.$$

The constant is different in the oaks and in the cedar from its nearly zero value in sea water; in plotting figure 26, we have subtracted the appropriate constant from the tree measurements in order to make all the measured points lie on the sea water curve; this subtraction has no effect on the slope of 8, of course.

We hope that analysis of stable isotopes in tree ring sequences will become an international undertaking, and that with the collaboration of many laboratories in many countries, the history of the climate in recent times, perhaps as far back as thirty thousand years, will be deduced, and that, from its determination, the likely trends of climate in the near future will be predicted.

It is a pleasure to thank Rudolph Black of the United States Advanced Research Projects Agency, who, in May 1971, funded our proposal that "temperature variations in past climates may be evaluated by measuring stable isotope ratios in natural data banks such as tree ring and varve sequences". We thank William Best of the U.S. Air Force Office of Scientific Research who monitored our study and Frank Eden of the U.S. National Science Foundation who subsequently provided further funds.

We thank our several collaborators, Patrick Payton and John Marshall 3rd who successfully carried out many of our difficult sample preparations, Bernd Becker of the Landwirtschaftliche Hochschule, Universität Hohenheim, Stuttgart, and V. Giertz-Siebenlist of the Tree Ring Laboratory, University of Munich, who provided authentically counted and dated chronological sequences of tree rings, as did K. Kigoshi of Gakushuin University, Tokoyo, and Henry Michael, University of Pennsylvania.

We used the measurements of organic carbon and uranium in a sea core, made by Paul R. Doose, Emil K. Kalil, and I. R. Kaplan of the University of California at Los Angeles, and we compared our stable isotope measurements with Hans Suess' measurements of radiocarbon in the bristlecone pine ring sequence prepared by Wesley Fergusson, (see fig. 27), Suess at the University of California at San Diego, and Fergusson at the University of Arizona, Tucson.

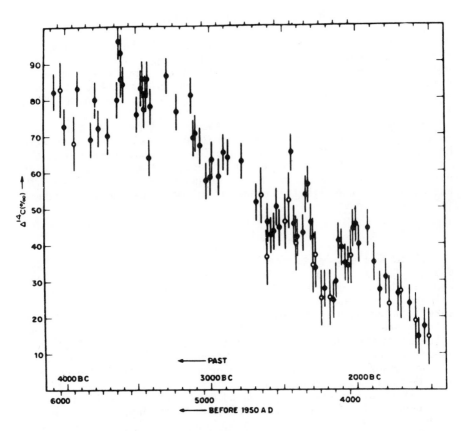

Figure 27.   Radiocarbon age vs. chronological age of tree rings vs. years 4000–
1500 B.C.

We thank Lawrence H. Levine for help with the UCLA PDP-11 computer and for preparing computer programs.

Most of all we thank Willard Frank Libby who at every setback advised and encouraged us and who took time and care to study our results.

References

[1] International Atomic Energy Agency, Environmental Isotope Data No. 1, 2, 3, 4, 5 (1953-1971), Vienna (1969).
[2] Craig, H., Science, 1961, 133, 1702-1703.
[3] Craig, H., Consiglio Nazionale delle Recherche, Spoleto, Italy, September 9-13, 1963, 17-53.
[4] Dansgaard, W., Tellus, 1964, 16, 436-467.
[5] Stewart, M. K., Friedman, I., J. Geophys. Res., 1975, 80, 3812-3818.
[6] Keeling, C. D., Tellus, 1960, 12, 200-203.
[7] Ergin, M., Harkness, D., Walton, A., Radiocarbon, 1970, 12, 495.
[8] Olsson, I., Stenberg, A., Radioactive Dating and Methods of Low Level Counting, IAEA Symposium, Monaco, March 2-10, 1967, 69-78.
[9] Olsson, I., Klasson, M., Radiocarbon, 1970, 12, 281-284.
[10] Vogel, J. C., Lerman, J. C., Radiocarbon, 1969, 11, 385.
[11] Gould, R. F., ed., Lignin Structure and Reactions, Advances in Chemistry Series, American Chemical Society Publications, 1966.
[12] Taube, H., Ann. Rev. Nucl. Sci., 1956, 6, 277.
[13] Sepall, O., Mason, S. G., Can. J. Chem., 1961, 39, 1934-1943.
[14] Grobecker, A. J., Climatic Impact Assessment Program (CIAP), Monograph 4, Department of Transportation, 1975, 4-51.
[15] DARPA, Defense Advanced Research Projects Agency, Contract Numbers F44620-73-0025, F44620-72-0029, monitored by Air Force Office of Scientific Research, 1971.
[16] Libby, L. M., Multiple Thermometry in Paleoclimate and Historic Climate, J. Geo. Res., 1972, 77, 4310-4317.
[17] IAEA, Environmental Isotope Data, No. 1-5, International Atomic Energy Agency, Vienna, Austria, 1969-1975.
[18] Gould, R. F., ed., Lignin Structure and Reactions, Advances in Chemistry Series, No. 59, American Chemical Society, Washington, D.C., 1966; and Rebello, A., "Composition of Wood", in Environmental Biogeochemistry, Nriagu, J., 2 vol., 1975, Ann Arbor Science Publishers, Ann Arbor, Michigan.
[19] Herzberg, G., Infrared and Raman Spectra of Polyatomic Molecules, Van Nostrand and Reinhold Co., New York, NY, 1945.
[20] Weast, R. C., ed., Handbook of Chem and Physics., Chem Rubber Co., Cleveland, OH, 45th Edition, F90 and F91.

[21] Huber, B., Ring Width Sequence in German Oaks, Ber. at. Bot. Gesell., 1935, 53, 711-719.

[22] Huber, Von K. M. B., Siebenlist, V. G., Unsere tausendjahrige Eichen-Jahrring-chronologie durch schnittliche 57(10-150)-fach belegt, Sitzberichte Abteilung I, Östereichische Akademie der Wissenschaft 178, Band I bis 4, Heft 37-42, Wien, 1969.

[23] Manley, G., The Mean Temperature of Central England 1698-1952, Quart. J. Royal Soc., 1953, 79, 242-261.

[24] Manley, G., Temperature Trends in England 1698-1957, Archiv. Meteorol. Geophis. Bioklmatol, Series B.A.K., Band 9, Heft 3/4, Vienna, 1959.

[25] Manley, G., Central Air Temperatures; Monthly Means, 1659-1973, J. Royal Meteorol. Soc., 1974, 100, 389-405.

[26] Libby, L. M., Pandolfi, L. J., Payton, P. N., Marshall, J., 3rd, Becker, B., Siebenlist, V. G., Isotopic Tree Thermometers, Nature, 1976, 261, 284-288, and see references therein.

[27] Epstein, S., Yapp, C. J., Climatic Implications of the D/H Ratio of Hydrogen in C-H Groups in Tree Cellulose, 1976, 30, 252-261.

[28] Libby, L. M., Pandolfi, L. J., Calibration of Two Isotope Thermometers in a European Oak Using Official Weather Records, Proceedings of International CLIMAP Conference, Norwich, England, 1973, 21-39.

[29] Libby, L. M., Pandolfi, L. J., Calibration of Isotope Thermometers in and Oak Tree Using Official Weather Records, Colloques Internationaux du Centre National de la Recherche Scientifique, Gif-sur-Yvette, France, June, 1973.

[30] Libby, L. M., Final Technical Report on Historical Climatology, DARPA order no. 1964-1, U. S. Air Force Office of Scientific Research, R&D Associates, Santa Monica, California, 1974.

[31] Libby, L. M., Pandolfi, L. J., Temperature Dependence of Isotope Ratios in Tree Rings, Proc. Nat. Acad. Sci., 1974, 71, 2482-2486.

[32] Libby, L. M., Pandolfi, L. J., Isotopic Tree Thermometers; Correlation with Radiocarbon, J Geo. Res., 1976, 81, 6377-6381.

[33] Rittenberg, D., Pontecorvo, L., Evolution of O_2 from Organic Compounds, Int. J. Appl. Radiat. Isotopes, 1956, 1, 208-214.

[34] Cohn, M., Urey, H. C., Exchange of Oxygen Isotopes Between CO_2 and H_2O, J. Am. Chem. Soc., 1938, 60, 679-687.

[35] World Weather Records, Europe, U. S. Dept. of Commerce, Environmental Services Administration, Washington, D.C., Vol. 2, 1966.

[36] Becker, B., Siebenlist, V. G., Tree Ring Sequence in Bavarian Fir, Flora, 1970, 159, 310-346.

[37] Daugherty, G. V., Acting Superintendent, United States Department of Interior, National Park Service, Three Rivers, California, collected the Sequoia gigantea and sent it to us, 1976.

[38] Michael, H., University Museum, University of Pennsylvania, counted the rings and dated them by comparison with the fiducial ring sequential pattern developed by the University of Arizona, 1976.

[39] Bergthorson, P., Ice Record in Rejkavik Harbor, Proc. Conf. Climate 11th to 16th Centuries, Aspen, Colorado, National Center for Atmospheric Research, Air Force Cambridge Research Laboratories, 1962.

[40] Blackman, R. B., Tukey, S. W., The Measurement of Power Spectra, Dover, NY, 1958; and see also Weast, Handbook of Chemistry and Physics, 54th Edition, Chemical Rubber Company, Cleveland, OH, 1973.

[41] Dansgaard, W., Johnson, S. J., Clausen, H. B., Langway, H. J., Ice Record of Oxygen Isotopes at Camp Century, Nature, 1971, 227, 482-483.

[42] Libby, L. M., Pandolfi, L. J., Climate Periods in Tree, Ice and Tides, Nature, 1977, 266, 415-417.

[43] Suess, H., Natural Radiocarbon, Endeavor, 1973, 32(115), 34-38.

[44] Kalil, E. K., Kaplan, I. R., Uranium and Organic Carbon Concentrations in a Deep Sea Core, PhD Dissertation, Dept. of Geochemistry, University of California at Los Angeles, 1976.

[45] Kalil, E. K., Kaplan, I. R., Uranium and Organic Carbon Concentrations in a Deep Sea Core, Marine Chemistry, in press, 1978.

[46] Doose, P. R., PhD Dissertation, Geochemistry Department, University of California at Los Angeles, 1978.

[47] Soutar, A., Isaacs, J. D., State of California Marine Research Committee, California Cooperative Oceanic Fisheries Investigation, 1969, 13, 63-70.

[48] Pandolfi, L. J., Kalil, E. K., Doose, P. R., Levine, L. H., Libby, L. M., Climate Periods in Trees and a Sea Sediment Core, 10th International Radiocarbon Conference, Bern and Heidelberg, August, 1979.

[49] Grobecker, A. J., ed. in chief, Monograph No. 1, Climatic Impact Assessment Program, Dept. of Transportation, Washington, D.C., 1975.

[50] Hurt, P. H., Stable Oxygen Isotopes in Tree Rings as an Indicator of Solar Variability, Masters Thesis, Chemistry Department, University of California at Los Angeles, 1978.

[51] Cohen, T. J., Lintz, P. R., Periodicities in Sunspot Numbers, Nature, 1974, 250, 398-399.

[52] Hill, J. R., Periodic Analysis of Sunspot Records, Nature, 1977, 266, 151-153.

[53] Berger, R., Libby, W. F., Exchange of CO_2 Between Atmosphere and Sea Water, Global Effects of Environmental Pollution, S. F. Singer, ed., D. Reidel Publishing Co., 1970.

[54] Libby, L. M., Globally Stored Organic Carbon and Radiocarbon Dates, J. Geophys. Res., 1973, 78, 7667-7670.

[55] Rafter, T. A., C^{14} Variations in Nature and the Effect on Radiocarbon Dating, N. Z. J. Sci. Technol., 1955, 37, 20-38.

[56] Kroopnick, P., Deuser, W. G., Craig, H., Carbon 13 Measurements on Dissolved Inorganic Carbon at the North Pacific (1969) Geosecs Station, J. Geophys. Res., 1970, 75, 7668-7671.

[57] Gordon, L. I., Williams, P. M., C^{13}/C^{12} Ratios in Dissolved and Particulate Matter in the Sea, Deep Sea Res., 1970, 17, 19-27.

[58] Maugh, T. H., Carbon Monoxide: Natural Sources Dwarf Man's Output, Science, 1972, 177, 338-339.

[59] Be, A. W., Bender, M. L., Eckelmann, W. R., Sackett, W. M., Temperature Dependence of Carbon Isotope Composition in Marine Plankton and Sediments, Science, 1965, 148, 235-237.

[60] Eadie, B. S., Carbon Fractionation in the Antarctic Marine Ecosystem (Abstract), Eos Trans. AGU, 1972, 53, 406.

[61] Degens, H. T., Deuser, W. G., Carbon Isotope Fractionation in the System CO_2 (gas)-CO_2 (aqueous)-HCO_3 (aqueous), Nature, 1967, 215, 1033-1035.

[62] Ladurie, E. L., Times of Feast; Times of Famine, Doubleday & Co., Garden City, New York, NY, 1971.

[63] Fritts, H. C., Growth Rings of Trees and Climate, Science, 1966, 154, 973-979.

[64] Franklin, B. A., New York Times, June 27, 1977, page 14.

[65] Landsberg, H., Time Series of Temperatures and Rainfall from Records in the Eastern U.S., Reduced to Philadelphia, University of Maryland, College Park, MD, 1975.

[66] Landsberg, W. H., Kaylor, R. E., Cyclic Analysis of Temperature and Rainfall Records for the Eastern U.S., J. Interdiscip. Cycle Res., 1976, 7, 237-243.

[67] Rinehart, J. S., Tidal Periods in Yellowstone Geysers, Science, 1972, 177, 346-347.

[68] Hurt, P. H., Libby, L. M., Pandolfi, L. J., Levine, L. H., Van Engel, W. A., Periodicities in Blue Crab Population of Chesapeake Bay, Climatic Change, 2, 75-78 (1979).

[69] Veenman, H., Zonen, De Landbouw in Brabants Westhoek in Het Midden van de Achttiende Eeuw, Agronomisch Historische Bijdragen, Wageningen, Netherlands, 1938.

[70] Samuelson, P. A., Economics, 8th Edition, McGraw Hill, New York, NY, 55-74, 1970.

[71] Meredith, H. O., Outline of the Economic History of England, Sir Isaac Pitman and Sons, Ltd., London, Chart B opposite p. 353, 1939.

[72] Steffen, G. F., Studien zur Geschichte der englischen Lohnarbeiter mit besonderer Berücksichtigung der Veränderungen Ihrer Lebenshältung, Hobbing & Buchle, Stuttgart, 3 Vols., 1901.

[73] Libby, L. M., Correlation of Historic Climate with Historic Prices and Wages, Indian J. of Meterol., 1977, 28(2).

[74] Craig, H., Science, 1961, 133, 1702.

[75] Stewart, M. K., Friedman, I., J. Geophys. Res., 1975, 80, 3812, and references quoted therein.

[76] Leventhal, J. H., Libby, W. F., J. Geophys. Res., 1968, 73, 2715.

[77] Libby, W. F., Libby, L. M., Pandolfi, L. J., The Slope of Eight, to be published, 1980.

RECEIVED August 10, 1981.

Glaciochemical Dating Techniques

MICHAEL M. HERRON

State University of New York–Buffalo, Department of Geological Sciences,
Amherst, NY 14226

Ice cores from Greenland and Antarctica contain valuable paleoenvironmental information from the past 100,000 years or more. Chemical analyses of the impurity content of polar snow and ice strata have proven useful in establishing and refining ice core chronologies through the identification of seasonal variations in concentrations and concentration event horizons. Measured concentration levels in polar precipitation range from about <1 ng/g to 50 ng/g necessitating the use of special contamination control procedures and analytical techniques such as INAA, IDMS, AAS, and ion chromatography [10][1]. The identification of an annual spring maximum in concentrations of crustally-derived Al, Ca, and Fe, of an annual winter maximum in concentrations of marine-derived Na, Mg, and Cl, and of a summer maximum for other species each allow ice core dating by counting annual layers. These, and other ice core chronologies can be improved by comparing ages with the known occurrence of event horizons. Sulfuric acid derived from the 1783 Laki, Iceland eruption resulted in an acid snow layer with a SO_4^{2-} concentration of 2200 ng/g which is more than an order of magnitude higher than in modern Greenland snow. The 1883 Krakatoa eruption resulted in Greenland concentrations of Cd, Zn, Pb, and Cu as much as a factor of 20 greater than average. Increased atmospheric turbidity during the last ice age resulted in prehistoric event horizons containing an order of magnitude more continental dust than in more recent ice.

---

[1]Figures in brackets refer to the literature references at the end of this paper.

The chemical composition of polar ice cores has provided valuable plaeoenvironmental information over time spans as great as several millenia [1-13]. Glaciochemical data have given insight into remote area baseline impurity concentrations and pollutant perturbations. Recently, chemical concentration data have also been used to construct and refine Greenland and Antarctic ice core chronologies. Dating is accomplished through the use of seasonal variations in impurity concentrations and glaciochemical horizons.

Analytical Methods

Surface samples were collected in snow pits under ultra-clean conditions described elsewhere [13] with the exception that samples for anion analysis were collected in polystyrene cups precleaned without the use of acids. Ice core samples were cleaned to remove surface contamination using the "dry-core" procedure involving rinsing and melting of exterior surfaces with ultra-pure water [13]. Shallow-depth firn cores are permeable and the dry-core rinsing is unsuitable. Therefore an inner core of 2.5 cm diameter was taken from intervals of the 7.6 cm diameter South Pole firn core using a specially-built precleaned stainless steel corer within a -15 °C cold room. Prior to this coring, exposed ends of core sections were shaved away with precleaned stainless steel chisels.

Aluminum concentrations were measured by graphite furnace atomic absorption spectrophotometry using a Perkin-Elmer Model 503 AA and HGA-2100 furnace within a class 100 clean room used exclusively for glaciochemical research [13]. Analytical precision at the 95 percent confidence level is ±10 percent. Concentrations of Na^+, Cl^-, NO_3^-, and SO_4^{2-} were measured by ion chromatography using a Dionex Model 10 IC with a 3 x 50 mm concentrator column, a 3 x 150 mm pre-column, a 3 x 500 mm separator column, and a 6 x 250 mm suppressor column. Sample volume was 10 mL. The anion eluent was 0.0029 M $NaHCO_3$ and 0.0023 M Na_2CO_3 and the cation eluent was 0.005 M HNO_3. Species were identified and quantified by comparison of retention times and peak heights with those of prepared standards. Analytical precision is ±5 percent except for low Cl^- concentrations where reproducibility at the 95 percent confidence level is ±20 percent.

Seasonal Variations in Greenland

The use of seasonal variations as a dating tool requires a glaciological regime of frequent, regular snowfall with minimal post-depositional modification. Large areas of the Greenland Ice Sheet are suitable for such stratigraphic techniques whereas much of the Antarctic receives only small amounts of snowfall which are then subject to intense erosion, redistribution, and wind-mixing

with older snow [14]. Consequently, the investigation of seasonal
variations in chemical impurity concentrations has been primarily
focused on samples from the Greenland Ice Sheet.

Verification of an annual singularity in concentration variations is possible for Greenland strata by direct comparison with
stable oxygen isotope values over the same ice core depth interval. Winter snow is isotopically lighter than summer snow thus
making it possible to determine the season of deposition from the
sinusoidal pattern of oxygen-18 delta values as a function of
depth [15]. Due to rapid isotopic smoothing by vapor phase diffusion, this method of seasonal identification is not feasible for
areas of low accumulation rates, including most of Antarctica
[16].

Seasonal concentration maxima have been found for a number of
chemical species in Greenland snow and ice [1,2,7,12,13,17,18].
Individual impurity sources, such as continental dust and sea
salts, generally display a single peak at a specific season of the
year. This makes species derived predominantly from a single
impurity source suitable for dating purposes. Examples of such
species, often called index species or index elements, are Al
which is dominantly derived from continental dust, and Na and Cl
which are usually of marine origin. On the other hand, species
such as Mg, with two or more prominent sources, may display
multiple peaks per year or have unresolvable peaks, and are therefore unsuitable for dating. Identification of the source of a
given species can be accomplished through multi-species chemical
analysis of samples, comparison of sample composition with known
source composition, comparison with mid-latitude and bi-polar
precipitation chemistry, and factor analysis of concentration
covariance, among other techniques [13].

Continental Dust - Continental dust, the windblown products
of crustal weathering processes, is the main source of the Al, Fe,
Si, and Ca, among other elements, in Greenland snow and ice [1,3,
10,11,13]. Verification of this origin has been accomplished by
scanning electron microscope studies and the agreement between the
relative abundances of these elements in Greenland snow and in
average crustal material [1,3,10,11,13,19,20]. Continental dust
is also the source of most insoluble microparticles [7].

Figure 1 shows concentrations of the index element Al and
oxygen isotope delta values as a function of depth in the Milcent,
Greenland ice core. A comparison of the two profiles reveals that
single spring peaks in Al concentration are readily discernible
and therefore quite useful for dating this interval. The annual
spring peak in continental dust concentrations, also observed in
mid-latitude precipitation, may be due to the low vernal vegetation cover coupled with "March winds" or to meteorological instability allowing dust to be carried higher in the troposphere.

The annual singularity in continental dust concentrations has
already been applied to cross-check the dating uncertainties of
O-isotope oscillations in several Greenland cores [8,9]. About

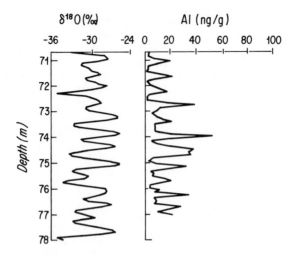

Figure 1. Stable oxygen isotope δ values (in per mil) and Al concentrations (in ng/g) as a function of depth in the Milcent, Greenland ice core. More negative δ values indicate winters and less negative values indicate summers. Al concentration peaks are in the spring.

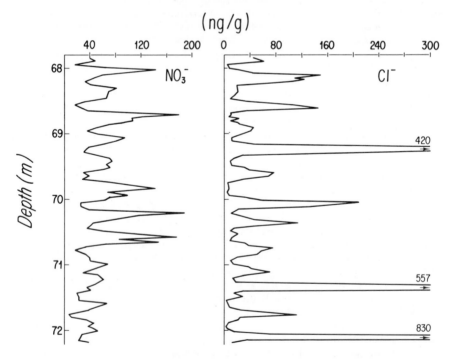

Figure 2. Concentration profiles of Cl^- and NO_3^- (in ng/g) in an ice core from northwest Greenland: Cl^- peaks demark winters and NO_3^- maxima are in summers. The three very high Cl^- concentrations may reflect early 19th century volcanism.

one year in seven is difficult to resolve in both the delta and Al or microparticle profiles [7,17]. However, since the uncertain years are randomly scattered, cross-checking one profile with the other leads to dating errors commonly of only a few percent.

Sea Salts - Most of the Na and Cl in both Greenland and Antarctica is of marine origin [1,13,21]. Near the ocean, sea salts may also account for most of the Mg, K, Ca, and $SO_4{}^{2-}$. Concentrations of Na and Cl display a maximum in winter Greenland precipitation which is coincident with the minimum oxygen isotope delta values [1,13,18]. The seasonal maximum in sea salt concentrations may be due to increased storminess over the ocean in winter or to an increased poleward latent heat flux during the polar night [13].

Figure 2 shows Cl^- concentrations as a function of depth in a core collected 10 km from Camp Century in northwest Greenland. (The three very high peaks may represent some early 19th century volcanic input which in this case did not disturb the seasonal pattern of Cl^- concentrations.) The clear winter peaks and summer minima demonstrate the utility of sea salts for dating purposes.

Nitrate - The origin of nitrate in polar snow is uncertain. A number of potential sources for Antarctic nitrate have been suggested by Parker et al. [22] including the tantalizing possibility that nitrate levels reflect solar activity which may modulate polar nitrogen-fixation processes. There are large reservoirs of nitrogen compounds in the atmosphere, however, and nitrate concentration variations may reflect changes in the magnitude of these reservoirs, or they may signify changes in the rates of nitrogen-fixation or nitrate or nitrate-precursor formation. The observation of a summer maximum in Greenland ice is in accord with the idea that photo-oxidation of a precursor is an important phase of polar nitrate deposition [13].

Regardless of the source(s) of polar nitrate and mechanisms of incorporation into snow, one point remains empirically clear. Nitrate is an excellent tool for dating Greenland snow and ice strata. Figure 2 demonstrates the clear summer nitrate peak coincident with minima in Cl^- concentrations in northwest Greenland. Herron [13] compared $NO_3{}^-$ concentrations with O-isotope delta values in the Milcent, central Greenland core and found a distinct summer maximum that is unaffected by explosive volcanic events, in contrast to the seasonal variations of many other species. Present work on the Dye 3 deep ice core from southern Greenland also shows clear $NO_3{}^-$ maxima which correlate with but are not necessarily coincident with summer melt features. Thus, $NO_3{}^-$ concentration variations appear to be useful in all Greenland latitudes and, when combined with continental dust and sea salt index species can be used to delineate summer, spring, and winter seasons of deposition.

The normally clear seasonal variations in certain chemical concentrations may be altered or rendered indecipherable by event perturbations. Following major explosive eruptions, such as that of Krakatoa in 1883, volcanism may be the dominant source of many species normally associated with continental dust or sea salts [13]. Volcanogenic species have a summer and/or winter maximum which may be related to the biannual exchange of mid-latitude and polar stratospheric air [23]. Following major eruptions Cl, Na, Al, and Ca peaks may thus give unreliable seasonal dating.

A second example of complications due to competing sources involves pollutants. Pollutant Pb and SO_4^{2-} concentrations in modern Camp Century snow show a winter maximum [1,2]. This is in contrast to the seasonal maximum in pre-1900 samples, which contain little or no pollutant component, where Pb has a spring maximum and SO_4^{2-} a late spring/summer maximum [12,13,18]. When pollutant and natural components are of approximately equal magnitude, as for SO_4^{2-} at Dye 3 in southern Greenland, seasonal variations may be obscured [12].

The fact that different impurity sources have maximum concentrations at different times of the year is important for ice core dating for two reasons. First, the occurrence of distinct winter, spring, and summer peaks means that it is unlikely that the peaks reflect changes in the efficiency of impurity incorporation mechanisms. If such mechanism efficiencies governed polar snow concentrations, a single annual maximum would be expected for all impurity sources. Instead, changes in snow concentrations reflect changes in atmospheric impurity concentrations. Second, the presence of maxima at different seasons means that dating accuracy may be increased by using several index species. When peak interpretation is uncertain for one seasonal indicator, additional species may be used to refine the dating.

Antarctic Seasonal Variations

In contrast to the favorable glaciological regime in Greenland, large areas of the Antarctic Ice Sheet may have such severe wind-redistribution of snow that short wavelength stratigraphy, including seasonal variations, is obscured or unreliable for dating. Partly in realization of this fact, less work has been done on Antarctic glaciochemical dating. In addition, since the oxygen isotope dating method is unusable in low accumulation zones [16], it is more difficult to verify an annual singularity as well as to determine the frequency of missing and multiple annual peaks. Thompson et al. [5] caution that conclusive proof of seasonal variations in microparticle concentrations is lacking.

In view of the likelihood of wind-mixing, particularly on the Antarctic plateau, it is not surprising that the concentration variances are lower than in Greenland. Figure 3 shows the concentration profiles of Na^+, Cl^-, NO_3^-, and SO_4^{2-} measured on the 1977 firn core from South Pole. The NO_3^- and SO_4^{2-} profiles have a very small variance, in marked contrast to the Camp Century Satellite profile shown in figure 2. Sea salt constituents Na^+ and Cl^-

have a detectable winter maximum at South Pole as in Greenland [11], but the concentration variance is much lower than in Greenland strata and the interpretation of winters indicated in figure 3 can only be regarded as tentative. Consequently, the potential dating errors are much greater for the Antarctic plateau than for Greenland.

The usefulness of continental dust concentration variations in dating snow from the Antarctic plateau is unclear. Boutron [11] found an insignificant variance in Al concentrations as a function of depth at South Pole, while Thompson and Thompson [24] found an annual peak in microparticle concentrations.

The glaciochemical dating potential is apparently much greater for the ice shelves of Antarctica than for the East Antarctic plateau. Seasonal variations in marine-derived Na concentrations have been found in Ross Ice Shelf strata with a winter maximum as in Greenland [25,26]. At Station J-9, 450 km from the edge of the ice shelf, Na concentrations vary sinusoidally with depth and the annual layering is easily decipherable. That the period of the concentration variations is one year has been verified by the agreement between snow accumulation rates determined glaciochemically, by ^{210}Pb dating, and by the depths of nuclear test debris horizons [26,28].

Sea salts appear to be less reliable seasonal indicators nearer the ocean on the Ross Ice Shelf. However, NO_3^- and SO_4^{2-} have easily interpretable seasonal variations and serve as useful substitutes. Figure 4 shows Cl^-, Na^+, NO_3^-, and SO_4^{2-} concentration variations in a two-meter deep pit at Station C-16, 350 km from the ice front. The high degree of covariance between Cl^- and Na^+ concentrations confirms their common marine origin. However, the concentration variations with depth are much less clear than the sinusoidal oscillations at J-9. At Station Q-13, only 70 km from the ice front, Na^+ and Cl^- concentrations are significantly higher than at C-16 (figure 5), but the variations are indecipherable in terms of seasonal variations. In contrast, seasonal variations are in NO_3^- and SO_4^{2-} concentrations are apparent at both sites. Verification that the peaks are seasonal is presently limited to a comparison of accumulation rates derived glaciochemically with those derived from the measured depth-snow density curve [29]. In both cases, differences are less than 5 percent. Additional verification will be possible when fission product studies at these sites are completed.

The loss of resolution of seasonal variations in sea salt concentrations near the ocean may be interpreted to reflect the presence of two marine sources with different seasonal maxima. A distant source with a winter maximum, as in the case in Greenland, dominates for more inland stations like J-9. Nearer the ice front, the local Ross Sea contribution is greatest in the summer when the sea ice cover is at a minimum. The contribution from the

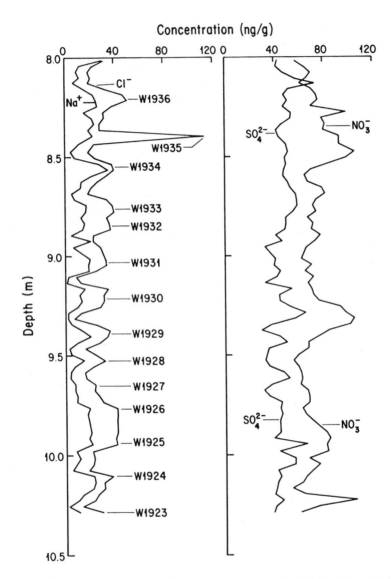

Figure 3.   Concentration profiles of Na⁺, Cl⁻, NO₃⁻, and SO₄²⁻ (in ng/g) in the 1977 South Pole firn core. Winter peaks in sea salt Na⁺ and Cl⁻ and approximate years of deposition are indicated. The small concentration variability probably reflects wind-redistribution of surface snow.

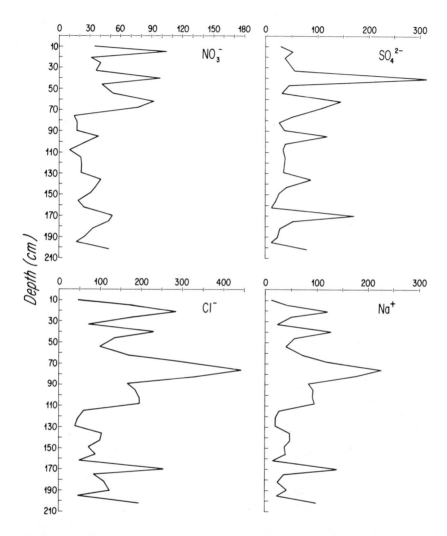

Figure 4. *Chemical concentrations of Na⁺, Cl⁻, NO₃⁻, and SO₄²⁻ (in ng/g) from a snow pit at Station C-16 on the Ross Ice Shelf, Antarctica. Annual accumulation rates derived from each of the profiles are in good agreement. The low accumulation rate and insufficiently detailed sampling resulted in some of the summer NO₃⁻ and SO₄²⁻ peaks being coincident with the winter sea salt peaks.*

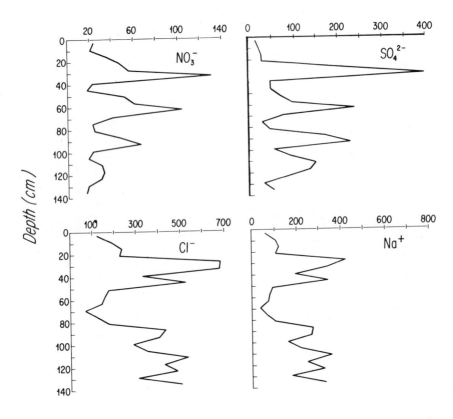

Figure 5.   Concentrations of NO_3^-, SO_4^{2-}, Cl^-, and Na^+ (in ng/g) in a snow pit from Station Q-13 on the Ross Ice Shelf, Antarctica. Station Q-13 is only about 70 km from the Ross Sea. Most of the Na^+ and Cl^- peaks, which are unusable for dating, occur in summer, coincident with the highly resolved NO_3^- and SO_4^{2-} peaks. This probably reflects the presence of local Ross Sea marine aerosols which are in greatest abundance during the summer minimum in sea ice cover.

local source results in higher sea salt concentration levels and a loss of clear seasonal variations at seaward stations. If this explanation is correct, it is reasonable to assume that sea salt dating will be applicable in regions of West Antarctica which are removed from local marine sources and where snow accumulation rates are greater than in East Antarctica.

Horizon Detection

Glaciochemical horizons are intervals of core with substantially higher or lower than average concentrations of certain chemical constituents. If a historical event of known age can be correlated with the event horizon in the core, the assigned age of that interval can be used to confirm the depth-age relationship which has been determined from seasonal variations or other dating methods. In addition, in deep ice where annual layers are too thin to count seasons reliably and dating is only possible by model calculations [15,30], these horizons provide check points for calculated ages.

The application of glaciochemical horizons to ice core dating is a recent and rapidly evolving development. A concerted search is under way to determine the impact of known events on the chemical composition of polar snow. So far, two types of universally applicable horizons have been detected in polar ice cores: volcanic eruptions and Late Wisconsin dust.

Some volcanic eruptions release large quantities of acids and trace metals into the stratosphere; these are then transported around the world. Following very large explosive eruptions sufficient acids and trace metals are deposited in Greenland snow to be detectable as glaciochemical events. In solid ice, the acids may be identifiable by elevated space charge [31]. In melted samples, volcanogenic acids are detectable by low pH, high specific conductance, and elevated SO_4^{2-}, Cl^-, and/or F^- concentrations [13]. Volcanogenic trace metals include Cd, Zn, Cu, Na, K, and Pb [13]. In contrast, volcanic aluminosilicate ash is quickly removed from the atmosphere and seldom makes a discernible impact on continental dust-derived Al or microparticle concentrations in Greenland [7,13].

An example of a volcanic glaciochemical event is provided by the 1783 Laki eruption in Iceland. Enormous quantities of sulfurous gases were vented from this volcano and oxidized to sulfuric acid. The sulfate concentration in snow deposited at Milcent, central Greenland rose from an average of 40 ng/g to over 2200 ng/g following the eruption and Cl^- concentrations reached 100 ng/g compared to a normal value of 20 ng/g (figure 6). This horizon, one of the strongest in recent Greenland ice, is detectable in all existing Greenland ice cores [9].

A second type of glaciochemical horizon, a long-term interval of anomalous concentrations rather than a sporadic peak, is demonstrated by Late Wisconsin dust. Greenland and Antarctic snow

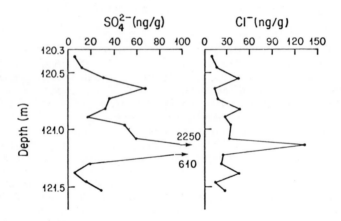

Figure 6. Concentrations of SO_4^{2-} and Cl^- (in ng/g) in Milcent, Greenland ice core samples deposited following the 1783 eruption of Laki in Iceland. The acid precipitation that followed this eruption left an unmistakable glaciochemical horizon all over Greenland.

that was deposited during the last third of the Wisconsinan
glaciation (ca. 25,000 to 10,000 BP) contains between 10 and 100
times as much continental dust as does more recent snow [5,6,8,19,
20]. The Late Wisconsin dust interval is as characteristic of
this time interval as are the strong shifts in oxygen isotope
content [30,32]. The changes to Holocene dust and O-isotope concentrations are both fairly sudden at about 10,000 BP. Late
Wisconsin dust in Greenland is more calcareous than dust in modern
snow, but under the electron microscope closely resembles crustal
weathering products that characterize present day dust [3,20].
The dust interval is detectable as elevated concentrations of Al,
Ca, Si, and microparticles in both hemispheres. Local volcanism
may have contributed to sporadically very high microparticle concentrations in Late Wisconsin ice from Byrd Station, Antarctica
[6]. However, the similarities in the temporal and quantitative
concentration perturbations in the Byrd Station and Camp Century,
Greenland cores suggest a common global phenomenon for this horizon [19].

A number of possible explanations for the Late Wisconsin dust
horizon have been offered [13]. Continental shelves, exposed by
lower glacial age sea level, may have provided additional wind-
erodable dust of a more calcareous nature than present day dust
[19]. Loess derived from glacially eroded continents, at least
some of which was highly calcareous [33], may have been another
important source. Reduced glacial age vegetative cover may have
resulted in more efficient erosion and/or transport to the polar
regions. Finally, dry deposition of aerosols in absence of precipitation may have been a more significant impurity incorporation
mechanism during the ice age, particularly if snow accumulation
rates were lower.

Deciphering the cause(s) of the Late Wisconsin dust interval
will be the subject of many future studies, but the interval
serves as a useful time marker at present. Koerner and Fisher
[34] have used elevated microparticle concentrations in the Devon
Island ice core as an indicator of Wisconsin-aged ice. Similarly
high continental dust concentrations are expected in the recently
recovered cores from the East Antarctic plateau.

Another type of glaciochemical horizon, limited in applications, is related to the outflow of ice from the polar highlands
of East Antarctica (elev. 2000-4000 m) and West Antarctica (1000-
1500 m) into the low elevation fringing ice shelves. According
to the concept of glaciochemical regimes, these three regions
can be characterized on the basis of sea salt content [26].
East Antarctic Na concentrations are about 10 ng/g (figure 3),
West Antarctic Na concentrations are 25-30 ng/g, and Ross Ice
Shelf Na concentrations (see figures 4 and 5) are greater than 30
ng/g. Underlying the snow that fell on the Ross Ice Shelf is ice
that flowed from the East and West Antarctic Ice Sheets. In two
cores that penetrated the ice shelf veneer, Na concentrations
characteristic of West Antarctica were found at depths marking the

transition horizon between the ice shelf and inland ice glaciochemical regimes [26,34]. Such information is valuable in
checking whether periodic ice sheet surges in flow velocity have
occurred in the recent past and in demarking previous boundaries
between West and East Antarctic drainage systems.

Other events, evidence of which has not yet been detected in
polar ice, may also eventually serve as useful time markers in the
future. Large meteors or meteor swarms that ablated in the upper
atmosphere may have left a significant chemical impact on succeed-
ing snowfalls. The same may be true for cometary collisions or
brushes with a comet's tail. As with ancient volcanic eruptions,
such events need not have been historically documented. If identified and accurately dated in one ice core, these events will
serve as useful time horizons in other cores.

In summary, a number of chemical constituents in polar snow
and ice have seasonal concentration variations that make them
suitable for dating ice cores by counting annual layers. In
Greenland, NO_3^-, Cl^-, Na^+, and Al or microparticles have usable
seasonal variations. Sea salt-derived Na and Cl are useful dating
species in East Antarctic snow although the variations are lower
than in Greenland and the dating errors are consequently higher.
Sea salts are satisfactory seasonal indicators in the inland
portion of the Ross Ice Shelf and presumably in West Antarctica.
Nitrate and sulfate have good seasonal variations on the Ross Ice
Shelf, even at the more seaward locations.

To date, seasonal microparticle analysis is the only glaciochemical technique which has been used for continuous dating of a
several hundred year old ice core. But the possibilities exist,
as demonstrated here, to apply other chemical measurements to
confirm or establish an ice core chronology. These applications
are of special importance where routine dating techniques are
inappropriate, as for very old ice where solid-state diffusion has
greatly smoothed the original oxygen isotopic signal [36], in low
accumulation areas where rapid vapor-phase diffusion renders the
oxygen isotope method inapplicable [16], and in the Antarctic
where an annual peak in microparticle concentrations has not yet
been conclusively proven.

Glaciochemical horizons not only provide checks on depth-age
relationships but are of great intrinsic interest. Continuous
chemical analysis of a well-dated core promises to greatly extend
and refine the record of explosive volcanism [13]. Each well-
dated eruption or other horizon then becomes of use as a time
marker for cores where the dating is less precise.

Literature Cited

[1] Murozumi, M., Chow, T. J., and Patterson, C., Geochim. Cosmochim. Acta, 1969, 33, 1247-1294.
[2] Koide, M. and Goldberg, E. D., J. Geophys. Res., 1971, 76, 6589-6596.
[3] Cragin, J. H., Herron, M. M., and Langway, C. C., Jr., USA CRREL Research Report 341, 1975.
[4] Weiss, H. V., Bertine, K. K., Koide, M., and Goldberg, E. C., Geochim. Cosmochim. Acta, 1975, 39, 1-10.
[5] Thompson, L. G., Hamiton, W. L., and Bull, C., J. Glaciol., 1975, 14, 433-444.
[6] Thompson, L. G., Int. Assoc. Hydrol. Sci. Publ. No. 118, 1977, 351-363.
[7] Hammer, C. U., Int. Assoc. Hydrol. Sci. Publ. No. 118, 1977, 297-301.
[8] Hammer, C. U., Int. Assoc. Hydrol. Sci. Publ. No. 118, 1977, 365-370.
[9] Hammer, C. U., Nature, 1977, 270, 482-486.
[10] Herron, M. M., Langway, C. C., Jr., Weiss, H. V., and Cragin, J. H., Geochim. Cosmochim. Acta., 1977, 41, 915-920.
[11] Boutron, C., These de Doctorate d' Etat, Universite Scientifique et Medicale de Grenoble, France, 1978.
[12] Busenberg, E. and Langway, C. C., Jr., J. Geophys. Res., 1979, 84, 1705-1709.
[13] Herron, M. M., Ph.D. Dissertation, State University of New York at Buffalo, 1980.
[14] Bull, C. B. B., "Research in the Antarctic", Quem, L. O., Ed., American Association for the Advancement of Science: Washington, D. C., 1971, 367-421.
[15] Dansgaard, W. and Johnsen, S. J., J. Glaciol., 1969, 8, 215-223.
[16] Johnsen, S. J., Int. Assoc. Hydrol. Sci Publ. No. 118, 1977, 210-219.
[17] Langway, C. C., Jr., Klouda, G. A., Herron, M. M., and Cragin, J. H., Int. Assoc. Hydrol. Sci. Publ. No. 118, 1977, 302-306.
[18] Klouda, G. A., M.A. Thesis, State University of New York at Buffalo, 1977.
[19] Cragin, J. H., Herron, M. M., Langway, C. C., Jr., and Klouda, G., "Polar Oceans", Dunbar, M. J., Ed., -Arctic Institute of North America: Calgary, Alberta, 1977, 617-631.
[20] Kumai, M., Int. Assoc. Hydrol. Sci. Publ. No. 118, 1977, 341-349.
[21] Boutron, C., Geochim. Cosmochim. Acta, 1979, 43, 1253-1258.
[22] Parker, B. C., Heiskill, L. E., Thompson, W. J., and Zeller, E. J., Nature, 271, 651-652.
[23] Lamb, H. H., Phil. Trans. Roy. Soc. London, 1970, 266, 425-533.

[24] Thompson, E. M. and Thompson, L. G., Ant. J. of the U.S., 1977, XII, 136-137.
[25] Warburton, J. A. and Linkletter, G. O., Int. Assoc. Hydrol. Sci. Publ. No. 118, 1977, 88-94.
[26] Herron, M. M. and Langway, C. C., Jr., J. Glaciol., 1979, 24, 345-357.
[27] Clausen, H. B. and Dansgaard, W., Int. Assoc. Hydrol. Sci. Publ. No. 118, 1977, 172-176.
[28] Koide, M., Michel, R., Goldberg, E. D., Herron, M. M., and Langway, C. C., Jr., Earth Planet. Sci. Let., 1979, 44, 205-223.
[29] Herron, M. M. and Langway, C. C., Jr., J. Glaciol., 1980, in press.
[30] Dansgaard, W., Johnsen, S. J., Moller, J., and Langway, C. C., Jr., Science, 1969, 166, 377-381.
[31] Hammer, C. U., Clausen, H. B., and Dansgaard, W., J. Glaciol., 1980, in press.
[32] Johnsen, S. J., Dansgaard, W., Clausen, H. B., and Langway, C. C., Jr., Nature, 1972, 235, 429-434.
[33] Flint, R. F., "Glacial and Quaternary Geology," John Wiley and Sons: New York, 1971, 880 p.
[34] Koerner, R. M. and Fisher, D., J. Glaciol., 1979, 23, 209-222.
[35] Langway, C. C., Jr., Herron, M. M., and Cragin, J. H., J. Glaciol., 1974, 69, 431-435.
[36] Hammer, C. U., Clausen, H. B., Dansgaard, W., Gundestrup, N., Johnsen, S. J., Reeh, N., J. Glaciol., 1978, 20, 3-26.

RECEIVED April 15, 1981.

Preliminary Studies on Dating Polar Ice by Carbon-14 and Radon-222

E. L. FIREMAN and T. L. NORRIS

Smithsonian Astrophysical Observatory, Cambridge, MA 02138

The Antarctic and Greenland ice sheets originate from snow and are vast storehouses of information on the history of the earth's atmosphere and ecology. At 60-m depth, snow at Byrd Station is compressed into ice and air is occluded in the form of gas bubbles under pressure [1,2][1]. The ice then slowly moves toward the continental margin, where most of it goes into the ocean. The travel distance for ice originating in central Antarctica is at least 2000 km. Since the average travel speed is a few meters per year at the surface and slower below the surface, samples of million-year-old ice may exist. The radioactive dating of polar ice would give important information on the history of the region and aid searches for ancient ice.

Interesting ice samples from Antarctica and Greenland have been and are being recovered. We studied samples of the Byrd core, which is a 12-cm-diameter core that extended to bedrock at 2100-m depth [1]. This core is presently kept at the Central Ice Core Storage Facility at S.U.N.Y. Buffalo (C. C. Langway, Jr., Curator). Its age-depth relationship has been calculated on the basis of rheological models [3,4,5], and comparisons of the $\delta^{18}O$ variations of the core with those in the Camp Century (Greenland) core. The age calculated for the bottom ice is between 50 x 10^3 and 100 x 10^3 years.

The $\delta^{18}O$ in Byrd core melted ice as a function of depth has been measured by mass spectrometry [4,5]. Since the $\delta^{18}O$ scale depends on the temperature of the ocean water that developed into snow flakes [6], accurate dating of the core itself is necessary to reveal the temperature history of the ocean surface water. Oeschger et al., [7] measured the ^{14}C contents of CO_2 extracted from 3 tons of ice melted in situ, at depths of 100, 175, 270, and 380 m near the Byrd site. Their ^{14}C ages for 270- and 380-m depths are 1300 ± 700 and 3000 ± 500 years, respectively.

---

[1]Figures in brackets refer to the literature references at the end of this paper.

0097-6156/82/0176-0319$05.00/0

It is necessary to measure ^{14}C in smaller ice samples to date specimens from the core itself. The uncertainties in the calculated age-depth relation increase with depth because it is difficult to measure ^{14}C at large depths. Another radioactive dating method is desirable for the bottom half of the core; [222]Rn, can be extracted and counted by essentially the same procedure as ^{14}C, and when combined with the measurement of [230]Th, is a promising dating method for the core bottom.

We analyzed ^{14}C and [222]Rn in Byrd core samples from 270-, 362-, 1070-, and 1496-m depths. We also analyzed surface and subsurface (5- to 25-cm depth) ice samples from the Allan Hills meteorite site [8], where more than 1000 meteorites have been recovered from an area of approximately 100 km^2 [9]. Allan Hills is a region near the continental margin where the ice movement is blocked by hills protruding above the surface and ice is rapidly dissipated by wind ablation. The terrestrial ages of more than a score of large meteorites, unmoveable by wind, have been measured and range from (11 ± 1) x 10^3 years to (700 ± 100) x 10^3 years [10, 11,12]. The recovery locations of these specimens are known relative to a network of stakes. A recovery location-terrestrial-age pattern is beginning to emerge that indicates a continued upflow of ice for ~700 x 10^3 years at one location and essentially stagnant ice for this period of time at another location [13]. The ages of the ice from these two locations should be determined by radiometric dating.

Figure 1 is a map of Antarctica showing the locations of Byrd Station, Allan Hills, and other well-marked sites.

Extraction and Purification Procedures

We extract the gas from the ice by an evacuation followed by a series of helium purges of the melt water. A crack-free section of the core is selected. The ice is rinsed with distilled water using the S.U.N.Y., Buffalo cleaning procedure to remove surface contaminants, carried to our laboratory, and placed in our extraction system. Figure 2 is a drawing of our gas extraction system. Up to 30 kg of ice can be lowered into the 6-inch-diameter glass cylinder. The cap with the large cold finger shown in figure 2 is bolted to the glass cylinder. Air is then removed from the cylinder by an evacuation to a pressure of 4 μm. The cylinder is then purged by filling it with He to 1.3 atm and then evacuating the He. This He purge procedure is continued until the ice takes on a glazed appearance showing that surface melting occurred; during this time 200-400 cc of melt water collects at the bottom of the cylinder. This procedure not only removes remnant air from the cylinder but also most of the air adsorbed on the ice surface.

The cylinder is then filled with He to 1.3 atm and closed off until the ice melts completely. The melting of 10-20 kg of ice takes between 8 and 16 hours.

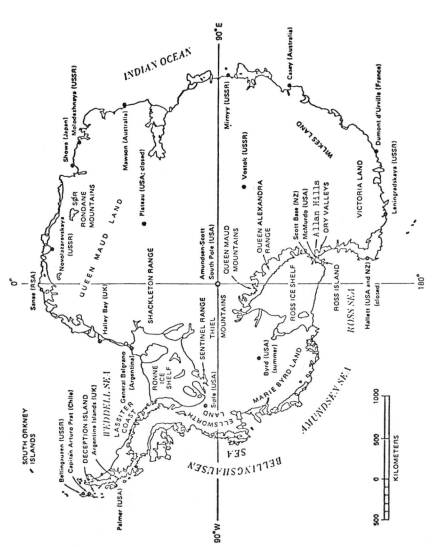

Figure 1.   Antarctica: selected stations and physical features.

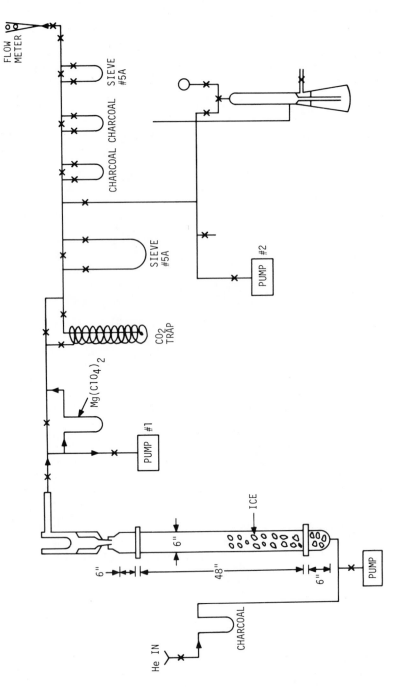

Figure 2. Extraction system.

The gas is removed from the melt water in three extractions:
(1) the gas in the volume above the melt water is collected; (2)
the melt water is purged with He; (3) the melt water is acidified
with H_2SO_4 to pH=1, heated to 55 °C, and then again He purged. We
recently started to acidify with HNO_3 which gives more reproducible Rn results. During the extractions, the cold finger above
the melt water is maintained at -70 °C to remove most of the H_2O
vapor (the remnant vapor is removed by a Mg (ClO_4) trap). The CO_2
is trapped on a fine glass spiral at -196 °C and the air (gas
minus CO_2) is collected by a molecular sieve at -196 °C. The CO_2
is removed from the CO_2 trap at room temperature, measured volumetrically, and stored. The air in the sieve is then removed at
100 °C measured volumetrically and stored.

Approximately 90 percent of the air and half the CO_2 is
collected from the volume above the melt water; approximately 10
percent of the air and half of the CO_2 is collected in the first
He purge. Very little air and CO_2 is collected in the second He
purge after acidifying and heating the water.

The CO_2 is accompanied by radon, which is difficult to
separate. The CO_2 is therefore often stored before counting its
^{14}C activity. A small amount of radon is in the bubble fraction;
larger amounts of radon are in the purged extractions. The CO_2 is
purified before counting by a series of freezing and unfreezing
steps.

After the third extraction, we measure the radon and its
yield by adding several cc STP of carrier Ar to the melt water
with He to 1.3-atm pressure and allow the radon to build up for
4 days. We also replace the spiral glass CO_2 trap with a charcoal
trap to insure the collection of the carrier Ar with the radon.
We then He-purge the water and collect the Rn with the carrier Ar
on the charcoal at liquid air temperature. The Ar plus Rn is
recovered from the charcoal at 300 °C; purified over hot Ti and
counted in a proportional counter.

^{222}Rn (3.8-day half-life) is the daughter of ^{226}Ra (1600-
year half-life). If ^{226}Ra is in solution, ^{222}Rn should accumulate
with a 3.8-day half-life after the He purge; furthermore, the loss
of ^{226}Ra by handling and storage the melt water should be
negligible. After removal of the melt water from the extraction
unit and storage, aliquots of the melt water are sometimes
returned for repetitive ^{222}Rn analyses. Aliquots of the melt
water can also be evaporated to dryness for ^{226}Ra and ^{230}Th
analyses by α particle-counting and neutron activation studies.
Measurements of ^{222}Rn purged from aliquots of the water reintro-
duced into the extraction unit after several months storage
showed a serious loss of ^{226}Ra, which we attribute to insoluble
$RaSO_4$ lost on the walls of the storage vessel. To eliminate this
problem, we now acidify with HNO_3 rather than with H_2SO_4.

Counter Description and Counting Procedure

Figure 3 shows a photograph and a drawing of one of our mini-counters. The counters are of Davis design [14]. Counters of 0.70 cm^3 volume were used for all the samples listed in Table 1 except the Allan Hills surface sample from stake 12, A.H.$_{12}$ sur., and frozen distilled water sample; for these 2.5-cm^3 volume counters were used. The efficiencies of the mini-counters were determined with a CO_2 standard having 2.15 dpm of ^{14}C per cm^3 (STP). This standard has been counted in a large counter of known efficiency. The mini-counter is evacuated overnight while heated to approximately 150 °C. The purified CO_2 together with an equal amount of purified Ar is put into the counter. The counter is removed from the filling line and placed in a low-level unit located in the basement of the Observatory Plate Stacks. The unit has a shield with 14-inch thickness of Pb, lined with 1-inch thickness of OHFC copper. The counter is placed in a copper can surrounded by a 2-inch thickness of Hg and a double layer of anticoincidence proportional counters of 42-inch length. The energy resolution of the counter is determined with an external ^{55}Fe source. Unless the 5.8-keV ^{55}Fe line has a resolution of better than 20 percent, the counter is rebaked and refilled with repurified CO_2 and Ar. The counts above 1.0 keV are recorded with a 100-channel analyzer and those above 7.5 keV with scalers. Superior backgrounds relative to efficiencies for ^{14}C are achieved by using the counts above 6.8 keV. For counts above this energy the background rate in the 0.70 cm^3 counters is approximately 2 counts day^{-1} with 1 atm pressure of CO_2 from a petroleum source and has an efficiency of 45 percent for ^{14}C. The 2.5-cm^3 counters give a background of 5 counts/day and 55 percent efficiency. Several months of counting together with periodic ^{55}Fe stability checks are necessary to obtain ^{14}C ages to $\sim10^4$ years with such small samples.

The counting of ^{222}Rn is relatively simple; no attempt was made to achieve backgrounds less than 0.1 counts/min. If the activity was higher than 0.2 counts/min, the sample was counted for approximately two weeks to ensure that the activity had the 3.8-day half-life. The problem with ^{222}Rn is that it is difficult to obtain reproducible results, particularly after the sample is removed from and reinserted into the extraction line. This type of reproducibility is necessary if ^{230}Th and ^{234}U are to be reliably determined on aliquots of melt water.

Results

Table 1 gives the amounts of gas minus CO_2 obtained in the first two extractions and in the heated water extraction, the percentage of CO_2 in the gas, the specific activity of the CO_2, ^{14}C age, and ^{222}Rn content of the ice. The amounts of gas minus

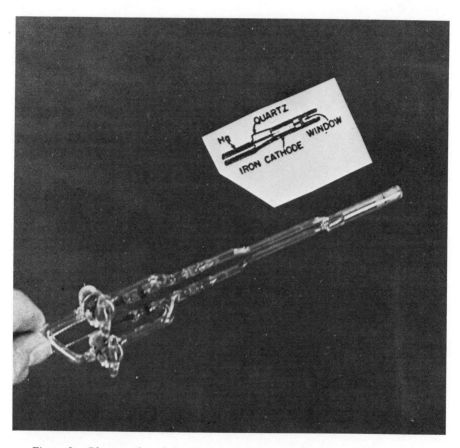

Figure 3. Photograph and drawing of a mini-counter used for the ^{14}C measurements.

Table 1. Amounts of Gas, Percentage of CO_2 in the Gas, Specific Activity of the CO_2, ^{14}C ages, and ^{222}Rn contents.

Sample (loc., depth, weight)	Extraction (temp, pH)	Gas (cm^3/kg)	CO_2 %	$^{14}C/CO_2$ (10^{-3} dpm/cm^3)	^{14}C age (10^3 yr)	^{222}Rn (dpm/kg)
Byrd 271 m 7.6 kg	24 °C, 5.5 / 55 °C, 1	66.8 / 0.1	0.0336	6.8 ± 1.0	1.0 ± 1.0	0.4 ± 0.1
Byrd 272 m 8.8 kg	24 °C, 5.5 / 55 °C, 1	57.5 / 0.1				
Byrd 362 m 6.2 kg	24 °C, 1 / 55 °C, 1	74.3 / 0.5	0.051	6.0 ± 0.5	2.0 ± 0.7	7.5 ± 0.4
Byrd 363 m 8.0 kg	24 °C, 1 / 55 °C, 1	69.3 / 1.3				
Byrd 1068 m 9.2 kg	24 °C, 5.5 / 55 °C, 1	117.8 / 0.2	0.0216	—	—	—
Byrd 1071 m 6.2 kg	24 °C, 5.5 / 55 °C, 1	101.2 / 7.5	0.0356	≤3.0	>8.0	<0.02
Byrd 1469 m 9.5 kg	24 °C, 5.5 / 55 °C, 1	115.1 / 0.4	0.0237	—	—	—
A.H.$_{-12}$, sur. 31.0 kg	24 °C, 1 / 55 °C, 1	47.0 / 0.9	0.187	27 ± 3	nuclear debris	—
A.H.$_{-18}$, sur. 16.6 kg	24 °C, 5.5 / 55 °C, 1	20.1 / 0.4	0.129	25 ± 6	nuclear debris	—
A.H., 5-25 cm 12.9 kg	24 °C, 5.5 / 55 °C, 1	46.0 / 1.5	0.054	7.0 ± 3.0	≤6	0.23 ± 0.02
Frozen water 7.8 kg	24 °C, 5.5 / 55 °C, 1	20.4 / 0.06	1.35	8.7 ± 0.5	0 ± 0.7	<0.02
Air		—	0.034	—	—	—

CO_2 ranged from 58 to 118 cm^3 STP/kg for the Byrd core and from 21 to 48 cm^3 STP/kg for the Allan Hills ice samples. Analyses for N_2, O_2, Ar, and $^{15}N/^{14}N$ were done by converting the O_2 to CO_2 over a hot carbon filament, measuring volumes, and studying the N_2 + Ar fraction with a mass spectrometer. For the Antarctic samples, the N_2, O_2, and Ar abundances and the $^{15}N/^{14}N$ ratios were the same as air; the deviations in Ar abundances and $^{15}N/^{14}N$ ratios outside of experimental error require further investigation. The gas from frozen distilled water had a very high CO_2 abundance (1.35 percent), high O_2 and Ar abundances (26.5 percent and 1.3 percent, respectively), and a low N_2 abundance (71.8 percent) which typifies dissolved air. The high CO_2 abundances in the Allan Hills surface samples from stakes 12 and 18 (0.187 and 0.129 percent, respectively) are attributed mainly to dissolved air because melting and refreezing occurs on Allan Hills surface ice. The CO_2 abundance in the subsurface Allan Hills ice (5- to 25-cm depth) is 0.054 percent; this value is similar to the 0.051 percent abundance in Byrd ice from 362- and 363-m depths, and is probably caused by a CO_2 contribution from carbonates. At 1068- and 1469-m depths in Byrd ice, the CO_2 abundances, 0.0216 and 0.0237 percent, are low; this result can be only explained by a low CO_2 abundance in the ancient atmosphere.

The pH of Antarctic ice is 5.5. The first three samples that were studied (362 and 363 m Byrd core and Allan Hills surface ice from stake 12) were acidified before purging.

The errors given for the specific ^{14}C activities are the sums of one standard deviation in the counting statistics for the CO_2 of the ice sample and the CO_2 of the petroleum background sample. CO_2 converted from a NBS oxalic acid standard was also counted in one 0.7-cm^3 and in one 2.5-cm^3 mini-counter and had specific activities of (8.0 ± 1.0) x 10^{-3} dpm/cm^3. The specific activities in the surface ice taken from Allan Hills at stakes 12 and 18 are three to four times higher than the oxalic acid standard and are attributed to nuclear particulate debris that has fallen on the ice surface in recent times. The Allan Hills ice from 5- to 25-cm depth from locale between stakes between 10 and 11 had a specific activity similar to oxalic acid CO_2. The large counting error for this sample results from the fact that its counting is not completed. The ^{14}C specific activities of the Byrd core samples with the exception of the sample from 271- to 272-m depth are lower than the oxalic acid standard and give ^{14}C ages consistent with ages calculated from rheological models.

The ^{222}Rn activities in the (362 + 363)m and in the 1068-m Byrd core samples were very high, 7.5 ± 0.4 and 12.5 ± 0.5 dpm/kg of ice; however, the activity was negligible in 1071-m Byrd sample. The ^{222}Rn activity is a direct measure of the ^{226}Ra present in the sample. The great difference between Byrd samples from 1068- and 1071-m depths indicates that there was a discontinued influx of ^{226}Ra and its parents, ^{230}Th and ^{234}U. These results favor the idea that there were large variations in the ^{226}Ra, ^{230}Th, and ^{234}U infall in the past due to volcanic eruption

and changing dust fallout patterns. There are similar amounts of
^{222}Rn in the Byrd samples from 271- and 1469-m depths even though
these samples have very different ages. The ^{222}Rn results show
that it is necessary to measure ^{230}Th in the same ice sample in
order to obtain a ^{230}Th-^{222}Rn age. Since very sensitive neutron
activation methods exist for the determination of ^{230}Th, the
^{230}Th-^{222}Rn transient equilibrium method [15] appears to be
promising for dating polar ice.

We thank J. C. DeFelice for his help in all phases of this
work, W. A. Cassidy and J. O. Annexstad for the Allan Hills ice,
and C. C. Langway, Jr., for the Byrd core ice. This research was
supported by NSF Grant DPP 78-05730.

References

[1] Gow, A. J., Ueda, H. T., Garfield, D. E., Antarctic Ice
    Sheets: Preliminary Results of First Core Hole to Bedrock,
    Science, 1968, 161, 1011-1013.
[2] Berner, W., Oeschger, H., Stauffer, B., Information of the
    CO_2 Cycle from Ice Core Studies, Berne preprint, 1980.
[3] Nye, J. F., Correction Factor for Accumulation Measured by
    the Thickness of the Annual Layers in an Ice Sheet, J.
    Glaciology, 1963, 4, 785-788.
[4] Epstein, S., Sharp, R. P., Gow, A. J., Antarctic Ice Sheet:
    Stable Isotope Analyses of Byrd Station Cores and Inter-
    hemispheric Climatic Implications, Science, 1970, 168,
    1570-1572.
[5] Johnson, S. J., Dansgaard, W., Clausen, H. B., Langway,
    C. C., Oxygen Isotope Profiles Through the Antarctic and
    Greenland Ice Sheets, Nature, 1972, 235, 429-434.
[6] Craig, H., Standard for Reporting Concentrations of Deuterium
    and Oxygen-18 in Natural Waters, Science, 1961, 133, 1833-
    1834.
[7] Oeschger, H., Stauffer, B., Bucher, P., Loosli, H. H.,
    Extraction of Gases and Dissolved and Particulate Matter
    From Ice in Deep Boreholes, Proc. Grenoble Symposium, 1975,
    IAHS, Publ. No. 118, 1977, 307-311.
[8] Cassidy, W. A., Olsen, E., Yanai, K., Antarctica: A Deep-
    Freeze Storehouse for Meteorites, Science, 1977, 198, 727-
    731.
[9] Cassidy, W. A., Antarctic Search for Meteorites During the
    1977-1978 Field Season, Antarctic J. of the U.S. XIII,
    1978, 39-40.
[10] Fireman, E. L., Carbon-14 Dating of Antarctic Meteorites and
    Antarctic Ice (Abstract), In: Lunar and Planetary Science
    XI, Lunar and Planetary Institute, Houston, 1980, 288-290.

[11] Nishiizumi, K., Arnold, J. R., Ages of Antarctic Meteorites (Abstract), In: Lunar and Planetary Science XI, Lunar and Planetary Institute, Houston, 1980, 815-817.

[12] Evans, J. C., Rancitelli, L. A., Reeves, J. H., ^{26}Al Content of Antarctic Meteorites: Implications for Terrestrial Ages and Bombardment History, Proc. Lunar Planet. Sci. Conf. 10th, 1979, 1061-1072.

[13] Fireman, E. L., ^{14}C and ^{39}Ar in ALHA Meteorites, Proc. Lunar Planet. Sci. Conf. 11th, 1980, 1215-1221.

[14] Davis, R., Harmer, D. S., Hoffman, K. C., Search for Neutrinos from the Sun, Phys. Rev. Lett., 1968, 20, 1205-1209.

[15] Friedlander, G., Kennedy, J. W., Miller, J. M., Nuclear and Radiochemistry, J. Wiley & Sons, Inc., 1964, 71-73.

RECEIVED August 11, 1981.

Dating Recent (200 Years) Events in Sediments from Lakes, Estuaries, and Deep Ocean Environments Using Lead-210

WILLIAM R. SCHELL

University of Washington, Laboratory of Radiation Ecology,
College of Fisheries, WH-10, Seattle, WA 98195

Sediments from lakes, continental margins and deep oceans record events caused by man and by natural processes and, by using radioactive tracers, their chronological record can be determined. The rhythmic pattern of sedimentation rates and benthic mixing processes has been determined using ^{210}Pb at lowland lakes of Washington State, at the Puget Sound Estuary and at two deep ocean sites — one 60 km off San Francisco at 1000 m and the other 350 km off New York City at 4000 m. The Pb input as a function of time is shown, for example, to increase from background to some 30 times background at Lake Washington (Seattle), and 13 times background at Sinclair Inlet (near a naval shipyard) of Puget Sound. The dates for maximum inputs of other trace metals into Sinclair Inlet were established; their relative enrichments above background were Cu..8.0 Zn..2.7, Ni..1.9, Cr..2.2. Man's effect on the Lake Washington watershed due to deforestation and urban development has resulted in three changes in sedimentation rates since 1850. The ^{210}Pb profiles in cores collected at the mouth of the Hudson Canyon (4000 m deep) show wide variations which may be due to sediment redistribution by deep currents and by biological mixing. An episodic event in the canyon may be registered synchronously, 100 years ago, at two stations; one at 2800 m and one at 4000 m deep. By measuring ^{239,240}Pu in deep sea cores together with ^{210}Pb, the past 30 years since the bomb testing can be identified since it is generally assumed that both ^{210}Pb and ^{239,240}Pu are associated with particulate matter. The deposition inventory for ^{210}Pb and ^{239,240}Pu shows that redistribution processes are active at the deep ocean stations and that ^{210}Pb serves as a time dependent tracer for both advective transport and biological mixing of sediments.

0097-6156/82/0176-0331$07.75/0

Recent events since the industrial revolution have significantly altered the environmental conditions existing at many
locations of our planet. Man has profoundly accelerated the
erosion of the land due to clearing of forests and construction of
roads; he has contaminated the environment by emission of pollut-
ing particles and gases through the processes of winning metals
from ores, by the production of energy and by the disposal of his
wastes. Natural changes in the environment are continually
occurring by the changing climatic conditions and by episodic
events such as volcanic and tectonic activity. Dating of these
events over the past 200 years can give some insight on what we
can expect in the future. This dating can be accomplished using
the rhythmical accumulation of sediments in available repositories
such as lakes, estuaries, atolls, and deep ocean environments
together with radioactive tracer chronometers. The purpose of
this paper is to outline the methods and to illustrate the use of
radioactive chronometers in solving environmental problems which
have occurred over the past 150 years.

Experimental Methods

Assumptions in the ^{210}Pb Method.   The method for using decay
products of naturally occurring ^{226}Ra for sediment chronologies
was first outlined by Goldberg (1963). The decay chain of ^{226}Ra
proceeds through several short half-life radionuclides to long
lived ^{210}Pb as:

^{226}Ra $\longrightarrow$ ^{222}Rn $\longrightarrow$ ^{218}Po $\longrightarrow$ ^{214}Pb $\longrightarrow$ ^{214}Bi $\longrightarrow$ ^{214}Po
(t =1620 yr)   (3.8 d)   (3.05 m)   (26.8 m)   (19.7 m)

$(1.6 \times 10^{-4} s)$

^{206}Pb $\longleftarrow$ ^{210}Po $\longleftarrow$ ^{210}Bi $\longleftarrow$ ^{210}Pb
(138 d)   (5.0 d)   (22.35 yr)
(Stable)

The basic principle for ^{210}Pb dating is that gaseous ^{222}Rn is
emitted to the atmosphere from the lithosphere, surface waters and
airborne dust and there decays to ^{210}Pb. After formation in the
troposphere, ^{210}Pb becomes attached to aerosol particles which
reside in the atmosphere for only 30 days or less depending on
season, latitude, frequency of rainfall, size and altitude of the
aerosols, Nevissi et al., [17][1], Schell [26], and Poet et al.,
[19].
This continuous source of ^{210}Pb provides a widespread flux to
land and water surfaces. The distribution of ^{210}Pb at coastal and

[1]Figures in brackets indicate the literature references at the end
of this paper.

marine stations fits a distribution pattern in precipitation similar to weapons-produced ^{90}Sr with mid-latitudinal peaks in the concentration as shown in figure 1, Schell [26]. The deposition of ^{210}Pb on the surface ocean is reflected by the atmospheric input distribution as also shown in figure 1. Similarly, at locations where prevailing winds are from over the ocean and where continental land masses are small, e.g., at islands and at west coasts of continents, a similar latitudinal input from precipitation could be expected. However, at the mid- to Eastern regions of continents, the atmospheric input of ^{210}Pb would be higher, more irregular and would depend strongly on local or regional terrestrial rock and soil composition. This is shown in the atmospheric ^{210}Pb data summarized by Rangarajan et al., [22].

Organic rich soil horizons are efficient in scavenging ^{210}Pb from the atmosphere which subsequently may be transported mainly by soil erosion processes, Lewis [16]. Upon entering riverine systems, the ^{210}Pb is rapidly removed from the water, Rama et al., [21]; it is also removed rapidly in lakes and estuarine systems since it follows the natural detrital particles. Thus, most of the ^{210}Pb entering lakes and estuaries comes directly from atmospheric input across the atmosphere-water interface and descends rapidly to the bottom with the suspended particulate matter. From a knowledge of the present ^{210}Pb input to a region, the age of a sediment horizon which has been isolated from the contemporary input can be determined. By measuring the ^{210}Pb concentrations in several different sediment horizons of a core, the rates of sediment accumulation can be calculated from the concentration measured and the radioactive decay rate. Because ^{210}Pb is also found in situ from the small but finite amounts of terrigeneous ^{226}Ra present in the sediment, this supported amount must be subtracted from the total concentration to obtain the unsupported ^{210}Pb; the unsupported ^{210}Pb represents the amount added to the water column from atmospheric sources.

Often the expected change in ^{210}Pb concentration with depth (obtained from the logarithmic plot of unsupported activity as a function of the overlying mass of dry sediment accumulated) shows variations which are outside the analytical errors expected from the measurement of radioactive decay. Possible reasons for these discrepancies are:

1) Disturbance or incomplete recovery of the upper layers of sediment by the coring device.
2) In situ mixing of the upper sediment layers due to burrowing of filter feeding macroorganisms, gas bubbles, etc.
3) Post depositional mobility by diffusion or by microorganism transport in pore water.
4) Inhomogeneity due to presence of wood chips, pebbles, shells, etc.
5) Actual changes in sedimentation rates.
6) ^{210}Po not yet in equilibrium with ^{210}Pb in the upper sediment layers.

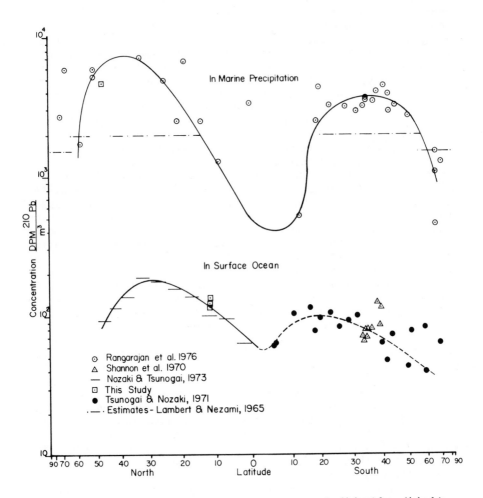

Figure 1. Global input of ^{210}Pb: (a) concentration in precipitation at marine stations; (b) concentrations in surface ocean water (26).

Sample Collection. Because of the importance placed on the upper layers of sediment, the devices used for collecting core samples must be designed so that neglibible mixing or losses occur in sampling. Hand collected cores, piston cores, and box cores are probably the most reliable collection devices; gravity core sampling is a convenient collection method but may not necessarily give reliable samples. For soft lake sediments a piston core designed by Donaldson [8] has been found satisfactory; for marine sediments, the box core designed by A. Soutar (Scripps Institution of Oceanography) has been used with success. A few gravity cores have been taken and used successfully when the water overlying the sediment is clear, indicating that no mixing has occurred. This latter method is the least desirable way to collect undisturbed cores.

Sample Processing. After collection in the field the sediment cores are cut into 0.5-, 1.0-, or 2.0-cm sections and placed into polyethylene bags. In the laboratory, the wet and dry weights are determined and a 2- to 3-g aliquant is spiked with 10.0 to 50.0 dpm of ^{208}Po (chemical yield tracer), leached and heated with concentrated HNO_3 and HCl for 4 hours, and centrifuged twice. The leachate is then evaporated to dryness and the residue converted completely to chlorides by repeatedly adding HCl and evaporating to dryness, Bruland [4] and Schell et al., [23]. Alternatively, after the ^{208}Po tracer addition and equilibration, the sediment sample is transferred into a quartz tube dehydrated and heated to 550-600 °C where the polonium is volatilized and condensed onto moist glass wool, Eakins and Morrison [11] and El-Daoushy [13]. The glass wool is then leached with HCl and the leachate evaporated to dryness. The residue from either method is dissolved in approximately 120 mL of 0.3 M HCl, and ascorbic acid is added to complexed iron in solution. The polonium isotopes are plated spontaneously at 65 °C onto a 2.2-cm diameter silver disk (one side coated with an insulating varnish, RedGlpt®) suspended in the solution. After plating the samples are measured for trace metals. The disks are rinsed, dried, and alpha-particle counted using Si surface-barrier detectors and pulse-height analysis for ^{208}Po and ^{210}Po. The ^{210}Po activity concentration is assummed to be identical to its parent ^{210}Pb concentration in sediment because of its short half life (138 days) and the resulting secular equilibrium. The supported ^{210}Pb concentration is determined by measuring ^{226}Ra concentrations in each core section, or by peeling the outer 3-mm layers of each core section, combining the outer layers, and gamma-ray counting of ^{214}Bi using Ge(Li) detectors and pulse-height analysis; the inner portion of each section is used for the ^{210}Po analysis. The Ge(Li) detector efficiency is determined by counting a known concentration of ^{226}Ra standard in the same geometry and density configuration as the sediment sample.

Results and Discussion

Lake Washington. The ^{210}Pb results from Lake Washington are based
on two replicate cores taken at 62 m using a piston corer near
Madison Park (the deepest part of the Lake, shown in figure 2).
Both cores contained a density anomaly at the 18-20 cm horizon
where a dense layer of silt was present. The origin of this layer
has been attributed to the erosion of littoral areas which
followed the lowering of the lake by about 2.4 m in 1916. At this
time, the Cedar River was diverted into the lake and a ship canal
and locks were constructed which connected the lake to Puget
Sound, Edmondson and Allison [12]. The data in figure 3 have been
obtained by averaging the ^{210}Pb and stable Pb values from both
cores after the silt band had been juxtaposed, Barnes et al., [2].
      The ^{210}Pb input from the atmosphere must have been constant
over the past 150 years due to the relative constancy in the
maritime climate (temperature and soil moisture influences the
radon emanation rate) and the resulting constancy in the input
source for ^{210}Pb. Therefore, the deviations from a single log-
linear relationship of the unsupported ^{210}Pb activity with the
dry mass of sediment accumulation must be due to some property
of the watershed. The three different relationships shown in
Figure 3 correspond to three sedimentation rates.
      The oldest of these sedimentation rates, Region I (150
$g \cdot m^{-2} \cdot yr^{-1}$ or 0.063 $cm \cdot yr^{-1}$) represents natural or the pre-
cultural sedimentation rates before 1889. Near this location in
the lake a volcanic ash layer overlain by 480 cm of sediment has
been found, Gould and Budinger [15]. Radiocarbon dating of
adjacent carbonaceous material determined that the ash layer was
deposited about 6700 years ago which would attribute it to the
Mt. Mazama (Crater Lake) eruption, Gould and Budinger [15] and
Powers and Wilcox [20]. Thus, the sedimentation rate during the
6600 years prior to 1890 (30 cm depth) was 0.068 $cm \cdot yr^{-1}$
{(480-30) cm/6600 yrs} and comparable to the rate determined by
the ^{210}Pb concentration profile below 30 cm.
      Between 1890 and 1902, the sedimentation rate identified as
Region II of figure 3 averaged 1800 $g \cdot m^{-2} \cdot yr^{-1}$ or 0.83 $cm \cdot yr^{-1}$,
i.e., greater than 10 times the pre-cultural rate. Historical
records and photographs show that by 1895 most of the land com-
prising the watershed had been logged and the suburbs of Seattle
had reached the lake shore. This rapid land development in the
watershed occurred about the turn of the century.
      Since 1916 the sedimentation rate, Region III of Figure 3,
has averaged 644 $g \cdot m^{-2} \cdot yr^{-1}$ or 0.3 $cm \cdot yr^{-1}$ or about 5 times the
pre-cultural rate. The diversion of the Cedar River (average
flow of 20 $m^3 \cdot s^{-1}$ into the lake in 1916 provided the water necessary to operate the ship and canal locks and contributes an
estimated 4-5 x 10^7 $kg \cdot yr^{-1}$ of allochthonous material, Crecelius
[7]. This riverine sediment input would contribute to the greater

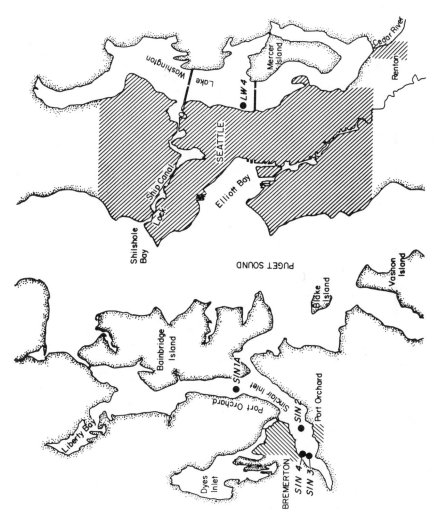

Figure 2.   Location of sampling stations at Lake Washington and Sinclair Inlet, Washington.

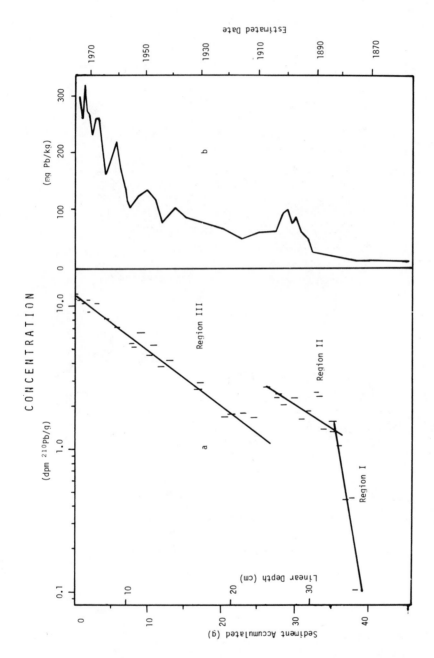

Figure 3.   Concentration profiles of (a) ²¹⁰Pb and (b) stable Pb in sediment cores from Lake Washington.

sedimentation rate at the Madison Park station even though its source is some 7 km south.

In addition to the chronometers of land clearing and construction, the input of stable Pb also serves as a cultural chronometer in fixing dates of sediment layers at Lake Washington. A major souce of Pb has been thought to be a smelter located about 35 km upwind of the lake. This installation produced lead bullion from its inception in 1890 until 1913 when it became a Cu smelter; current emissions are 40 percent PbO, Crecilius and Piper [7]. However, recent results indicate that coal burning may have been a major source for Pb between 1890 to 1910 (Barnes, 1981, personal communication). Since 1923, when Pb additives were combined with internal combustion engine fuel, the increases in the Pb concentrations in Lake Washington sediments closely parallel increases in local gasoline consumption, Schell and Barnes [24].

The stable Pb concentration profile shown in figure 3 gives a precultural background of less than 10 mg·kg until about 1885 followed by a rapid increase during the 1890's. The decline in Pb concentrations from about 1902-1916 appears to be the result of dilution by the silt from land erosion rather than from a reduction in external sources. From 1930-1973, when the cores were collected, the levels of Pb increased about threefold; the increase from pre-cultural levels being about 30 times. These data are consistent with the chronology developed by ^{210}Pb measurements and illustrate the sensitivity and accuracy possible in determining man's local pollution as recorded by lake sediments. If man had not changed the watershed, there would be about 26 cm less sediment deposited at the deepest part of Lake Washington.

Sinclair Inlet.   Sinclair Inlet is located at 47°33'N, 122°38'W, and is a fjord-like estuarine inlet separated from the central basin of Puget Sound by Bainbridge Island (figure 2). It is approximately 1.5 km wide, and has no major freshwater input from rivers. The city of Bremerton (39,000 people), Port Orchard (4,000 people) and the major U. S. Naval Shipyard on Puget Sound are located on its shoreline. Most of the inlet has an average depth of 10 m except at the mouth where the depth is 16 m. The shipyard is the major construction and maintenance facility for the U. S. Navy in the Northwest and has been in operation since 1891, with the facility's most extensive use being World War II. The considerable industrial activity and the presence of a growing population at the inlet suggest that pollution, particularly by trace metals, might be significant. By combining information on the sedimentation rates and the input of trace metals, the pollution history may be determined from the sediment cores.

A series of gravity cores was collected inside the inlet and at a control station located some 5 km north of the inlet in Port Orchard Passage, as shown in figure 2. The sedimentation rates as determined by ^{210}Pb are shown in figure 4a where a least squares fit of the data gives an average sedimentation rate of

1200-1800 $g \cdot m^{-2} \cdot yr^{-1}$ or 0.2 to 0.3 $cm \cdot yr^{-1}$ at the four stations. The ^{210}Pb concentration profile at Station SIN 4 illustrates a non-systematical accumulation of sediments due to disturbance either by biological processes or by man's activities. Since this station is located near the shipyards at a depth of 10 m the disturbance of sediment could be due to the large ships' propellers and/or to biological activity. Stations SIN 1A (30.5 m), SIN 2 (13.7 m) and SIN 3 (10.0 m) appear to give average sediment accumulation rates and the ^{210}Pb concentration profiles can be used to give dates. Consequently, it is possible that by measuring more sections of the cores, the fine structure of the sedimentation history could be determined for the inlet. However, because of the limitations imposed on the study, only the average sedimentation history was determined.

The use of the sedimentation rates from the ^{210}Pb measurements has permitted the fixing of time of input of the trace metals Cu, Pb, Zn, Ni, and Cr. Figures 4b-4e also show the history of trace metals input at the four stations. The "control" station SIN 1A (figure 4b) shows that an increase from pre-cultural levels has occurred and that the time of this small increase corresponds to the large increases of trace metals in Sinclair Inlet. The circulation system transports a fraction of the metal pollutants to the "control" station. The average metals concentration of sediment at the four stations before the impact of man on Sinclair Inlet (1880) (in ppm) is: Pb—10, Cu—10, Zn—52, Ni—26, Cr—32. Release of metals into the inlet from mans' activities has increased these environmental levels by the following factors: Pb—13, Cu—8, Zn—5.7, Ni—1.9, Cr—2.2. The highest level of Pb was deposited in the late 1930's to mid 1940's at Station 4 (nearest to the shipyard) corresponding to the time just prior to and during World War II.

The elements deposited within the sediment matrix show that mobilization processes may be occurring in the upper layers. At Station SIN 3, figure 4d for example, the element deposited ($\mu g \cdot cm^{-2}$) in the topmost layers decreases, often much more than in the concentration ($\mu g \cdot g^{-1}$). This may be due to organic matter decomposition and/or to environmental chemical reactions of solubility and precipitation of the given element. The metal must have been removed rapidly from the water column since the sediment concentration is shown to decrease rapidly with distance from the shipyard (Stations SIN 3 and SIN 2). Lead may not be mobilized significantly after deposition since any diffusion in the pore water would tend to "smooth" the concentration profile with time.

The maximum deposition of Zn, Cu, Ni, and Cr appear to be at about the mid 1960's, whereas the maximum for Pb appears to be just following World War II. The shape of the deposition profile curve for Zn, Ni, Cr, and possibly Cu indicates that these elements may be migrating downward in the sediments, possibly in the pore water. The decrease in deposition at the sediment-water interface indicates diffusion of the metals out of the sediment into the overlying water column. Diffusion resulting from bacterial activity and/or chemical solubility reactions would tend to smooth the deposition profile with time giving a maximum at the surface where the loss occurs. By a more detailed study of the sediment geochemistry, (oxidation-reduction reactions and pH), significant new information may be obtained on the mobilization and diffusion coefficients of trace metals in these natural sediment systems where the time of input can be determined.

Bikini Atoll Lagoon.   Bikini Atoll is located in the Central Pacific Ocean at 11°N, 122°E and was used extensively for nuclear weapons testing from 1946 to 1958. In biogeochemical studies of the radionuclides present in the ecosystem, one of the concerns was the transport of radioactive coral debris from the localized bomb crater areas to deposition at a wide area of the lagoon floor, Schell et al., [28]. Transport of radionuclides by the circulation system of the lagoon and subsequent uptake by the abundant life forms indigenous to the atoll would be greater if re-distribution rather than burial of the contaminanted debris were occurring.

Figure 5 shows the profile of radionuclide concentration distribution with depth in a gravity core collected near the Bravo Crater at Station B-2, Schell and Watters [25]. The radionuclide concentration in the top 12 cm appears to be significantly different from the lower layers Schell et al., [28]. The unsupported ^{210}Pb concentration measured in the upper 12 cm gave an effective sedimentation rate of 0.58 cm·yr^{-1} (correlation 0.98) obtained by calculating a linear regression of the values. Thus, the age of the 11 cm section was calculated to be 1955 (11 cm/.58 cm yr = 19 yr), a date consistent with the period of testing at Bikini, Schell et al., [28]. The implication of these data is that two different processes were responsible for the radionuclide profile in the 40 cm of sediment sampled at this station (B-2). Rapid accumulation of sediments containing no unsupported ^{210}Pb but now containing significant levels of bomb-produced radionuclides occurred at the depth interval from 40 cm to 11 cm; above 11 cm slow accumulation is occurring from re-distribution of coraline debris. The rapid accumulation was probably part of the excavation of the island where the bomb was detonated. (A crater, 40 m deep with a diameter of 500 m, now exists from the Bravo detonation in 1954). The radionuclide profile observed below 11 cm may be due to diffusion and/or microorganism transport of small particles containing the several radionuclides.

The ^{210}Pb tracer data have shown: 1) the year of the detonation which affected this station, and 2) that material is being eroded in one area and transported for deposition at another area by the currents in the lagoon; and 3) that this has occurred at a constant rate between 1954 and 1972.

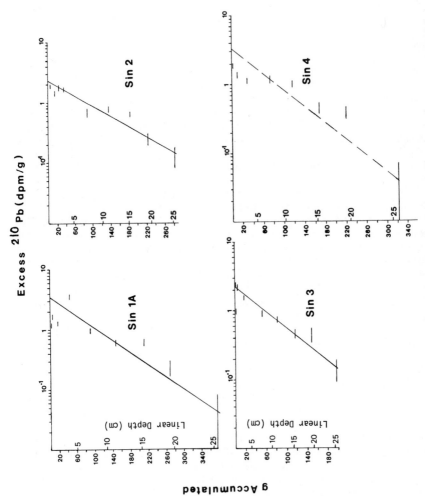

Figure 4a.   Sedimentation rates at Stations SIN 1A (control station), SIN 2, SIN 3, and SIN 4 in Sinclair Inlet, Washington.

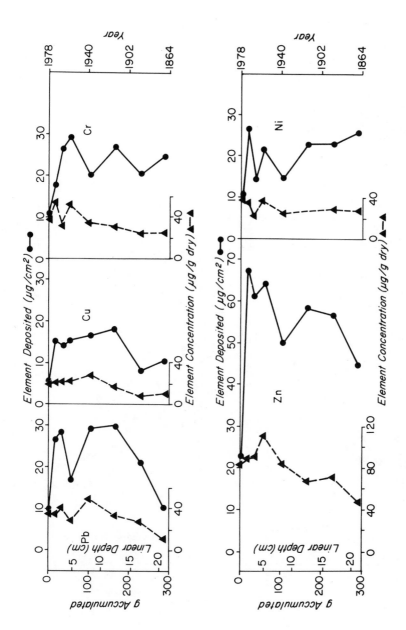

Figure 4b.   Concentration profile of the trace metals Pb, Cu, Cr, Zn, and Ni and the estimated dates of sediment core sections from Sinclair Inlet, Washington (control Station SIN 1A).

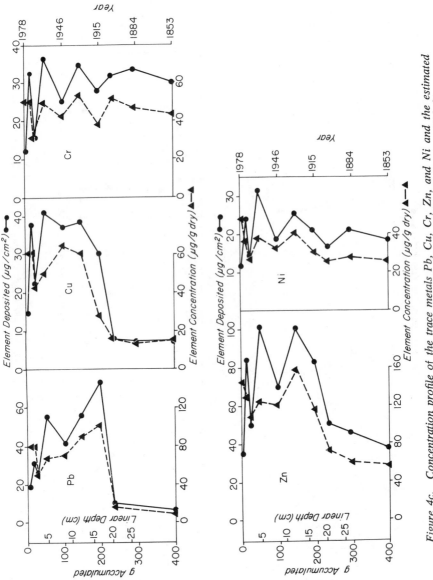

Figure 4c. Concentration profile of the trace metals Pb, Cu, Cr, Zn, and Ni and the estimated dates of sediment core sections from Sinclair Inlet, Washington (Station SIN 2).

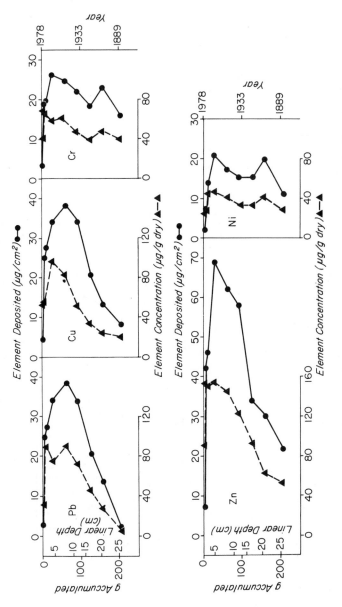

Figure 4d. Concentration profile of the trace metals Pb, Cu, Cr, Zn, and Ni and the estimated dates of sediment core sections from Sinclair Inlet, Washington (Station SIN 3).

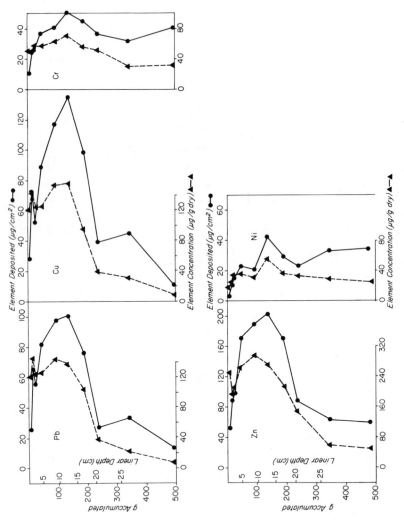

Figure 4e. Concentration profile of the trace metals Pb, Cu, Cr, Zn, and Ni and the estimated dates of sediment core sections from Sinclair Inlet, Washington (Station SIN 4).

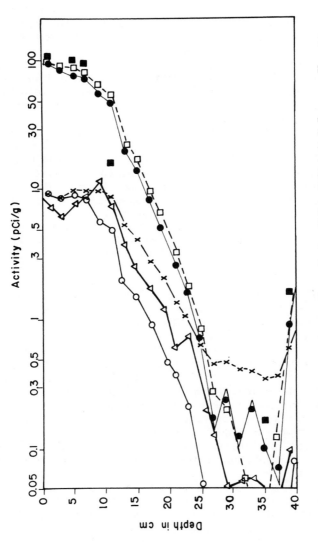

Figure 5.   Radionuclide concentration profile at Station B-2 in Bikini Atoll Lagoon. Key: ×, ^{60}Co; ○, ^{137}Cs; □, ^{155}Eu; △, ^{207}Bi; ●, ^{241}Am; *and* ■, $^{239,240}Pu$.

Deep Ocean. As part of two projects to evaluate possible radio-nuclide leakage from waste disposal canisters deposited at depths of 900 to 4000 m, sedimentation mixing, and dating of layers have been attempted using the ^{210}Pb methods. One site is located in the eastern Pacific Ocean near the Farallon Islands at 37°30'N, 123°17'W, 60 km off San Francisco (figure 6a) and the other site is located in the western Atlantic Ocean at the mouth of the Hudson Canyon at 38°30'N, 72°06'W, 350 km off New York City (figure 6b). Because of the low sedimentation rates expected at the outer continental shelf (2 - 10 cm/10^3 yr), ^{210}Pb measurements were not expected to yield information on sedimentation rates at these deep stations. However, ^{210}Pb was expected to yield informa-tion on the depth of biological mixing (bioturbation) which would occur within the time period of 200 years.

Box cores were collected at two stations at the Farallon Islands site and at seven stations at the Hudson Canyon site. Cores were also collected by manned submersibles; the PYCES IV in the Pacific and the DMV ALVIN in the Atlantic. After each box core was taken, the water overlying the sediment was carefully removed by a siphon and several sub-cores of 66.7 mm diameter were collected for the various participants in the project. The submersible cores were collected by the mechanical arm of the submersible which pushed a 66.7 mm core liner into the sediments. On deck the cores were sectioned into 2 cm fractions for gamma radionuclide measurements and into 1 cm fractions for ^{210}Pb mea-surements.

The results of the unsupported ^{210}Pb measurements are shown in figure 7 for Station 13A (Farallon Island site) at a depth of 1045 m, Schell and Sugai [27]. The least squares fit of the data gives a sedimentation rate of 789 ± 45 g·m^{-2}·yr^{-1} or .112 ± .011 cm·yr^{-1}. The depth of physical mixing is indicated by the unsupported ^{234}Th concentrations which was found in the upper 4 cm layers. Since ^{234}Th has a half life of 24 days, its presence in the core sections indicates rapid mixing in the upper 4 cm, i.e., within 3 months, Aller and Cochran [1]. Below this depth, apparent mixing is not observed by the ^{234}Th tracer concentra-tions. Also shown in figure 7 is the concentration profile of ^{210}Pb at Station 39 Farallon Island site at a depth of 1470 m. Two effective sedimentation rates are shown; below 7 cm the rate is 274 ± 11 g·m^{-2}·yr^{-1} or 0.035 ± .002 cm·yr^{-1} and above 7 cm the rate is 1717 ± 26 g·m^{-2}·yr^{-1} or 0.28 ± .01 cm·yr^{-1}. Table 1 shows the sedimentation rates and least squares fit of ^{210}Pb values for the two sections of the core. The upper layers may be due to bioturbation but the lower layers in both these cores should be representative of the real sedimentation rates which are much higher than the expected rates of 2-10 cm·10^{-3}·yr^{-1}. An alterna-tive explanation is that different mixing processes are occurring in the two sections of the sediments. Further verification is needed.

Table 1. Sedimentation Rates and Curve Fitting of ^{210}Pb Measurements in Cores Collected at the U. S. Radioactive Waste Disposal Sites Near the Farallon Islands 60 km off San Francisco and at the Hudson Canyon, 350 km off New York City.

Station	Water Depth (m)	Core Depth Range (cm)	Sedimentation Rate (cm·yr ^{1}±SD)	Equation of Line
			Farallon Islands Site	
13A	1043	0 - 18	0.112 ± 0.011	log y = 1.806 ± .140 - (0.120 ± .012) x
39	1469	0 - 5	0.278 ± 0.008	log y = 1.563 ± .005 - (0.048 ± .001) x
		5 - 12	0.035 ± 0.002	log y = 3.552 ± .187 - (0.386 ± .018) x
			Hudson Canyon Site	
3	4015	0 - 8	3.550 ± 5.340	log y = 1.615 ± .032 - (0.004 ± .006) x
		8 - 20	0.105 ± 0.027	log y = 2.847 ± .578 - (0.129 ± .036) x
4	3950	0 - 8	0.199 ± 0.058	log y = 1.689 ± .118 - (0.068 ± .020) x
		8 - 14	0.049 ± 0.003	log y = 3.471 ± .170 - (0.278 ± .015) x
5	3945	0 - 4	0.023 ± 0.001	log y = 1.906 ± .072 - (0.589 ± .031) x
7	3885	0 - 2	0.010	log y = 2.223 - 1.406 x
8	3995	0 - 6	0.055 ± 0.007	log y = 1.889 ± .147 - (0.244 ± .029) x
		6 - 17	0.153 ± 0.040	log y = 1.657 ± .268 - (0.088 ± .023) x
15	3655	0 - 10	0.127 ± 0.020	log y = 1.244 ± .089 - (0.107 ± .016) x
813	3970	0 - 6	0.401 ± 0.735	log y = 0.935 ± .241 - (0.037 ± .061) x
		6 - 13	0.043 ± 0.004	log y = 3.326 ± 2.880 - (0.318 ± .027) x
816	2812	0 - 8	0.074 ± 0.007	log y = 2.007 ± .102 - (0.182 ± .017) x
		8 - 17	0.127 ± 0.025	log y = 1.538 ± .244 - (0.094 ± .018) x

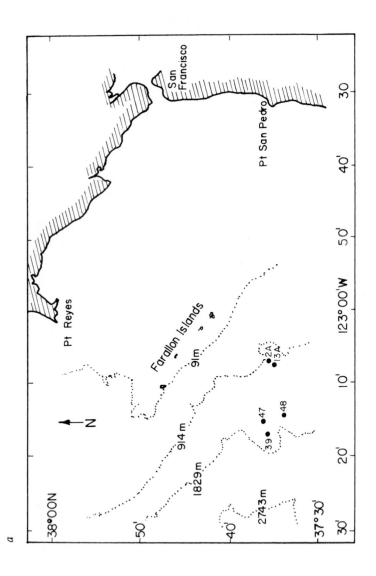

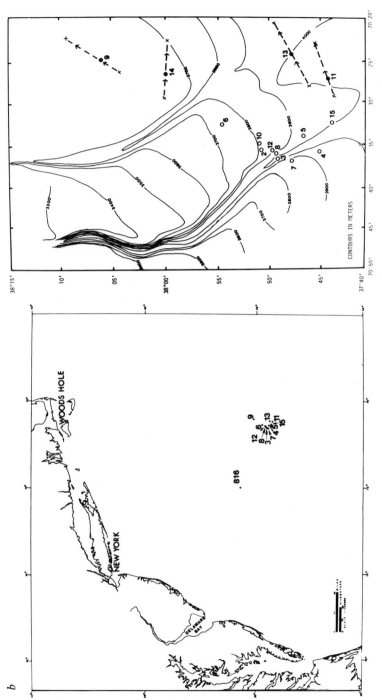

Figure 6. Location of sampling stations at nuclear waste disposal sites: (a—facing page) Pacific and (b) Atlantic.

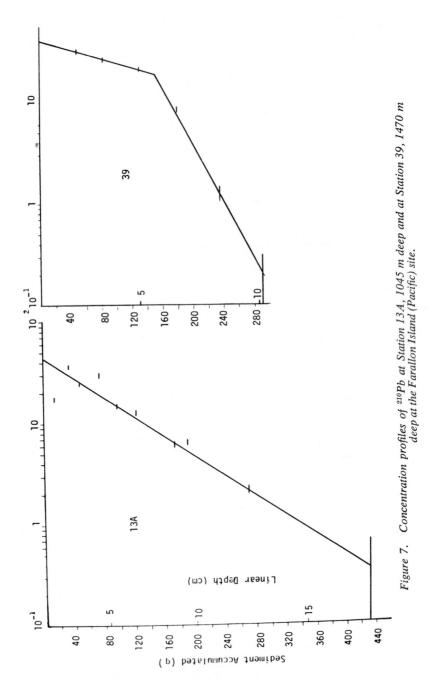

Figure 7.   Concentration profiles of ^{210}Pb at Station 13A, 1045 m deep and at Station 39, 1470 m deep at the Farallon Island (Pacific) site.

One of the most interesting areas where ^{210}Pb techniques have been used is at the nuclear waste disposal site, 350 km off New York at the mouth of the Hudson Canyon (figure 6b). Measurements of the concentrations of $^{239,240}Pu$ and other radionuclides which could possibly be leaking from disposal canisters was the primary mission of the study. Figures 8a and 8b show the ^{210}Pb concentration profile with depth for the seven box cores collected at a depth of 4000 m. Table 1 also shows the sedimentation rates and least squares fit of the ^{210}Pb values for different portions of the depth profile at the Atlantic site. The sedimentation rate is shown to vary from 0.010 to 0.199 cm·yr^{-1} at the several stations. It is apparent that the ^{210}Pb measurements give sedimentation rates which are five to ten times greater than those reported for the sediments of the nearby continental shelf and rise, Turekian [29] and Doyle et al., [9]. In question is the possibility that these high sedimentation rates are due to biological mixing of ^{210}Pb and may not be representative of real sediment accumulation rates. The interpretation of the data from this unique set of box cores from the deep ocean area can give significant new information on processes occurring in this most difficult sampling environment.

The sedimentation history of the upper Hudson Canyon has been measured at the 430 m depth level by Drake et al., [10]. The average sedimentation rate over the past 6,300 years, as determined by ^{14}C dating of a piston core, has been an average of 70-80 cm/10^3 yrs; the past 2,000 years appear to have a sedimentation rate of about 150 cm/10^3 years, Drake et al., [10].

In the outer Hudson Submarine Canyon, active erosion of the canyon walls and floor has been observed at depths of 2900 m and 3600 m by Cacchione et al., [5]. Visual observations and shallow cores collected using the ALVIN submersible revealed that bottom currents and benthic organisms have reworked the surface sediments of the canyon floor. Cacchione et al., [5] speculates that the active erosion by strong bottom currents is probably an episodic process because current measurements obtained by the ALVIN and by two moored current meters during the four-day dive period were down-canyon and consistently weak (speed less than 15 cm·s^{-1}). The steady seaward flow, if representative of longer durations, would produce a very significant material flux to deeper zones. An extremely thin cover of unconsolidated floculent sediment was observed on each dive by Cacchione which was in marked contrast to the thick accumulations of fine grained sediments observed by Drake [10] in the shallower sections of the Hudson Canyon. Abundant salp bodies were observed to be rolling along the bottom down-canyon. Continued down-canyon transport of these organisms could contribute large quantities of organic matter as a food source and a cementing agent to the deep sea.

Episodic events occur periodically and may be restricted to given small areas. These events may result from gradual erosion which ultimately reaches the slope instability of the canyon walls

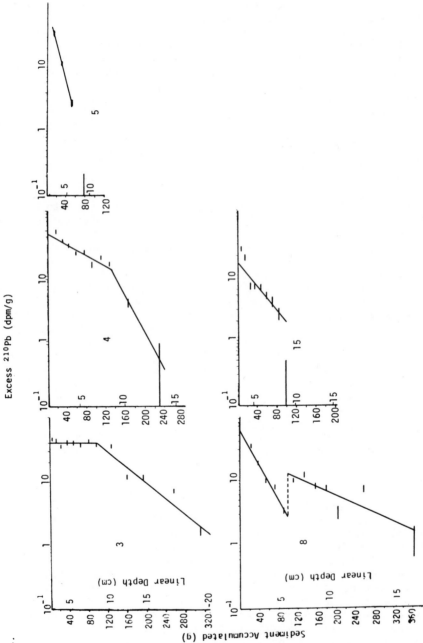

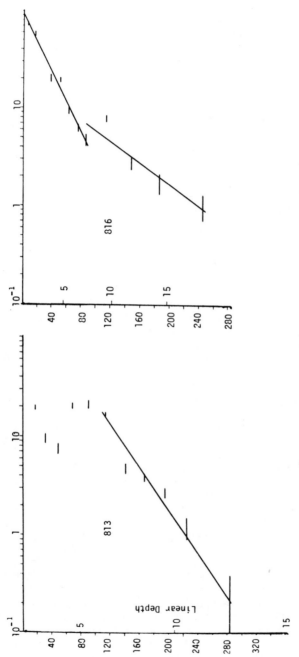

Figure 8.   *Concentration profiles of* ^{210}Pb *for 3800-m deep stations at the Hudson Canyon Nuclear Waste Disposal Site (Atlantic): (a—facing page) Stations 3, 4, 5, 8, and 15; and (b) Stations 813 and 816.*

and a slide occurs. The material is then transported down-canyon. The episodic events are shown at Stations 8 and 816 (Figure 8a and b) where approximately 100 ^{210}Pb years ago an event occurred which decreased the sedimentation rates at Station 8 from 153 to 55 cm/10^3 yrs and at Station 816 from 127 to 74 cm/10^3 yrs.

Deposition Inventory. To determine whether disposal site sediments contained ^{239,240}Pu in excess of that attributable to atmospheric fallout over the past 30 years, ^{239,240}Pu inventory to 1978 was calculated for each core. The total amount of ^{239,240}Pu found at each 2 cm interval was added together and divided by the area of the core, 34.9 cm^2, Schell and Sugai [27]. Values for the outer Hudson Canyon (Table 2) are compared with estimates made by Noshkin and Bowen [18] on the integrated deposition from fallout at this latitude. A wide range of values (0.09 - 0.4 mCi·km^{-2}) were obtained at the several stations.

The ^{210}Pb inventory in the sediments was calculated in a similar way to the ^{239,240}Pu method described previously, but 1 cm increments were used in determining the integrated deposition. The deposition rate of the ^{210}Pb for each core was determined as the product of the integrated deposition and the radioactive decay constant, ($\lambda = .0311 \cdot yr^{-1}$). Since the unsupported ^{210}Pb ($t_{1/2}$ of 22.35 yrs) would decay to background levels in 200 yrs, the time of mixing of the upper sediment layers can be estimated. The value for the deposition rate was expected to be constant, or reasonably constant for the local area at a depth of 4000 m. However, as shown in Table 2, variations of between 0.5 to 5.9 dpm·cm^{-2}·yr^{-1} were observed.

Benninger et al., [3] found values of about 1 dpm·cm^{-2}·yr for the deposition rate of ^{210}Pb in soils and sediments near New Haven, Connecticut and Long Island. Recent measurements in Puget Sound at Sinclair Inlet by W. R. Schell (unpublished data), indicated that the deposition rate was about 0.35 dpm·cm^{-2}·yr. Thus, to explain the very high and variable values for the ^{210}Pb deposition rate at the deep ocean stations, one must propose a mechanism whereby material from the topmost layers of sediments near the Atlantic Disposal Site is transported and re-deposited at these stations.

By combining the findings of Cacchione, Drake and the results reported here, a coherent model can be proposed to explain the deposition inventory of the radionuclides. The down-canyon current transports large quantities of sediment toward the radioactive waste disposal site at 4000 m. Within the upper canyon, fine material is transported the furthest. Near the mouth of the canyon, sediment erosion of the walls occurs due to the down-canyon currents meeting a proposed opposing on-shore bottom current. The eroded material from the walls is transported and the finer material is deposited in eddies formed where the two currents meet.

Table 2.   Deposition of ^{210}Pb and $^{239,240}Pu$ in Sediment Cores
Collected at the U. S. Radioactive Disposal Site Near
the Farallon Islands, 60 km off San Francisco and at the
Hudson Canyon, 350 km off New York City.

No.	Station Location	Depth (m)	^{210}Pb Deposition Rate $dpm \cdot cm^{-2}yr$	$^{239,240}Pu$ Integrated Deposition $mCi \cdot km^{-2}$ (30 yr)
			Farallon Island Box Cores	
13A	37°38.10'N 123°08.00'W	1043	2.9	4.3
39	37°38.10'N 123°08.00'W	1469	3.5	1.3
			Hudson Canyon Box Cores	
3	37°49.30'N 70°36.71'W	4015	5.9	.40
4	37°45.03'N 70°35.75'W	3950	3.8	.24
5	37°46.75'N 70°34.01'W	3945	0.7	.12
6	37°54.65'N 70°32.69'W	3740	—	.20
7	37°48.10'N 70°37.11'W	3885	0.5	.09
8	37°49.79'N 70°36.13'W	3995	2.9	.31
15	37°04.24'N 70°26.38'W	3655	0.6	.11
	39°02'N 42°36'W	4810	—	0.17 ± .02[a]
	"	4810	—	(0.40)[b]

[a]Collection made in 1969, Noshkin and Bowen (1973).

[b]Extrapolated to 1978 from Noshkin and Bowen (1973) estimate of
fallout deposited at 4000 m on their sinking rate model.

If material which contains nutrients, ^{210}Pb and $^{239,240}Pu$ on particulate matter is raining down from the water column to the sediment and is initially deposited on ridges at the ocean bottom, currents may subsequently transport part of this material and deposit it at other locations. The low deposition rate found at Stations 5, 7, and 15 (Table 2) may be due to the initial deposition on ridges at the bottom where only a small fraction of the material is retained in the topmost layers. The remainder of the material is transported to settling zones of the submerged valley such as at Stations 3, 4, 8, 813, and 816.

At Station 3, a mixed zone occurs to a depth of 12 cm as indicated by the ^{210}Pb profile. At this station, the highest deposition rate of ^{210}Pb is found compared to the other eight sampling stations. If this ^{210}Pb profile represents biological mixing, the station should be rich in benthic organisms because of the deposition of particulate matter including nutrients from a wide area of bottom. This station also has the highest integrated deposition of $^{239,240}Pu$ at 0.4 mC·/km^{-2} and one would expect that this higher concentration could be from the waste disposal canisters in this area. However, it is possible that the $^{239,240}Pu$ and ^{137}Cs from atmospheric fallout, like ^{210}Pb, is being concentrated at this area and is mixed by biological organisms to a depth of 12 cm. Thus, the high levels of $^{239,240}Pu$, ^{137}Cs, and ^{210}Pb require that physical or biological mixing occurs and that additional material is being transported advectively and deposited at this station. The high concentrations found may not necessarily be due to leakage from waste disposal canisters.

According to Dr. D. Rice (California State University) (personal communication), the biological species diversity shows that at the stations which have high ^{210}Pb and $^{239,240}Pu$ values for sediment deposition, namely Stations 8, 4, and 3, there were 25, 20, and 15 species, respectively; at the low sediment deposition Stations 15, 7, and 5, there were 6, 6, and 10 species, respectively. These biological data do not disagree with the advective mechanism for sediment transport proposed to explain the high sediment deposition data for ^{210}Pb, ^{137}Cs, and $^{239,240}Pu$. The organisms would be more concentrated in regions where the available particulate matter, including nutrients, would be deposited. Thus, the higher levels of radionuclides at certain stations may not be due to leakage of radioactive wastes from the canisters but to redistribution of particulate matter deposited from the water column.

The $^{239,240}Pu$ can be accounted for as being from nuclear weapons-produced fallout over the past 30 years, and not from the waste disposal canisters. The mechanism must be by transport and redistribution of the fine floculent sediment at the sediment-water interface. Because of the significant amount of sediment which is being transported down-canyon, any leakage from the canisters would probably be scavenged locally by the environmental sediment material.

Conclusions

Dating of recent (200 year) events in a wide variety of fresh and salt water environments has been made by measuring natural levels of unsupported ^{210}Pb concentrations in sections of sediment core samples. If the atmospheric ^{210}Pb input has been determined, a single or few layers can be used to find the age of deposition at a given depth in shallow waters. However, a series of measurements of core sections can better define the sedimentation rates. Where man's activities or natural events have altered the deposition history, the ^{210}Pb profile has been used to determine the magnitude of the changes in the sedimentary regime and to define the time when these changes occurred.

Increases in Pb concentrations of 30 times have been measured in Lake Washington caused mainly through the burning of leaded fuel in internal combustion engines. Significant increases in the concentrations of Zn (6 times), Pb (13 times), and Cu (8 times) have been introduced by man's activity near Sinclair Inlet, probably as a result of shipyard's activities.

By measuring other elements in the same core profiles, their diffusion as compared to ^{210}Pb have been observed. Biological mixing and possible sedimentation rates have been determined from the concentration profiles, assuming that the ^{210}Pb tracer is fixed strongly to the sediment particles at the deep ocean stations. The ^{210}Pb tracer of sediments has been shown to compare times when anthropogenic events occurred with times obtained from the radioactive chronology in fresh, estuarine, and marine waters.

These studies introduce and utilize a natural biogeochemical tracer of sedimentary processes. The use of ^{210}Pb to trace deep ocean currents where flocculent material is transported in one region and deposited in another may provide valuable information on deep sea transport. However, more work is required to verify these initial findings. Additional work also is needed to evaluate mechanisms responsible for the differences in diffusion of certain elements in sediments compared to lead and to identify if two types of biological mixing may indeed be responsible for the high sedimentation rates found in the deep ocean.

I would like to acknowledge the assistance given by Robert S. Barnes for the Lake Washington Studies, by the students in 1979, 1980 Fisheries 477, 478 classes for the Sinclair Inlet studies and by Susan Sugai and Ahmad Nevissi for the radiochemical measurements in the ocean studies.

Literature Cited

[1] Aller, R. C., and Cochran, J. K., $^{234}Th/^{238}U$ disequilibrium in nearshore sediment: particle reworking and diagenic time scales, Earth and Planet. Sci. Lett., 29, 37-50 (1976).

[2] Barnes, R. S., Birch, P. B., Spiridakis, D. C., and Schell, W. R., Changes in the sedimentation history of lakes using ^{210}Pb as a tracer of sinking particulate matter, In: Isotope Hydrology - 1978, IAEA-STI/PUB/493, 875-8981 (1978).

[3] Benninger, K. E., Lewis, D. M., and Turekian, K. K., The use of natural Pb-210 as a heavy metal tracer in the river - estuarine system, Marine Chem., 12, 202-210 (1975).

[4] Bruland, K. W., Lead-210 Geochronology in the Coastal Marine Environment, PhD Dissertation, University of California, San Diego (1974).

[5] Cacchione, D. A., Rowe, G. T., and Malakoff, A., In: Sedimentation in Submarine Canyons, F. Stanley and G. Kelling, eds., Dowden, Hutchinson & Ross, Stroudsburg, Pa., Chapt. 4, (1977).

[6] Crecelius, E. A., Arsenic geochemical cycle in Lake Washington, Limnol. Oceanog., 20, 441-451 (1975).

[7] Crecelius, E. A., Piper, D. Z., Particulate lead contamination recorded in sedimentary cores from Lake Washington, Seattle, Environ. Sci. Technol., 6, 274-278 (1973).

[8] Donaldson, J. R., The Phosphorous Budget of Iliaamna Lake, Alaska, as Related to the Cyclic Abundance of Sockeye Salmon, PhD Dissertation, University of Washington, Seattle (1967).

[9] Doyle, L. J., Woo, C. C., and Pilkey, O. H., Sediment flux through the inter canyon slope areas: U. S. Atlantic Continental Margin, Abst., Geol. Soc., AW Prog., 8, 8-843 (1976).

[10] Drake, D. E., Hatcher, P., and Keller, G., Suspended particulate matter and mud deposition in the upper Hudson submarine canyon, F. Stanley and G. Kelling, eds., Dowden, Hutchinson & Ross, Stroudsburg, Pa., Chapt. 3 (1977).

[11] Eakins, J. D., and Morrison, R. T., United Kingdom Atomic Energy, Auth, Harwell, AERE-R8475 (1976).

[12] Edmondson, W. T., and Allison, D. E., Recording densitometry of X radiographs for the study of cryptic laminations in the sediments of Lake Washington, Limnol. Oceanog., 15, 138-144 (1970).

[13] El-Daoushy, M. F. A. F., The determination of ^{210}Pb and ^{226}Ra in lake sediments and dating application, Uppsala, Sweden report UUIP-979 (1976).

[14] Goldberg, E. D., Geochronology with lead-210, In: Radioactive Dating IAEA STI/PUB, 68, 121-131 (1963).

[15] Gould, H. R., and Budinger, T. F., Control of sedimentation and bottom configuration by convection currents, Lake Washington, J. Mar. Res., 17, 183-198 (1958).

[16] Lewis, D. M., The use of ^{210}Pb as a heavy metal tracer in the Susquehana River system, Geochem. et Cocmochin. Acta, 41, 1557-1564 (1977).

[17] Nevissi, A., Beck, J. N., and Kuroda, P. K., Long lived radon daughters as atmospheric radioactive tracers, Health Phys. 27, 181-188 (1974).

[18] Noshkin, V. E., and Bowen, V. T., Concentrations and distributions of long lived fallout radionuclides in open ocean sediments, In: Radioactive Contamination of the Marine Environment, IAEA - STI/PUB, 313, 631-686 (1973).

[19] Poet, S. E., Moore, H. E., and Martell, E. A., Lead-210, Bismuth-210, and polonium-210 in the atmosphere: accurate radio measurement and application to aerosol residence time determination, J. Geophys. Res., 77, 6515-6527 (1972).

[20] Powers, H. A., Wilcox, R. E., Volcanic ash from Mount Mazama (Crater Lake) and from Glacier Park, Science, 144, 1334-1336 (1964).

[21] Rama, M. Koide, and Goldberg, E. D., Pb-210 in Natural Waters, Science, 134, 98-99 (1961).

[22] Rangarajan, C. S., Gopalakrishnan, S., and Eapen, C. D., Global variation of lead-210 in surface air and precipitation, USAERDA, Health and Safety Laboratory Environmental Quarterly Report, HASL-298, UC-11, I-63-82 (1976).

[23] Schell, W. R., Jokelo, T., and Eagle, R., Natural ^{210}Pb and ^{210}Po in a marine environment, IAEA-STI/PUB, 313, 701-711 (1973).

[24] Schell, W. R., and Barnes, R. S., Lead and Mercury in the aquatic environment of Western Washington State, In: Aqueous Environmental Chemistry of Metals, A. J. Rubin, ed., Ann Arbor Science, Ann Arbor, Michigan, 129-165 (1974).

[25] Schell, W. R., and Watters, R., Plutonium in aqueous systems Health Phys., 29, 589-597 (1975).

[26] Schell, W. R., Concentrations, Physical-Chemical states and mean residence times of ^{210}Pb and ^{210}Po in marine and estuarine waters, Geochem. Cosmochim. Acta, 41, 1019-1031 (1977).

[27] Schell, W. R., and Sugai, S., Radionuclides at the U. S. Radioactive waste disposal site near the Farallon Islands, Health Physics, 39:475-496 (1980).

[28] Schell, W. R., Lowman, F. G., and Marshall, R. P., Geochemistry of the Transuric elements at Bikini Atoll, In: Transuranic Elements in the Environment, W. C. Hanson, ed., DOE/TIC 22800, 541-577 (1980).

[29] Turekian, K. K., Oceans, Prentice Hall, Englewood Cliffs, New York, 120 (1968).

RECEIVED July 23, 1981.

Deep-Sea Sedimentation:
Processes and Chronology

S. KRISHNASWAMI and D. LAL

Physical Research Laboratory, Navrangpura Ahmedabad 380 009, India

The application of naturally produced and artificially injected radionuclides to determine the chronology of marine sedimentary processes is discussed in this review, with particular reference to models describing their transport to the sea floor and distribution within the sediment pile. The particle-reactive radionuclides belonging to the U-Th series, cosmic-ray produced and artificially injected nuclides have proven useful in this pursuit. Studies on the radiogenic descendants of ^{238}U and ^{232}Th (^{234}Th, ^{230}Th, ^{210}Pb and ^{228}Th) in sea water reveal that the particle reactive radionuclides are removed to sediments from overlying water in short time scales, 10-100 years; however, their scavenging mechanisms have not yet been clearly identified. The removal most likely occurs through a combination of two processes, adsorption onto sinking particles and scavenging at the sediment-water interface. Evaluation of the relative significance of these two processes has been one of the intriguing problems in marine geochemistry in recent years.

Radionuclide profiles in well preserved sediment cores indicate that their distribution in the sediment pile, particularly near the sediment-water interface, is dominated by particle reworking processes, much unlike the commonly assumed undisturbed grain-by-grain deposition of open ocean sediments. The 'proper' interpretation of tracer profiles in sediments, therefore, rests on a knowledge of the processes affecting the tracer distribution and their relationship with time. The availability of several radionuclides with different half-lives and source functions and mathematical models to describe the sedimentary processes considerably helps in resolving at least some of the problems. The state-of-the-art in these studies is reviewed with a case study of the application of deep-sea sediments to resolve changes in cosmic ray production of radioisotopes in the past.

0097-6156/82/0176-0363$06.50/0

Deep-sea sediments have long been considered as an ideal repository for the retrieval of records pertaining to earth's past history. Major events such as global climatic changes, temporal variations in the meteorite and cosmic dust influx and volcanic activity are some of the key problems in earth sciences, which may be illuminated through the study of records preserved in deep-sea sediments. The introduction of 'time' dimension in these studies is central for their understanding in a chronological frame. As in several important discoveries in astrophysics and earth sciences, the development of a method for determining the chronology of ocean sediments was incidental and was a consequence of the studies on the origin of penetrating radiations observed by ionization chambers on board ships. These studies indicated the presence of significant concentrations of radium in deep-sea sediments [1][1], an observation which paved the way for development of the 'radium method' for dating deep-sea sediments [2,3]. In subsequent years several major advances were made both in the understanding of the marine geochemistry of radionuclides and in the techniques for their measurement. These advances helped in establishing by about 1960, radioactive geochronometry of marine deposits as a routine exercise in pursuit of an 'age' [4,5,6,7]. However, during the last few years' precise measurements of radionuclides in carefully collected and well preserved sediment cores have revealed that their distribution in the sediment pile is a resultant of several complex processes [8,9,10], much unlike the simple particle by particle accumulation model envisaged earlier. These findings have emphasized the necessity to decouple the effects of various processes on the radionuclide distribution in the sediment pile before an age-depth relationship can be derived.

In this article we plan to focus on two aspects (i) the transport of radionuclides to the ocean floor and the processes which govern their distribution in deep-sea sediments and (ii) the application of deep-sea sediments to retrieve historical records of large scale phenomena, e.g. long term changes in the rate of production of nuclides by cosmic rays. Even while discussing these aspects, our emphasis will be mainly on the processes rather than on the details of the chronometric method.

Delivery of Radionuclides to the Ocean

The radionuclides which are commonly used for determining the chronology of deep-sea sedimentary processes are given in Table 1. The principal pathways of introduction of these nuclides into the

---

[1]Figures in brackets indicate the literature references at the end of this paper.

Table 1. Common Chronometers of Deep-Sea Sedimentary Process.

Nuclide	Half-Life (year)	Mode of Input to the Ocean	Average Input Rate to Oceans (atoms/cm²·min)	Average Deposition Rate on the Sea Floor (atoms/cm²·min)	References
Cosmic-Ray Produced Nuclides					
^{32}Si	~300	atmospheric deposition	0.01	<0.0001	69, 76
^{14}C	5.73×10^3	air-sea exchange	210[a]	8	77
^{10}Be	1.6×10^6	atmospheric deposition	1.0	1.0	78
U-Th Series Nuclides					
^{210}Pb	22.3	atmospheric deposition + in situ production from $\underline{^{226}Ra}$	60-150	6-60	32, 69, 74
^{231}Pa	3.25×10^4	in situ production from $\underline{^{235}U}$	40	40	5, 7, 10, 79
^{230}Th	7.52×10^4	in situ production from $\underline{^{234}U}$	1000	1000	5, 7, 10, 79
Bomb-Produced Nuclides					
^{239,240}Pu	2.44×10^4 6.6×10^3	atmospheric deposition	--	--	

[a]This number is in excess of its overhead production, most of the ^{14}C depositing on the land surface eventually reaches the ocean surface.

oceans are: atmospheric deposition, river run off, and in situ production in the ocean water.

The atmospheric deposition predominates for nuclides which originate in the atmosphere (except for nuclides in gaseous phase such as ^{14}C). These include nuclides which are produced in the atmosphere either through the interaction of cosmic rays with atmospheric nuclei (^{10}Be, ^{26}Al, and ^{32}Si) or through the decay of radioactive parents ($^{222}Rn \rightarrow {}^{210}Pb$) and those injected into the atmosphere through nuclear weapon tests (^{55}Fe and $^{239,240}Pu$). Both the cosmic ray produced and artifically injected nuclides (except ^{14}C, Table 1) exhibit seasonal and latitudinal variations in their deposition, with peak fallout in the mid-latitides in spring [11,12]. The deposition pattern of cosmic ray produced nuclides is identical in both the hemispheres because of hemispherically symmetric source functions. The situation is quite different in case of the fallout pattern of the artifically injected nuclides which exhibits a much higher deposition in the northern hemisphere, since the weapon tests were conducted in this hemisphere.

^{14}C is introduced into the oceans mainly through the exchange of $^{14}CO_2$ at the air-sea interface. ^{14}C once produced in the atmosphere gets quickly oxidized to $^{14}CO_2$ and enters the exchangeable carbon system.

River runoff and in situ production are the major sources of U-Th series nuclides (Table 1) to the oceans. The concentrations of the various U-Th series nuclides in rivers vary considerably and depend upon several factors prime among them being their chemical reactivity [13], the chemistry of river water, and the nature of the river bed.

Release of uranium (^{238}U) to river waters follows the general trend of chemical weathering of rocks as can be inferred from the strong positive correlation between the abundances of ^{238}U and total dissolved salts (TDS) in rivers (figure 1 [14,15,16,17]). The ^{238}U-TDS relationship yields an average annual ^{238}U flux of 1.05 x 10^{10} g [17] to the ocean corresponding to mean ^{238}U concentration of about 0.3 µg/liter of river water.

^{234}U is the granddaughter of ^{238}U. The $^{234}U/^{238}U$ activity ratio in surface waters is generally >1 and average about 1.2 on a global scale [14,15,17]. The source of the 'excess' ^{234}U activity in river waters is known to be rocks through which the rivers flow, however the exact mechanism of production of this 'excess' is still not well understood. Processes like preferential leaching [18,19], in situ decay of recoiled ^{234}Th [20] and leaching through alpha recoil tracks [21] have been suggested as possible mechanisms to produce the commonly observed supra-equilibrium values of ^{234}U in surface waters. Whatever may be the mechanism of production of the excess ^{234}U activity in surface waters, it serves as a handle to probe into the geochemistry of uranium isotopes in the marine environment and as a dating tool for selected marine deposits [22,23].

As is common with all materials introduced into the oceans through river runoff, uranium isotopes also enter the open ocean through the fresh and salt water mixing zone. The behaviour of uranium isotopes in this domain was studied recently [17,24,25, 26]. These studies reveal that the distribution of uranium isotopes in estuaries beyond ~0.1 g Cl⁻/L are controlled only by the mixing of fresh and salt water (figure 2) and that these regions of estuaries (i.e., >0.1 g Cl⁻/L) are neither a source nor a sink for ^{234}U and ^{238}U. The geochemistry of uranium isotopes in the very low chlorosity region ($\leq$0.1) is presently uncertain.

The ^{238}U concentration and $^{234}U/^{238}U$ activity ratio in open ocean waters are nearly uniform and center around 3.3 ± 0.2 µg/L and 1.14 ± 0.03 respectively [27].

Three other nuclides of the U-Th series which serve as chronometers for deep-sea sedimentary processes are ^{210}Pb, ^{230}Th and ^{231}Pa (Table 1). All these nuclides are being produced continuously in the rocks and soils of the earth's crust through radioactive decay of their parents and are expected to be mobilized by weathering processes and transported to the oceans through surface and subsurface waters. However, measurements of ^{210}Pb and ^{234}Th in continental waters [28,29,30,31] indicate that these nuclides are chemically very reactive in natural waters, and hence are removed from soluble to suspended phases in very short time scales, 1-2 days. These nuclides tend to associate themselves with soil particles and hence are likely to be mobilized only through soil erosion and subsequent transport of the solids to the oceans. This leads to the natural conclusion that the dissolved supply of chemically reactive nuclides to the oceans via rivers is negligible.

The dominant source of ^{230}Th and ^{231}Pa to the oceans is their in situ production in the water column. The production rates of ^{230}Th and ^{231}Pa in the open ocean waters are nearly uniform because of the homogeneous distribution of their parents, ^{234}U and ^{235}U respectively. ^{210}Pb is introduced into the oceans through atmospheric deposition [32] and through in situ production in the water column. The production rate in the water column exhibits both lateral and vertical variations because of the changing concentrations of ^{226}Ra. The production essentially follows the deep water movement, with lowest values occurring in the North Atlantic and highest in the North Pacific.

Table 1 contains 'average' supply rate of these nuclides to the oceans.

Delivery of Radionuclides to the Ocean Floor

Particles present in the oceans play the role of 'scavengers' for the removal of radionuclides from the water column to the sediments. The type of particles executing the transfer may vary

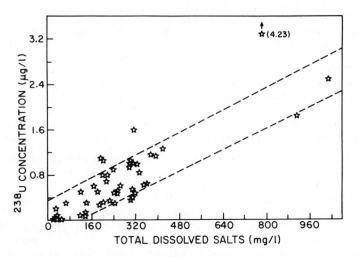

Figure 1.   Scatter diagram of ^{238}U vs. total dissolved salt (TDS) abundances in the aqueous phases of selected world rivers (data from Ref. 17).   The covariation between ^{238}U and TDS abundances is clearly evident.

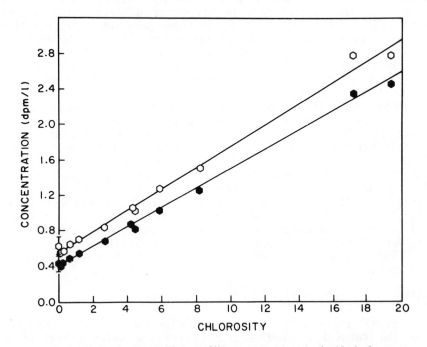

Figure 2.   The distribution of ^{238}U and ^{234}U concentrations in the Narbada estuary as a function of chlorosity (data from Ref. 16).   The results strongly suggest that the concentration profiles of U isotopes in estuaries are governed mainly by the mixing of fresh and salt water. Key: $\bigcirc$, ^{234}U; and $\bullet$, ^{238}U.

from nuclide to nuclide and seem to depend upon the involvement of the radionuclides in the marine biological and geochemical cycles.

[14]C and [32]Si are two nuclides for which the principal pathways of transport to the ocean floor are sufficiently known. These nuclides once introduced into the ocean waters equilibrate with their stable counterparts (mainly HCO_3^- for ^{14}C and H_4SiO_4 for [32]Si) and follow their cycles in the ocean [33]. Both carbon and silicon are essential ingredients of the hard parts ($CaCO_3$ and SiO_2) of the plant and animal populations in sea water. During their growth these organisms incorporate in their shells both ^{14}C and [32]Si essentially in the same ^{14}C/C and [32]Si/Si ratios as in the ambient waters they grow. After the death of the organisms, their calcareous and siliceous skeletal remains descend to the ocean floor [33]. A fraction of these shells dissolves in deep waters during their downward journey, and that which survives, deposits on the 'ocean floor'. Here again, further solution of the biogenic skeletons occur, and those which outlive the dissolution effects become a part of the sediments.

The sequestering agents and details of scavenging processes responsible for the removal of other chronometric tracers (Table 1) from sea water to sediments are not well understood. For [10]Be, [26]Al, ^{210}Pb, [230]Th, and [231]Pa, the scavengers are speculated to be either clay particles or (Fe, Mn) hydrous oxides. This speculation follows from the observation that these nuclides are relatively more abundant in slowly accumulating fine grained clay sediments rich in Fe and Mn. This is also strengthened from the finding that planktonic organisms subscribe only very little towards the downward transport of these nuclides from the mixed layer as demonstrated through the measurements of ^{210}Pb and [234]Th [34,35].

The removal of uranium isotopes to sediments is most likely brought about by their incorporation in organic matter [36].

The extent and rate of removal of radionuclide tracers from sea water can be ascertained through a knowledge of their 'residence time' in the water column. The residence time, τ, of a tracer due to its physical removal from one reservoir to another is defined [37,38,39,40,41] as:

$$\tau = M/\phi \qquad (1)$$

where M is the tracer content of the reservoir (gm) and ϕ its physical removal rate (gm/yr). Implicit in this definition of τ is that the tracer distribution in the reservoir is in steady state (i.e., the total influx of tracer into the reservoir equals total efflux). The residence time of tracers in oceans is derived following two types of models, the box or compartment model and the dynamic model.

In the box-model the aqueous domain of the oceans is treated as a box in which the tracer distribution is assumed to be homogeneous and in a steady state. The material balance equation for a radionuclide tracer in this domain is given by:

$$P = \lambda C + \phi' \qquad (2)$$

where P is its flux into sea water reservoir (atoms/cms$^2 \cdot$sec), C its inventory in sea water (atoms/cm^2), λ its radioactive decay constant, and ϕ' its outgoing flux (atoms/cm$^2 \cdot$sec). The physical removal, ϕ', is usually assumed to be a first order process, $\phi' = \psi C$, analogous to radioactive decay. In such a case the residence time, τ_ψ, due to physical removal is $\tau_\psi = 1/\psi$, analogous to the radioactive mean life, $\tau_\lambda = 1/\lambda$. Rearranging relation (2) yields:

$$\tau_\psi = \frac{\tau_\lambda}{[(P/\lambda C) - 1]} \qquad (3)$$

In case of nuclides for which the input into the box is only due to their <u>in situ</u> production from radioactive decay of their parents, equation (3) modifies to:

$$\tau_\psi = [\frac{R}{(1-R)}] \tau_\lambda \qquad (4)$$

where R is the daughter-parent activity ratio in the box. It is a common practice to compute the box-model residence times of tracers separately for the mixed layer and the deep ocean, with a view to estimate removal times and reaction rates in these domains. Studies of radioactive disequilibrium between ^{234}Th:^{238}U; ^{210}Po:^{210}Pb and ^{228}Th:^{228}Ra in the mixed layer have yielded residence times in the range of 0.3-1.0 yrs for Th and Po isotopes [42,43,44,45,46,47]. Calculations made similarly for the oceanic residence times of ^{230}Th, ^{231}Pa, and ^{10}Be yield values between 10-200 years [48,49,50] several orders of magnitude lower than their radioactive mean-lives. The short residence time leads us to infer that these nuclides reach the sediment within short time after their injection in sea water and that most of their oceanic inventory is contained in sediments. Both these inferences have been experimentally confirmed which make the application of these nuclides as chronometers on a stronger footing.

In the second model, the distribution and removal rates of tracers in the ocean are characterized through a one dimensional, (vertical) diffusion-advection equation. In this model, which ignores all horizontal processes, the equation governing the distribution of tracer in the soluble phase is [51,52,53,54]:

$$\frac{\partial}{\partial z} (K \frac{\partial C}{\partial z} - \omega C) + P - \lambda C \pm J(z) = \frac{\partial C}{\partial t} \qquad (5)$$

where C is the dissolved concentration, K the vertical eddy diffusion coefficient, ω the vertical advection velocity, λ the radioactive decay constant, P its production rate, and J (z) its in situ introduction or removal rate (non-conservative tracer case [42,52]. Equation (5) is solved for the depth profile of the dissolved phase of the tracer assuming steady state ($\frac{\partial C}{\partial t}$ = 0) and K and ω to be depth independent. Analogous to that in box-models, it has also been common to assume that the removal rate, J(z), is a first order process i.e., J(z) = ψC where ψ is a first order scavenging rate constant. The removal is speculated to occur through an irreversible in situ scavenging or absorption process on to settling particulate matter. In this case, the scavenging residence time of the nuclide would be:

$$\tau_\psi = \frac{\int C(z)\ dz}{\int J(z)\ dz} = 1/\psi \qquad (6)$$

The value of ψ (and hence τ_ψ) is deduced by curve fitting of the dissolved profiles of the radionuclides. This model was first applied to the ^{210}Pb profiles in the Pacific which yielded a scavenging residence time of 54 ± 4 years for ^{210}Pb [53]. The estimated scavenging residence time in effect represents the characteristic time for the transfer of ^{210}Pb from solution to suspended phase by a process mimicking radioactive decay.

The vertical scavenging model also allows one to predict the distribution of particulate radionuclide profiles. Following Craig et al. [53] the particulate phase activity would be given by the solution of the equation:

$$\frac{\partial}{\partial z} \{K \frac{\partial \chi}{\partial z} - (\omega+S)\ \chi\} + J(z) - \lambda\chi = \frac{\partial \chi}{\partial t} \qquad (7)$$

where χ is the concentration of radionuclide in particulate phase and S the settling velocity of particles labelled with the radionuclides. Equation (7) is also solved assuming steady-state ($\frac{\partial \chi}{\partial t}$ = 0), K, ω, and S to be constant with depth and J(z) = ψC. These assumptions of convenience and simplications (neglecting effects of K and ω compared to S) reduce equation (7) to:

$$S \frac{d\chi}{dz} + \lambda\chi - \psi C = 0 \qquad (8)$$

from which the depth profiles of particulate phase activity, $\chi\lambda$, can be derived in terms of dissolved concentration. Equation (8) predicts a downward increase in the particulate phase activity for nuclides whose source functions in the water column are either uniform (e.g. ^{234}Th, ^{230}Th, ^{231}Pa) or increase with depth (e.g. ^{210}Pb); the magnitude of increase would depend on the settling

velocity of particles. Attempts to check on this, as a test for
the vertical scavenging model have yielded inconsistent results.
The data of Somayajulu and Craig [55] and Krishnaswami et al.,
[56,57] on particulate ^{210}Pb and [230]Th profiles (figure 3) are in
agreement with the predictions of the model, whereas the particulate ^{210}Pb profiles in the Antarctic [55] and in the Atlantic [45,
figure 4] do not show an unambiguous increasing trend as expected
from the vertical scavenging model. The latter result, if it has
to be explained in terms of the vertical scavenging model would
require too large (> km/yr) particle sinking rates which are
uncommon in the marine environment. These findings coupled with
other observations such as (i) the ^{210}Pb removal is rapid in
regions of high particulate concentration, the surface waters and
near sediment-water interface and (ii) the lack of sufficient
surface groups in sinking particulate matter [58] to produce the
observed scavenging effects, have led to the development of two
alternate models. The first model, the settling model [59]
assumes that the transfer rate (ψ) is very fast and that the
residence time of reactive nuclides such as ^{210}Pb is controlled
principally by the settling velocities of particles carrying them.
These assumptions are in contrast to those of the model of Craig
et al., [53] and imply quite different partitioning of [210]Pb
between soluble and suspended phases. In the second model, the
horizontal diffusional transport model [45,58], it is suggested
that the removal of radionuclides, particularly ^{210}Pb is a dual
process: adsorption onto sinking particles and removal at
sediment-water interface, the latter process accounting for most
of the removal. The application of this model to other nuclides
such as [230]Th, which exhibit gross deficiency throughout the water
column and on a world ocean scale is yet to be demonstrated. Also
the observation that the standing crops (Σ dpm/cm^2) of [230]Th and
[210]Pb in several cores are similar to that expected from their
deficiencies in the overlying water column is difficult to be
explained by the horizontal transport model. Instead this
observation seems to favour the vertical scavenging model.

It is amply clear from the foregoing discussion that several
of the naturally produced and artificially injected radionuclides
are removed from sea water to sediments on short time scales, but
the questions of their removal sites and principal scavenging
mechanism are still wide open. It is likely that a systematic
study of the interrelation of particulate phase activities with
various components and properties of suspended solids in the ocean
may help resolve some of these questions.

Recently, the importance of large particle formation and
their subsequent rapid transport to abyssal depths of the oceans
have been invoked as a mechanism to explain the distribution of
several oceanic variables [60]. These studies have shown that
there exists a direct transport path between sea surface and sea
floor and that the sea floor can no longer be considered remote
from the surface. This transport path may become significant for

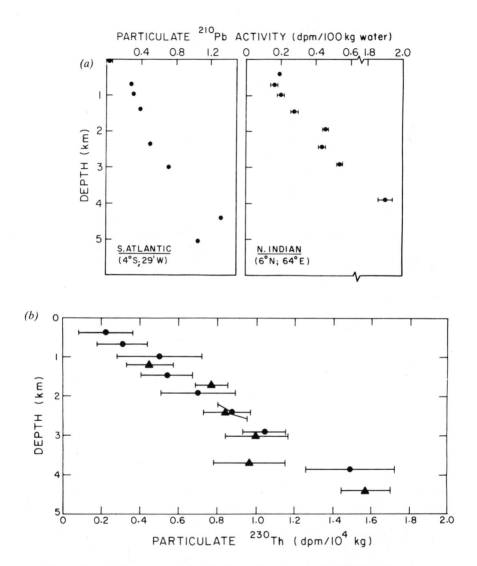

Figure 3.   (a) The concentration depth profiles of particulate ^{210}Pb in the sea water column (data from Refs. 55 and 57). The increasing trend with depth supports the in situ vertical scavenging model. (b) The concentration profiles of particulate [230]Th in two profiles from the Indian Ocean (data from Ref. 57). These results, similar to the ^{210}Pb results in Figure 3a, also strongly favor the in situ scavenging model for nuclide removal from the sea water column. Key: ●, Station 418; and ▲, Station 440.

tracers introduced at the ocean surface [36,60,61,62]. Perhaps
the study of radionuclide fluxes mediated through large and small
particles may also help resolve some of the questions raised in
the foregoing discussions.

Distribution of Radionuclides in Sediments

Once the radionuclides reach the sediments they are subject
to several processes, prime among them being sedimentation, mixing, radioactive decay and production, and chemical diagenesis.
This makes the distribution profiles of radionuclides observed
in the sediment column a residuum of these multiple processes,
rather than a reflection of their delivery pattern to the ocean
floor. Therefore, the application of these nuclides as chronometric tracers of sedimentary processes requires a knowledge of
the processes affecting their distribution and their relationship
with time. Mathematical models describing some of these processes and their effects on the radionuclide profiles have been
reviewed recently [8,9,10] and hence are not discussed in detail
here. However, for the sake of completeness they are presented
briefly below.

All models available to date rest on two basic assumptions,
viz. (i) the flux of the radionuclide to the sediment water interface has remained constant with time (or if the flux has varied,
then its time variations are known) and (ii) the tracers and
sediment particles are not independently mobile in the sediment
pile. If these conditions are met then the temporal variations in
the radionuclide concentration at depth, z, below the moving sediment-water interface would be given by [63,64,65,66]:

$$\frac{\partial}{\partial t}(\rho A) = \frac{\partial}{\partial z}\left\{ K\frac{\partial}{\partial z}(\rho A)\right\} - S\frac{\partial}{\partial z}(\rho A) - \lambda\rho A \qquad (9)$$

where K is the mixing coefficient (cm^2/time), S the sediment
accumulation rate (cm/time), A the radionuclide concentration
(dpm/g) at depth z from the sediment-water interface, ρ the in
situ density of sediment (g/cm^3) and λ the radioactive decay
constant of the nuclide ($time^{-1}$).

The first term on the right hand side of equation (9) represents the effects of particle reworking on the radionuclide distribution. The reworking results in a vertical mixing of particles
in the sediment pile and is brought about by physical and biological processes, such as action of bottom currents and burrowing and
feeding activities of benthic organisms. It is customary to treat
the particle reworking process as a diffusion analogue as in
equation (9). The major justification for this analogy is that
the mixing is a random process and that the sample size integrates
several of these events both spatially and temporally. Attempts
to test this speculation through laboratory experiments [67] have
confirmed that the mode of mixing of sediment solids by organisms
can be described in terms of an eddy diffusion model.

Equation (9) is quite generalized and allows for variations in the mixing rate, K, and *in situ* density, ρ, with depth in determining the activity-time relationship of radionuclide profiles. However, in all commonly used models, equation (9) is further simplified using assumptions such as K, ρ and S to be constant and $\frac{\partial A}{\partial t} = 0$ (the steady state assumption) and that the sediments consist of two layers, the upper mixed layer with a constant mixing coefficient and the lower layer free of particle reworking processes. The equations governing the depth distribution of radionuclide in the two layers are:

$$K \frac{d^2A}{dz^2} - S \frac{dA}{dz} - \lambda A = 0 \text{ for } 0<Z<Z_m \tag{10}$$

$$- S \frac{dA}{dz} - \lambda A = 0 \text{ for } Z>Z_m \tag{11}$$

where Z_m is the thickness of the mixed layer.

Solution of equation (10) which involves sedimentation in the presence of mixing and that of equation (11) which contains the sedimentation term only, are exponential in nature. The major conclusion which arises from this is that the logarithmic nature of the activity-depth profiles by itself is not a guarantee for undisturbed particle by particle sediment accumulation, as has often been assumed. The effects of mixing and sedimentation on the radionuclide distribution in the sediment column have to be resolved to obtain pertinent information on the sediment accumulation rates. (It is pertinent to mention here that recently Guinasso and Schink [65] have developed a detailed mathematical model to calculate the depth profiles of a non-radioactive transient tracer pulse deposited on the sediment surface. Their model is yet to be applied in detail for radionuclides.)

The depth profiles of radionuclides in the two layers of sediment are qualitatively described in figure 5. It is evident from figure 5 that sediment mixing, if present, reduces the concentration gradient of the radionuclide, which in effect yields a higher sediment accumulation rate. The radionuclide profile in the mixed layer will depend upon the intensity of mixing and the half-life of the nuclide. The same apparent effect is obtained even in case of a transient tracer, wherein also reworking mixes the tracer to deeper depths (older sediments) resulting in an overestimation of sediment accumulation rates. Thus if mixing and sedimentation govern the radionuclide distribution, then the sediment accumulation rate computed from the log A vs. depth plot, without correcting for mixing effects, will be an upper limit.

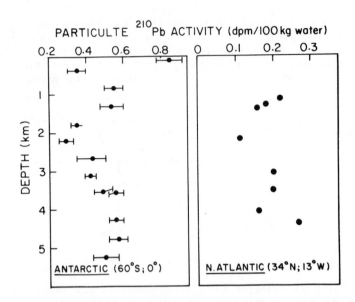

Figure 4. Particulate ^{210}*Pb distribution in profiles from the Antarctic and the Atlantic (data from Refs. 45 and 55). The near constancy of particulate* ^{210}*Pb abundance with depth is difficult to explain by the vertical scavenging model of Craig et al. (53).*

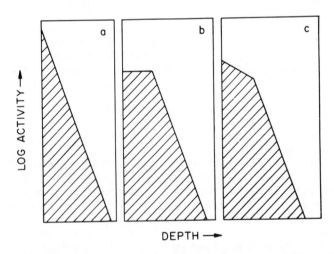

Figure 5. Qualitative diagram depicting the effect of mixing on radionuclide distribution in a sediment core.

(a) Hypothetical case of no mixing, i.e., the nature of the profile is only due to sedimentation; (b) intense mixing in the surficial layers, resulting in a well-defined mixed zone; and (c) slower mixing compared to (b) near the interface. These types of profiles also can result from a changing rate of sediment accumulation with time.

A quantitative and fairly easy method to obtain particle reworking rates in deep sea sediments became possible after the elegant work of Nozaki et al., [68] based on ^{210}Pb distribution in them. The radioactive half-life of ^{210}Pb is too short (22.6 yrs) to produce measurable depth profiles in deep sea sediments based on sedimentation alone since its activity would be limited to the top $\lesssim$ 1 mm layer. In such a case its depth profile predominantly records the effects of particle reworking and its distribution can be approximated as:

$$A = A_0 \exp\left\{- z \left(\frac{\lambda}{K}\right)^{1/2}\right\} \tag{12}$$

In figure 6 are presented some of the available ^{210}Pb profiles in deep sea sediments. The data available to-date yield values ranging between (20-400) $cm^2/10^3$ year for particle mixing coefficients, K [69]. The data also show that the mixing coefficients do not exhibit any systematic trend either with the sediment type or with sedimentation rate [69].

Another nuclide which has begun to find application to study particle mixing rates in pelagic sediments is cosmic ray produced [32]Si (half life $\sim$ 300 yrs). First attempts to evaluate particle reworking rates based on [32]Si in Antarctic and Pacific siliceous and clay-siliceous oozes has been reported by Demaster [69]. The ^{210}Pb and [32]Si based particle mixing rates in these cores are mutually compatible.

The short-lived nuclides, though yield valuable information on the mixing coefficients, they may provide only a lower limit on the mixing depth. This is because the mixing may be operative much deeper than the depth to which the radionuclide can be detected unambiguously. Estimates on the mixing depth are therefore best obtained through a study of the distribution of longer lived nuclides, e.g., [230]Th, [231]Pa, ^{14}C etc. Figures 7 and 8 show typical profiles of ^{14}C and [230]Th in carefully collected deep sea sediments. These results clearly show the well mixed surface layer followed by an exponentially decreasing activity profile. The thickness of the well mixed surface layer in the various types of pelagic sediments is $\sim$ 10 cm [69,70]. The logarithmic activity profiles below the surface mixed layer are usually ascribed to sedimentation, and the accumulation rates are calculated based on the slopes of these lines. However, as mentioned earlier, it is plausible that even at these depths there can be particle mixing and the nature of the profile could be resultant of these two processes. (Indeed, the recent report of Thomson and Wilson [71] shows that burrow like structures are present even up to depths of 1.5 m in deep sea cores). In such cases, the best method to obtain reliable sediment accumulation rates would be to measure the concentration depth profiles of at least two radionuclides with comparable half-lives (e.g., [230]Th, [231]Pa, ^{14}C) and solve for both K and S. Alternatively, it would

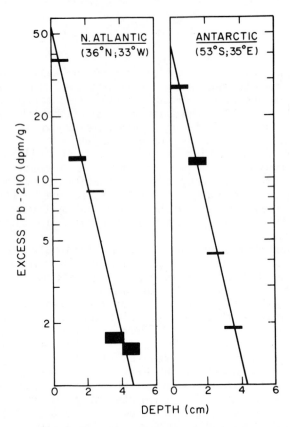

Figure 6. The ^{210}Pb excess profiles in two deep-sea sediment cores (data from Refs. 68 and 69). The data yield particle mixing coefficients in the range of ~ 100 cm²/ky.

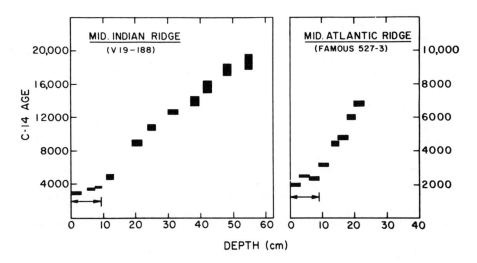

Figure 7. Depth profiles of [14]*C in two cores (data from Refs. 68 and 70). The near constant age up to about 8 cm from the core top is attributable to mixing. The thickness of the mixed layer is indicated by the arrows.*

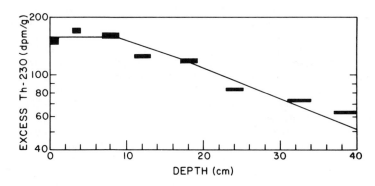

Figure 8. The [230]*Th excess profile in a box core from the DOMES area (data from Ref. 74). The mixed-layer thickness in the core is about 8 cm.*

suffice to measure the distribution of a very long lived nuclide (e.g., ^{10}Be) whose profile would be overwhelmingly due to sedimentation.

Deep Sea Sediments and Historical Records

The advent of new techniques to collect undisturbed sediment cores, with well preserved sediment – water interface has brought into sharper focus the various deep sea sedimentary processes, their rates and their effects on the preserved records. As mentioned earlier, recent studies have shown that the record contained in sediments is not a direct reflection of the delivery pattern of a substance to the ocean floor as has so far been assumed; the record is modified as a result of several complex physical, chemical and biological processes. Therefore, information on the temporal variations in the tracer input to oceans, if sought, has to be deciphered from the 'sediment-residuum'. In the following we consider one specific example of retrieval of information from the sediment pile: the application of deep sea sediments to obtain historical records of cosmic ray intensity variations.

Records of cosmic ray history can be derived through the study of either extraterrestrial (e.g., moon and meteorites) or terrestrial samples (e.g., tree rings, sediments and glaciers). The record preserved in the terrestrial samples is 'differential' in character, whereas the record contained in the extraterrestrial samples are 'integral' ones. The high frequency variations in the cosmic ray intensity are heavily attenuated in the meteorites and only those variations with periods comparable to the mean-life of the radionuclide can be unambiguously studied using these samples. This is in contrast to the terrestrial records, where a resolution as fine as $\sim$ 10 years can be obtained, the lower limit being controlled by the transfer/mixing rates in the geospheres. In the oceans, the controlling factors are: (i) the residence time of the nuclide in the water column and (ii) the particle reworking at the sediment-water surface. The effects due to these two parameters are evaluated below.

The global average production rate of any nuclide, $Q(t)$, at any time, t, will be primarily dependent on the cosmic ray intensity, $I(t)$. If the intensity varies sinusoidally with a period $T(\omega = 2\pi/T)$, $Q(t)$ will also vary sinusoidally. The standing crop of a nuclide in the sea water column for a production function, $Q(t) = Q_0 (1 + a \cos \omega t)$, a being the amplitude, can be deduced to be:

$$N(t) = \frac{Q_0}{\Lambda_1} \left\{ 1 + \frac{a \cos (\omega t - \alpha)}{\{1 + (\omega/\Lambda_1)^2\}^{1/2}} \right\} \qquad (13)$$

where $\alpha = \tan^{-1}(\omega/\Lambda_1)$, Λ_1 being the total removal rate constant, i.e., the sum of radioactive decay and the physical removal rates $(\Lambda_1 = \lambda + \psi)$. Implicit in the derivation of equation (13) is that the removal of the nuclide to the ocean floor is first order; i.e., ψN. For ^{10}Be and ^{26}Al, which are long-lived and have short residence time in sea water, $\Lambda_1 \simeq \psi$, their deposition flux $(N\psi)$ to the ocean floor $Q_d(t)$ is expected to vary approximately as

$$Q_d(t) = Q_o \left\{ 1 + \frac{a \cos(\omega t - \alpha)}{\{1 + (\omega/\Lambda_1)^2\}^{1/2}} \right\} \quad (14)$$

The amplitude attenuation factor, $\{1 + (\omega/\Lambda_1)^2\}^{1/2}$ for nuclides satisfying relation [14], for various values of Λ_1 and T are presented in figure 9. It is obvious from figure 9 that the attenuation is minimal when $\Lambda_1 > \omega$, i.e., when the removal residence time of the nuclide from sea water is less than the period in the variation of cosmic ray intensity.

Once these nuclides deposit on the ocean floor they are likely to be subjected to particle mixing processes. In the following we discuss attenuation due to a simple case of mixing, in which the sedimentary particles are mixed to a constant depth, L, from the sediment-water interface [72,73]. For such a case the temporal variation in the standing crop (atoms/cm^2) C, of the radionuclide in the mixed layer is given by:

$$\frac{dC}{dT} = -C\left(\lambda + \frac{S}{L}\right) + Q_d(t) \quad (15)$$

where λ is the radioactive decay constant, S the sediment accumulation rate (cm/time), L is the mixed layer thickness (cm) and $Q_d(t)$ is the deposition flux on the ocean floor. Solving equation (15) by substituting for $Q_d(t)$ from relation (14) and assuming S and L to be constant, we obtain:

$$C(t) = \frac{Q_o}{\Lambda_2} \left\{ 1 + \frac{a \cos(\omega t - \alpha - \beta)}{(1 + \omega^2/\Lambda_1^2)^{1/2} \; (1 + \omega^2/\Lambda_2^2)^{1/2}} \right\} \quad (16)$$

where Λ_2 is the sum of removal rate constants from the mixed layer, $\Lambda_2 = (\lambda + \frac{S}{L})$, and $\beta = \tan^{-1}(\omega/\Lambda_2)$. For ^{10}Be, ^{26}Al, $\Lambda_2 \simeq (S/L)$, and therefore the flux of these nuclides out of the mixed layer of the sediments into the historical layer, $Q_s(t)$ would be

$$Q_s(t) = Q_o \left\{ 1 + \frac{a \cos(\omega t - \alpha - \beta)}{\{1 + (\omega/\Lambda_1)^2\}^{1/2} \; \{1 + (\omega/\Lambda_2)^2\}^{1/2}} \right\} \quad (17)$$

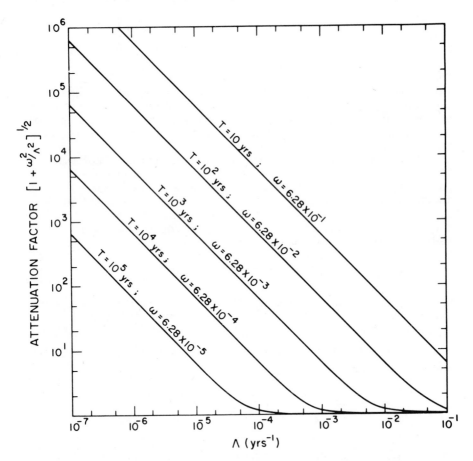

Figure 9.   Calculated attenuation factors for various values of T *(period in the sinusoidal cosmic ray intensity variations) and* Λ *(the total rate constant).   See section on deep-sea sediments and historical records for discussion.*

The forms of equations 14 and 17 describing the deposition of the tracer on the ocean surface, ocean floor and into the historical layers of the sediments are all similar. However, the amplitude variations in the historical layers of sediment are attenuated considerably compared to variations in deposition on the ocean surface, i.e., at input, the attenuation being governed by the effective residence times of nuclides in sea water and in the mixed layer of the sediments.

In Table 2 we present the expected attenuation factors of ^{10}Be. A residence time of 200 years ($\psi = 5 \times 10^{-3}$ yr^{-1}) for ^{10}Be in sea water and a mixed layer thickness of 10 cm have been assumed in the calculations.

Table 2.   Calculated Attenuation Factors for ^{10}Be at Deposition on Ocean Floor and Within Sediments.

Period in Q(t)	At Deposition	Attenuation Factor Within Sediments S=0.2 cm/ky	S=2 cm/ky
200 yrs.	6.4	1570	157
7000 yrs.	1.02	45	4.6
10^5 yrs.	1.0	3.3	1.05

The calculated values in Table 2 show that for ^{10}Be even in fast depositing sediments (S=2 cm/10^3 yrs.) the imposed signal is appreciably attenuated for Q(t) < 10^5 yrs. The major attenuation in this case is due to particle mixing which constitutes an apparent decay constant, S/L, 2×10^{-4} yr^{-1} for L=10 cm. The obvious conclusion which emerges from these simple calculations is that for studying the periodicities of cosmic ray intensity, one should select samples from regions of minimal or zero particle reworking, e.g., the anoxic sediments depositing in the Arabian Sea, Black Sea, etc. The ^{10}Be studies in pelagic sediments from oxic basins, can at best yield information on the long-term ($\sim 10^5$ yrs.) changes in Q(t).

(In the preceding discussion, we have calculated the attenuation factors for ^{10}Be for three periods, 200, 7×10^3 and 10^5 years. Of these the 200 and 7000 year periods are well established and have been ascribed to solar cycle variations and earth's magnetic field excursions, respectively. For detailed calculations on the effect of these variations on the production rates of isotopes by cosmic rays reference is made to Castagnoli and Lal, 75.)

Short-term changes in the cosmic ray intensity (T $\sim$ 200 yrs.) can be best studied through the analyses of ^{32}Si and ^{14}C in relatively fast accumulating sediments (S > 10 cm/10^3 yr.). Both these nuclides are introduced in surface waters, from where they are transported to sediments by sinking remains of organisms within 1-2 years. Hence these nuclides are expected to provide high resolution data on cosmic ray intensity variations.

Synthesis

Since the discovery of ^{226}Ra enrichment in deep sea sediments nearly a century ago, several new methods based on U-Th series nuclides and cosmic ray produced isotopes have been developed to 'date' deep-sea sediments. Recent advances in sample collection, analysis and modeling have radically revolutionized the prevailing concepts of radionuclide transport to the oceans, their behavior in the water column and their distribution in sediments. Studies of radionuclide distributions have yielded far more information than relative chronology; they have contributed significantly to the general understanding of various processes occurring in the marine environment. The aim of this article has been to present an overall view of the modern ideas on the transport of radionuclides to sediments and their distribution within the sediment pile.

One of the more significant advances of recent times has been to bring into sharper focus the effect of particle reworking processes on the distribution of properties in the sediment column and their effects on the historical records. These studies have demonstrated that sediments act as a 'low pass filter' and that the preserved records are a smudgy version of their delivery pattern on the ocean floor. Fortunately there exists an explicit interrelation between the imposed and the preserved signals and one can be derived from the other through appropriate mathematical modeling. In spite of all the complexities governing the transport and distribution of tracers in the sediments, these deposits are still one of the best available storehouses to retrieve earth's historical records.

Literature Cited

[1] Joly, J., Phil. Mag., 6, 196 (1908).
[2] Petterson, H., Das Verhaltnis Thorium Zu Uran, In: den Gesteinen Und im Meer. Sitzleer Akad, Wiss. Wien, Math-naturw K. 127, Mitt. Inst. Radium forsch wien Nr., 400 a (1937).
[3] Piggot, C. S., Urry, W. D., Am. J. Sci., 240, 1-12 (1942).
[4] Broecker, W. S., Isotope Geochemistry and the Pleistocene Climatic Record, In: Wright, H. E. and Grey, D. G., ed., The Quartenary of the United States, Princeton University Press, New Jersey (1965).
[5] Goldberg, E. D., Bruland, K., Radioactive Geochronologies, In: Goldberg, E. D., ed., The Sea, Wiley Interscience, New York, 5, 451-489 (1974).

[6] Burton, J. D., Radioactive nuclides in the marine environment, In: Riley, J. P., Skirrow, G., eds., Chemical Oceanography, Academic Press, London, 3, 92-191 (1975).

[7] Ku, T. L., Ann. Rev. Earth Planet. Sci., 4, 347-379 (1976).

[8] Krishnaswami, S., Lal, D., Radionuclide Limnochronology, In: Lerman, A., ed., Lakes — Chemistry, Geology, Physics, Springer-Verlag, New York, 153-173 (1978).

[9] Robbins, J. A., Geochemical and Geophysical Applications of Radioactive Lead, In: Nriagu, J. O., ed., Biogeochemistry of Lead, Elsevier Scientific Publishers, Netherlands, 285-393 (1978).

[10] Turekian, K. K., Cochran, J. K., Determination of marine chronologies using natural radionuclides, In: Riley, J. P., Chester, R., eds., Chemical Oceanography, Academic Press, London, 7, 313-360 (1978).

[11] Lal, D., Peters, B., Cosmic ray produced radioactivity of the earth, In: Handbuch der physik, Springer-Verlag, Berlin, 46, 551-612 (1967).

[12] Joseph, A. B., Gustafson, P. F., Russel, I. R., Schuert, E. A., Volchok, H. C., Tamplin, A., Sources of Radioactivity and their characteristics in Radioactivity in the marine environment, U. S. National Academy of Sciences, Washington, D.C., 6-41 (1971).

[13] Goldberg, E. D., Minor elements in seawater, In: Riley, J. P., and Skirrow, G., eds., Chemical Oceanography, Academic Press, London, 1, 163-196 (1965).

[14] Bhat, S. G., Krishnaswami, S., Proc. Ind. Acad. Sci., 70, 1-17 (1969).

[15] Turekian, K. K., Chan, L. H., The Marine Geochemistry of the Uranium Isotopes, ^{230}Th and ^{231}Pa, In: Activation Analysis in Geochemistry and Cosmochemistry, Universitetsforlaget, Oslo, 311-320 (1971).

[16] Borole, D. V., Radiometric and trace elemental investigations on Indian estuaries and adjacent seas, PhD Thesis, Gujarat University, Ahmedabad (1980).

[17] Borole, D. V., Krishnaswami, S., Somayajulu, B. L. K., submitted to Geochim. Cosmochim. Acta 1980.

[18] Rosholt, J. N., Shields, W. P., Garner, E. L., Science, 139, 224 (1963).

[19] Hussain, N., Krishnaswami, S., Geochim. Cosmochim. Acta, 44, 1287-1292 (1980).

[20] Kigoshi, K., Science, 173, 47-48 (1971).

[21] Fleischer, R. L., Science, 207, 979-981 (1980).

[22] Thurber, D. L., Broecker, W. S., Blanchard, R. L., Portraz, H. A., Science, 149, 55-58 (1965).

[23] Veeh, H. H., J. Geophys. Res., 71, 3379-3386 (1966).

[24] Borole, D. V., Krishnaswami, S., Somayajulu, B. L. K., Est. Coast. Mar. Sci., 5, 743-754 (1977).

[25] Martin, J. M., Nijampurkar, V. N., Salvadori, F., Uranium and Thorium isotopes behaviour in estuarine systems, In: Bio-geochemistry of Estuarine Sediments, Proceedings of Workshop held in Melreux, Belgium, Unesco Pub., 111-127 (1976).
[26] Martin, J. M., Meybeck, M., Pusset, M., Neth. J. Sea Research, 12, 338-344 (1978).
[27] Ku, T. L., Knauss, K. G., Mathieu, G. G., Deep Sea Res., 24, 1005-1017 (1977).
[28] Rama, Koide, M., Goldberg, E. D., Science, 134, 98-99 (1961).
[29] Goldberg, E. D., Geochronology with ^{210}Pb, In: Radioactive dating, International Atomic Energy Agency, Vienna, 121-131 (1963).
[30] Benninger, L. K., Lewis, D. M., Turekian, K. K., The use of natural ^{210}Pb as a heavy metal tracer in river estuarine system, In: Church, T. M., ed., Marine Chemistry in the Coastal Environment, Am. Chem. Soc. Symp., 18, 222-255 (1975).
[31] Hussain, N., Krishnaswami, S., unpublished results.
[32] Turekian, K. K., Nozaki, Y., Benninger, L. K., Ann. Rev. Earth Planet. Sci., 5, 227-255 (1977).
[33] Broecker, W. S., Chemical Oceanography, Harcourt Brace Jovanovich, Inc., New York (1974).
[34] Turekian, K. K., Geochim. Cosmochim. Acta, 41, 1131-1138 (1977).
[35] Krishnaswami, S., Lal, D., Marine Particles, to appear in Ocean Handbook, Horne, R. A., Hood, D. W., ed., Marcel Dekker, Inc. (1980).
[36] Brewer, P., Nozaki, Y., Spencer, D. W., Fleer, A., Sediment Trap Experiments in the Deep North Atlantic: Isotopic and Elemental Fluxes, J. Mar. Res., 38, 703- (1980).
[37] Barth, T. F. W., Theoretical Petrology, John Wiley and Sons, New York (1952).
[38] Craig, H., Tellus, 9, 1-17 (1957).
[39] Li, Y. H., Geochim. Cosmochim. Acta, 41, 555-556 (1977).
[40] Nozaki, Y., Chikyukagaku, Geochemistry, 12, 27-36 (1978).
[41] Whitefield, M., Mar. Chem., 8, 101-124 (1979).
[42] Bhat, S. G., Krishnaswami, S., Lal, D., Rama, Moore, W. S., Earth. Planet. Sci. Lett., 5, 483 (1969).
[43] Broecker, W. S., Kaufman, A., Trier, R. M., Earth Planet. Sci. Lett., 20, 35 (1973).
[44] Matsumoto, E., Geochim. Cosmochim. Acta, 39, 205 (1975).
[45] Bacon, M. P., Spencer, D. W., Brewer, P. G., Earth Planet. Sci. Lett., 32, 227 (1976).
[46] Nozaki, Y., Thompson, J., Turekian, K. K., Earth Planet. Sci. Lett., 32, 304 (1976).
[47] Knauss, K. G., Ku, T. L., Moore, W. S., Earth Planet. Sci. Lett., 39, 235-249 (1978).
[48] Moore, W. S., Sackett, W., J. Geophys. Res., 69, 5401 (1964).

[49] Somayajulu, B. L. K., Goldberg, E., Earth Planet. Sci. Lett., 1, 102 (1966).
[50] Raisbeck, G. M., Yiou, F., Fruneau, M., Loiseaux, J. M., Lieuvin, M., Earth Planet. Sci. Lett., 43, 237-240 (1979).
[51] Lal, D., J. Oceanogr. Soc. Japan, 20, 600-614 (1962).
[52] Craig, H., J. Geophys. Res., 74, 5491-5509 (1969).
[53] Craig, H., Krishnaswami, S., Somayajulu, B. L. K., Earth Planet. Sci. Lett., 17, 295-305 (1973).
[54] Veronis, G., Use of tracers in circulation studies, In: The Sea, Wiley Interscience, New York, 6, 169-188 (1976).
[55] Somayajulu, B. L. K., Craig, H., Earth Planet. Sci. Lett., 32, 268-276 (1976).
[56] Krishnaswami, S., Lal, D., Somayajulu, B. L. K., Weiss, R. F., Craig, H., Earth Planet. Sci. Lett., 32, 420-429 (1976).
[57] Krishnaswami, S., Sarin, M. M., Somayajulu, B. L. K., Earth Planet. Sci. Lett. (in press).
[58] Brewer, P. G., Hao, W. M., Oceanic Elemental Scavenging, In: Chemical Modeling in Aqueous Systems, American Chemical Society, 261-274 (1979).
[59] Tsunogai, S., Nozaki, Y., Minagawa, M., J. Oceanogr. Soc. Japan, 30, 251-259 (1974).
[60] Spencer, D. W., Honjo, S., Brewer, P. G., Oceanus, 21, 20-25 (1978).
[61] Spencer, D. W., Brewer, P. G., Fleer, A., Honjo, S., Krishnaswami, S., Nozaki, Y., J. Marine. Res., 36, 493 (1978).
[62] Cherry, R. D., Higgo, J. J. W., Fowler, S. W., Nature, 274, 246-248 (1978).
[63] Goldberg, E. D., Koide, M., Geochim. Cosmochim. Acta, 26, 417-450 (1962).
[64] Sarma, T. P., Dating of Marine Sediments by Ionium and Protactinium Methods, PhD Thesis, Carnegie Institute of Technology, Pittsburgh, Pennsylvania (1965).
[65] Guinasso, N. L., Schink, D. R., J. Geophys. Res., 80, 3032-3043 (1975).
[66] Nozaki, Y., J. Geol. Soc. Japan, 8, 699-706 (1977).
[67] Robbins, J. A., McCall, P. L., Fisher, J. B., Krezoski, J. R., Earth Planet. Sci. Lett., 42, 277-287 (1979).
[68] Nozaki, J., Cochran, J. K., Turekian, K. K., Kellar, G., Earth Planet. Sci. Lett., 34, 167-173 (1977).
[69] Demaster, D. J., The marine silica and ^{32}Si budgets, PhD Thesis, Yale University, New Haven, Connecticut, 260 (1979).
[70] Peng, T. H., Broecker, W. S., Kipphut, G., Shackleton, N., Benthic mixing in deep sea cores as determined by ^{14}C dating and its implications regarding climate stratigraphy and fate of fossil fuel CO_2, In: Anderson, W. R., Malahoff, A., eds., Fate of Fossil Fuel CO_2 in the Ocean, Mar. Sci. Ser, 6, 355-373 (1976).

[71] Thomson, J., Wilson, T. R. S., Deep Sea Res., 27, 197-202 (1980).
[72] Berger, W. H., Heath, G. R., J. Mar. Res., 26(2), 134-143 (1968).
[73] Berger, W. H., Johnson, R. F., Killingley, J. S., Nature, 269, 661-663 (1977).
[74] Cochran, J. K., The Geochemistry of ^{226}Ra and ^{228}Ra in Marine Deposits, PhD Thesis, Yale University, New Haven, 260 (1979).
[75] Castagnoli, C., Lal, D., Radiocarbon, 22, 133-158, 1980, in press.
[76] Kharkar, D. P., Nijampurkar, V. N., Lal, D., Geochim. Cosmochim. Acta, 30, 621-631 (1966).
[77] Lal, D., Suess, H., Ann. Rev. Nucl. Sci., 18, 407-434 (1968).
[78] Somayajulu, B. L. K., Geochim. Cosmochim. Acta, 41, 909 (1977).
[79] Krishnaswami, S., Geochim. Cosmochim. Acta, 40, 425 (1976).

RECEIVED August 21, 1981.

V. CHEMICAL EVOLUTION, EXTINCTION, AND ARCHAEOLOGY

The Antiquity of Carbon

CYRIL PONNAMPERUMA and ELAINE FRIEBELE

University of Maryland, Laboratory of Chemical Evolution, College Park, MD 20742

In the geochemical approach to the study of the origin of life, a major objective is to answer the question, "When did life on Earth begin?" Organic molecules, as old as the solar system, have been identified in carbonaceous chondrites. The oldest known sediments of the Earth, dated at 3.8×10^9 years, contain compounds of biological significance. In the absence of direct methods for dating the earliest organic matter, an array of criteria have been used to establish the syngenicity of the most primitive organic compounds.

In the study of the origin of life on earth, the element carbon is essential. Carbon is a required component of the fundamental molecules of life: amino acids, bases, and sugars. In addition, a large variety of carbon compounds is necessary in the complex biochemical cycles of living organisms. The physical and chemical nature and geometry of the carbon atom make it well suited to form the vast array of molecules involved in the chemistry of life.

How long has carbon been available to the processes of chemical evolution in the cosmos? About 18 billion years ago, the universe came into being with the explosion of the primeval concentration of matter. The lighter elements up to beryllium appeared almost instantly. The rest of the elements in the periodic table up to iron were formed in nuclear fusion reactions occurring during the condensation of massive stars. The heavier elements resulted from supernovae explosions. Thus, most of the elements, including the carbon in all living organisms today, were formed in the development of an earlier generation of stars before the formation of the solar system 4.6 billion years ago. We are indeed the stuff of which stars are made.

Our knowledge of the chemical composition of interstellar space is growing rapidly due to recent advances in observation techniques using detection at ultraviolet, infrared, and radio wavelengths. Studies of interstellar clouds, which are composed

0097-6156/82/0176-0391$05.00/0

predominately of hydrogen and helium, show that the heavier elements are depleted by 1/3 to 1/10 relative to their total abundance in the galaxy [1][1]. These depleted elements are found either in molecular form, as charged ions, or in interstellar dust grains.

Interstellar molecules are grouped by their location in three types of clouds: diffuse, dark, and black clouds [1]. The diffuse clouds are characterized by low gas density, predominately atomic hydrogen, and diatomic carbon compounds such as CH, CH^+, CN, and CO. Since ultraviolet light penetrates diffuse clouds, the photolifetimes of these molecules are one hundred to one million years. Dark and black clouds have higher gas densities and molecular hydrogen. Carbon is mainly in the form of CO, the second most abundant intersellar molecule. $(CO:H_2 = 3 \times 10^{-5})$ [1]. Other observed organic molecules within these clouds include H_2CO, HCN, HCO^+, HCC, C_3N, C_4H. The high density and temperature of black interstellar clouds facilitates a richer chemistry in which molecules such as dimethyl ether and ethyl alcohol are formed [1]. Figure 1 summarizes the carbon compounds which have been found in interstellar space and their abundances relative to hydrogen. Note that the carbon compounds decrease in abundance with increasing complexity.

There is no doubt that carbon plays a significant role in interstellar chemistry. Of the known interstellar molecules, 75 percent contain carbon, with CO binding about 10 percent of the total cosmic carbon. Many of the organic compounds of space (free radicals, molecular ions, etc.) are different from familiar terrestrial molecules. $C \equiv C$ is common in interstellar molecules, while $C=C$ is rare [1]. However, the presence of molecules such as HCN, formaldehyde, and cyanoacetylene in interstellar clouds, as well as energy sources such as UV irradiation, cosmic rays, and association of $H \rightarrow H_2$, make it possible for the reactions of chemical evolution producing biologically significant monomers to occur in intersteller space. Cyanides and formaldehydes are involved in the formation of amino acids. A probable pathway for purine formation is by rearrangement of HCN. Pyrimidines can be formed by reaction of cyanoacetylene and urea, and sugars are condensation produces of formaldehyde. These chemical products which are essential to life have not been observed in interstellar space, but as the technology of observation of interstellar molecules improves, it may become possible to detect more complex molecules such as heterocyclic compounds, if they exist. Many of the molecules detected in interstellar space, including those mentioned above, resemble products and precursor molecules found in prebiotic synthesis experiments.

---

[1]Figures in brackets refer to the literature references at the end of this paper.

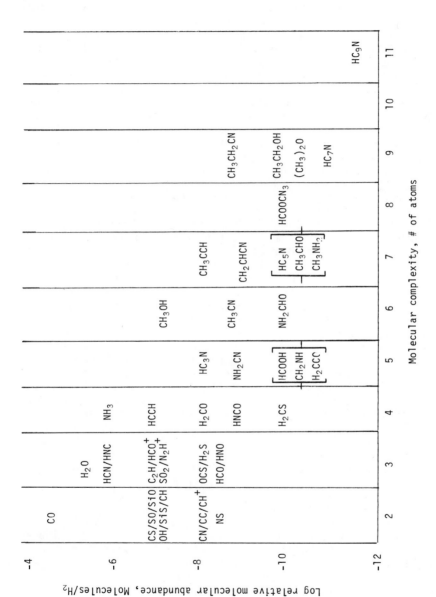

Figure 1.   Abundances of interstellar molecules relative to H_2 (1).

The similarities in products and pathways between interstellar molecules and terrestrial laboratory experiments imply a unity of physical and chemical laws in the universe. Given certain conditions and appropriate energy sources, the same chemical pathways will be followed to create certain products from the elements. That is not to say that life, even in primitive form, could be supported in interstellar space. The significant precursor molecules found in interstellar space are at extremely low concentrations, but if they were transported to planetary atmospheres, perhaps by comets, they might then react in the proper environment and evolve into self-replicating systems.

Is the existence of these important precursor carbon compounds limited to our galaxy? Perhaps not. Observations of interstellar clouds in other galaxies have shown the following molecules to exist in abundances similar to those in our galaxy: CO, HCN, OH, H_2O, CO, H_2 [1]. The continuing study of extragalactic chemistry may add to our knowledge of processes leading to the origin of life in our galaxy, and enlighten us as to the probability of the evolution of life beyond the Milky Way.

Another important area of study which is concerned with carbon compounds of extraterrestrial origin is that of meteorite analysis. The meteorites which fall to the Earth's surface present an opportunity to study at close range organic molecules imported from extraterrestrial sources. The origin of meteorites is not fully understood. It is generally thought that they share a common parent body with asteroids; however, their derivation from the dense nuclei of comets cannot be ruled out. Astronomical evidence favors a model whereby comets, asteroids, meteors, and planets in our solar system were formed by accretion from a common parent body; solidification of chondrite parent bodies is placed at 4.6×10^9 years ago by radionuclide decay studies [2]. Analysis of carbonaceous chondrites, which were probably derived from volatile retention areas of meteorite parent bodies and consequently have a higher concentration of carbon, has yielded many carbon compounds of interest in chemical evolution. Significant findings were made in studies of the Murchison meteorite, a type II carbonaceous chondrite, which fell in Australia in 1969. Gas chromatographic analysis of benzene-methanol extractions of the pulverized meteorite fragments revealed a mixture of alkanes with maximum abundance at C_{16} and C_{18} [3]. In addition to aliphatic hydrocarbons, 14 aromatic compounds were also extracted and identified by gas chromatography/mass spectrometry [4]. These compounds were not those generally associated with microorganisms. Of greatest significance was the detection of protein as well as non-protein amino acids. The D and L isomers were present in nearly equal abundances [3,5]. The presence of a wide variety of amino acids some of which are found in protein and others which are not has been confirmed by ion exchange chromatography, gas chromatography, and GC/MS [3,5]. Determination of $\delta^{13}C$ values for various carbon fractions showed that the meteorite, in contrast to terrestrial samples, is enriched in ^{13}C, minimizing the possibility of terrestrial contamination [3].

Essentially the same amino acids, and nearly equal quantities of D and L enantiomers, were detected in the Murray meteorite, another type II carbonaceous chondrite [6]. Recent expeditions to Antarctica have returned with a large number of meteorites, many of which are carbonaceous chondrites. These may have been protected from terrestrial contamination by the pristine Antarctic ice. Careful analysis of two of these, the Yamato (74662) and the Allan Hills (77306), both type II carbonaceous chondrites, by ion exchange chromatography, gas chromatography, and GC/MS, have detected a wide variety of both protein and non-protein amino acids in approximately equal D and L abundances [9,10]. Fifteen amino acids were detected in the Yamato meteorite and twenty in the Allan Hills, the most abundant being glycine and alanine. The amino acid content of the Yamato meteorite is comparable with that of the Murchison and Murray, but the Allan Hills contains 1/5 to 1/10 that quantity. Unlike earlier meteorites from other locations, the quantities of amino acids in the exterior and interior portions of the Yamato and Allan Hills meteorites are almost identical [9,10]. Thus, these samples may have been preserved without contamination since their fall in the blue ice of Antarctica, which is 250,000 years old in the region of collection.

Several factors indicate that the amino acids detected in all of these carbonaceous chondrites are indigenous and that they must have originated abiotically. First, the presence of protein and non-protein amino acids, with approximately equal quantities of D and L enantiomers points to a nonbiological origin and precludes terrestrial contamination. In addition, the non-extractable fraction of the Murchison is significantly heavier in ^{13}C than terrestrial samples. Finally, the relative abundances of some compounds detected resemble those of products formed in prebiotic synthesis experiments. The aliphatic hydrocarbons are randomly distributed in chain length, and the C_2, C_3, and C_4 amino acids have the highest concentrations (i.e., the most easily synthesized amino acids with the least number of possible structures are most abundant) [4].

Hydroxypyrimidines have been detected in the Murchison, Murray, and Orgueil carbonaceous chondrites in abundances similar to those of amino acids [7]. Earlier analyses of the Orgueil meteorite by thin layer chromatography of organic extracts indicated the presence of melamine, ammeline, adenine, and guanine [8]. Although these could not be confirmed by Folsome, et al., [7] using GC/MS, recent studies by Schwartz [11] and by Hayatsu, et al., [12] have shown that these constituents of the nucleic acids may indeed exist in the carbonaceous chondrites.

Another recent interesting finding is that previously unknown organic polymers or "amorphous carbon," which are noble gas carriers in meteorites, are actually carbynes. Five different carbynes have been identified in the Murchison and Allende carbonaceous chondrites [13,14]. Rare gas studies and thermal testing of the Allende carbynes indicate that they were probably formed at low temperature from the solar nebulae. Greater than 80 percent of the total carbon in the Allende meteorite was initially present as carbynes [14]. In contrast, the thermal stability of the Murchison carbynes suggest that they were formed as high temperature condensates of a solar gas. The anomalous nature of neon and xenon isotope abundances in the Murchison might be explained by a presolar origin of the meteorite involving a red giant, nova, or supernova [13].

The careful study of at least five different carbonaceous chondrites establishes the fact that these meteorites contain carbon compounds of extraterrestrial origin and of great significance in chemical evolution. Their presence confirms that the chemical reaction paths producing biologically important monomer molecules occur in the far reaches of our solar system.

After examining the chemistry of carbon in interstellar space and in meteorites which are samples from other parts of our solar system, we must turn to the carbon on our planet Earth. Of course, carbon is abundant in both the biosphere and lithosphere today. But how far back in time has carbon been utilized and recycled in the processes of life? The oldest fossil evidence of primitive life on Earth is found in stromatolitic structures found in ancient sediments. Stromatolites are biogenic structures commonly found in carbonate rocks that form as a result of carbonate precipitation by primitive life forms such as blue-green algae or bacteria followed by binding or trapping in sediment [15]. Although older stromatolitic fossils do not contain microstructures recognizable as cells, they are usually evaluated as possible biogenic structures by criteria such as depositional environment, surrounding geology, and similarities between fossil morphological features and those in present algal mat communities.

Very compelling evidence for the existence of life about 2.7 billion years ago has been found in Archean rocks at Slave Province and Steep Rock Lake, Canada, and also at Bulawago and Belingwe, Rhodesia [16-19]. Probable stromatolites found in the Pongola Supergroup, South Africa, are slightly older at 2.9-3 billion years old [20]. Knowl and Barghoorn reported the presence of microstructures resembling dividing yeast cells in stromatolite-like structures in Barberton, South African sediments 3.4 x 10[9] years old [19]. But almost as old as the most ancient rocks known are the stromatolitic structures of the Pilbara Block in Western Australia dated at 3.4-3.5 billion years. Dunlop et al., [21] first reported matlike structures similar to stromatolites in the Warrawoona Group of the Pilbara Block, but these were isolated, and a biogenic origin was uncertain. Two more recent reports describe macrostructures existing in cherts of the Pilbara Block; these are layers and conical structures which closely resemble younger stromatolites attributed to algal mat communities [22,23]. Thus, there is convincing evidence, at least on the macrostructural scale, for a benthic microbiota existing 3.4-3.5 billion years ago.

Since the number of well preserved microfossils in metamorphosed sediments which can be identified as biological cells is small, the examination of organic compounds in ancient rocks is a method that can add significantly to our knowledge of biological activity during the Precambrian era. Most organic molecules which survive weathering in Precambrian rocks are either destroyed or rendered unrecognizable by the high temperatures of metamorphism. The large geologic time span also reduces the temperature required for drastic alteration of these molecules [24]. Measurement of $^{13}C/^{12}C$ of organic carbon fractions is an important method that is independent of molecular form and gives clues to possible sources of the carbon. Carbon in biological organic matter has a significantly lower $^{13}C/^{12}C$ ratio, due to fractionation by photosynthesis, than carbonate minerals and atmospheric CO_2. $\delta^{13}C$ values in the reduced carbon fraction of Precambrian stromatolites has been determined at -18.8 to -29.8, while $\delta^{13}C$ of carbonate carbon from the same rock source is -0.7 to +2.4 [24]. $\delta^{13}C$ values in rocks of marine origin are generally -20 to -30 per mil, but $\delta^{13}C$ is even lower (-34 to -37) in some of the older Precambrian rocks. Some explanations proposed are that the isotopic ratio of carbonate available for photosynthesis could have been different in Precambrian times; that during diagenesis mobile compounds with higher amounts of ^{13}C were released; or that differences in the mode of fixing inorganic carbon existed in Precambrian times [24].

Molecular fossils have been successfully identified in younger Precambrian rocks and linked to certain classes of biological source material. In organic analyses of ancient sediments the cleaned, pulverized rocks are treated with organic solvents to extract a soluble fraction containing the less complex and more easily identifiable compounds. However, this fraction is more subject to contamination since it is not locked within the rock matrix. Normal alkanes have been identified in extracts of the 3 billion year old Fig Tree Shale. These alkanes have a probable biological origin in cellular lipids. The odd and even-numbered alkanes are evenly distributed, a characteristic of alkanes from ancient rocks. It is uncertain, however, whether these compounds were present at the time of deposition or derived from a later source [24].

The isoprenoid hycrocarbons pristane and phytane (derived from the phytol side chain of chlorophyll), as well as porphyrins, have been detected in organic extracts of the Nonesuch Shale of 1.1 billion year age [23]. Their presence points to the existence of photosynthetic pigments in the Precambrian era, but it is also possible that these extractable substances could have been contributed to the rock at a later time. However, in this instance contamination appears to be less likely on account of the large abundance of organic material in this shale.

Data showing a discrepancy between the $\delta^{13}C$ values of carbon in kerogen and in the extractable organic fractions of Precambrian rocks suggests that extractable compounds might have come from younger source materials leaching into the rocks [24]. Therefore, although kerogen, or the insoluble organic matter in rocks, is much more complex material and presents problems in analysis, it is much less likely to be contaminated by younger compounds than the soluble material. Kerogen is isolated by destruction of the rock matrix by HCl and HF treatment. Analysis then proceeds by splitting the complex molecules into smaller fragments by pyrolysis, or strong oxidation or reduction. The smaller molecules may be studied by gas chromatography or GC/MS. Generally the greater degree of metamorphism the rock has undergone, the larger will be the ratio of aromatic to aliphatic hydrocarbons, due to dehydrogenation of original aliphatic structures. Analysis of the Fig Tree Shale kerogen has shown the presence of benzene, toluene, and other substituted benzenes [24] as well as the normal alkanes C_{12}-C_{19} [25]. The ultimate effect of metamorphism on kerogen is graphitization [24].

Recently, researchers have detected 2,5-dimethylfuran and 2-methylfuran and normal alkanes in kerogen of the 2.7×10^9 year old Belingwe, Rhodesia stromatolites, by the method of pyrolysis/ GC/MS [26]. They concluded that although furans could probably be derived from many compounds, their probable origin is in bacterial and algal sugars, and that the alkanes are either products of decarboxylation of fatty acids or unaltered constituents of ancient organisms.

In our laboratory, we have undertaken studies of the Isua supracrustal rocks of West Greenland, the oldest rocks known, dated at more than 3,750 million years. These are highly metamorphosed sediments containing graphite. If morphological or chemical evidence of biological activity is found in these rocks, it would constitute a historical record for the existence of life less than a billion years after the formation of the Earth. Since as many as four periods of recrystallization have occurred in the rocks after sediment deposition, the preservation of any biological cells within the rock is highly improbable. Structures within the rock reported to be microfossils [27,28] have been described as mineral filled fluid inclusions [29], but they are more likely to be weathered dolomite grains possibly primary in nature [30].

Organic geochemical studies must therefore provide any discoveries about the existence of life 3.8 billion years ago. Yet, any chemical evidence left by cells existing at that time might have been destroyed or at least considerably altered by the extensive metamorphism. A laboratory procedure of cleaning the rocks by mechanical means, then by solvent extractions and crushing is followed until no organic compounds are detected by gas chromatography [31]. This ensures that all superficial contamination is removed from the rocks. Then the rock matrix is dissolved by HCl and HF. Usually no organics are found in these solutions [31].   The acid insoluble residue is then treated with further solvent extractions until no organics are detected.   The final residue consists mainly of graphite, which is then analyzed by pyrolysis (at 250-1000°) gas chromatography/mass spectrometry.   Ion fragments from the analyses are mostly small ($\leq C_2$) with some larger fragments (m/e = 175) at less than ng/g concentrations.   A benzene fragment has also been identified (31).   These studies suggest that some of the Isua graphite was derived from organic compounds during metamorphism, but is is impossible to say from the scant evidence left, whether these compounds are of abiotic or of biological origin.   The carbon isotope values for the graphite present the best evidence for life 3.8 billion years ago. $\delta^{13}C$ values are -26.9 to -5.9, suggesting, even with the possible effects of metamorphism, that photosynthetic fractionation may have taken place [32,33].   Although the complex metamorphic history of the oldest rocks known on Earth has made it difficult for researchers to determine whether life existed 3.8 billion years ago, some of the Isua graphite appears to be derived from organic carbon deposited with sediments of that time.   A definitive answer to this question is not yet forthcoming from the present studies.

The history of carbon in our universe has been of great importance in the evolution of life on Earth and quite possibly in other galaxies as well.   Elemental carbon was formed sometime after the galaxy 10 to 15 billion years ago.   Observations of interstellar carbon compounds matching the important percursor molecules of prebiotic synthesis confirm that reactions of chemical evolution have occurred beyond our planet since the formation of the solar system.   The presence of hydrocarbons and amino acids in meteorites as old as the solar system supports this hypothesis.   On the Earth, the oldest organic carbon is contained in Isua, West Greenland graphite formed from sediments deposited 3.8 billion years ago.   The oldest known biological carbon is contained in Western Australian stromatolites 3.4 to 3.5 billion years old.   Further analytical studies of meteorites and ancient sediments and further-reaching observations of interstellar chemistry can help us to understand the carbon chemistry of life, how it began, and the chances of life based upon carbon existing elsewhere in our solar system and beyond.

References

[1] Gammon, Richard H. Chem. Eng. News, 56, 1978, 21.
[2] McCall, G. J. "Meteorites and Their Origins," John Wiley and Sons, New York, 1973, p. 34-35.
[3] Kvenvolden, K., Lawless, J., Pering, K., Peterson, E., Flores, J., Ponnamperuma, C., Kaplan, I. R., Moore, C., Nature, 228, 1970, 923.
[4] Ponnamperuma, C., Ann. N. Y. Acad. Sci., 194, 1972, 56.

[5] Kvenvolden, K., Lawless, J. G., Ponnamperuma, C., Proc. Natl.
     Acad. Sci., 68, 1971, 486.
[6] Lawless, J. G., Kvenvolden, K. A., Peterson, E.,
     Ponnamperuma, C., Moore, C., Science, 173, 1971, 626.
[7] Folsome, C. E., Lawless, J. G., Romiez, M., Ponnamperuma, C.,
     Geochim. Cosmochim. Acta, 37, 1973, 455.
[8] Hayatsu, R., Studier, M., Oda, A., Fuse, K., Anders, E.,
     Geochim. Cosmochim. Acta, 32, 1968, 175.
[9] Shimoyama, A., Ponnamperuma, C., Yanai, K., Nature, 282,
     1979, 394.
[10] Kotra, R. K., Shimoyama, A., Ponnamperuma, C., Hare, P. E.,
     J. Mol. Evol., 13, 1979, 179.
[11] Stoks, P. E., Schwartz, A. W., Geochim. Cosmochim. Acta, 45,
     1981, 563.
[12] Hayatsu, R., Studier, M. H., Moore, L. P., Anders, E.,
     Geochim. Cosmochim. Acta, 39, 1975, 471.
[13] Whittaker, A. G., Watts, E. J., Lewis, R. S., Anders, E.,
     Science, 209, 1980, 1512.
[14] Hayatsu, R., Scott, R. G., Studier, M. H., Lewis, R. S.,
     Anders, E., Science, 209, 1980, 1515.
[15] Henderson, J. B., in "Chemical Evolution of the Early Pre-
     cambrian", Ponnamperuma, C., ed., Academic Press, New York,
     1977.
[16] Henderson, J. B., Can. J. Earth Sci., 12, 1975, 1619.
[17] Joliffe, A. W., Econ. Geol., 50, 1955, 373.
[18] Macgregor, A. M., Trans. Geol. Soc. S. Afr., 43, 1941, 9.
[19] Schopf, J. W., Oehler, D. Z., Horodyski, R. J., Kvenvolden,
     K. A., J. Paleont. 45, 1971, 477.
[20] Mason, T. R., Von Brunn, V., Nature, 266, 1977, 47.
[21] Knowl, A. H., Barghoorn, E. S., Science, 198, 1977, 396.
[22] Dunlop, J. S. R., Muir, M. D., Milne, V. A., Groves, D. I.,
     Nature, 274, 1978, 676.
[23] Lowe, D. R., Nature, 284, 1980, 441.
[24] Walter, M. R., Buick, R., Dunlop, J. S. R., Nature, 284,
     1980, 443.
[25] Hoering, T. C., in "Researches in Geochemistry", Abelson,
     P. H., ed., John Wiley and Sons, New York, 1967, p. 87-111.
[26] Scott, W. M., Modzeleski, V. E., Nagy, B., Nature, 225,
     1970, 1129.
[27] Sklarew, D. S., Nagy, B., Proc. Natl. Acad. Sci., 76, 1979,
     10.
[28] Pflug, H. D., Naturwissenschaften, 65, 1978, 611.
[29] Pflug, H. D., Nature, 280, 1979, 483.
[30] Barghoorn, E. S., personal communication.
[31] Walters, C., Shimoyama, A., and Ponnamperuma, C., Proc. 6th
     Meeting Intern. Soc. Study Origins of Life, 1980, Jerusalem,
     Israel.
[32] Oehler, D. A., Smith, J. W., Precambrian Res., 5, 1977, 221.
[33] Schidlowski, M., Appel, P. W. V., Eichmann, R., Junge, C. E.,
     Geochim. Cosmochim. Acta, 43, 1979, 189.

RECEIVED August 4, 1981.

Results of a Dating Attempt:
Chemical and Physical Measurements Relevant to the Cause of the Cretaceous–Tertiary Extinctions[1]

FRANK ASARO and HELEN V. MICHEL—University of California—Berkeley
Berkeley, CA 94720

LUIS W. ALVAREZ—University of California—Berkeley, Space Sciences Laboratory,
Lawrence Berkeley Laboratory, Berkeley, CA 94720

WALTER ALVAREZ—University of California—Berkeley, Department of
Geology and Geophysics, Berkeley, CA 94720

The historical background is presented for the asteroid-impact theory that is based on the iridium anomaly found in rocks frm the Cretaceous-Tertiary boundary. Recent measurements of Ir, Pt, and Au abundances from such rocks in Denmark have shown that the element abundance ratios are different from mantle-derived sources and agree with values for chondritic meteorites within one standard deviation of the measurement errors (7-10%). Rare-earth patterns for these rocks are presented, and their potential significance with respect to the asteroid-impact site are discussed. Future directions of experimental studies necessary to prove aspects of the asteroid-impact theory are also discussed.

In Gubbio, Italy, a 1 cm layer of clay between extensive limestone formations marks the boundary between the Cretaceous and Tertiary Periods. This clay layer was known to have been deposited about 65 million years ago when many life forms became extinct, but the length of time associated with the deposition was not known. In an attempt to measure this time with normally deposited meteoritic material as a clock, extensive measurements of iridium abundances (and those of many other elements) were made on the Gubbio rocks. Neutron activation analysis was the principal tool used in these studies. About 50 elements were searched for in materials like the earth's crust, about 40 were detected and about 30 were measured with useful precision [26-28][2].

---

[1]Incorporated in paper are data and references that were acquired after the Symposium and before September 1, 1980.

[2]Figures in brackets refer to the literature references at the end of this paper.

0097-6156/82/0176-0401$05.00/0

We were not able to determine exactly the length of time associated with the deposition of the clay layer. Instead the laboratory studies on the chemical and physical nature of the Cretaceous-Tertiary boundary led to the theory that an asteroid collision with the earth was responsible for the extinction of many forms of life including the dinosaurs.

Experimental Research

Our research to date has demonstrated that deep-sea lime-stones exposed in Italy [1-3], Denmark [1], New Zealand [4], and northern Spain [6] show ususual increases in the abundance of iridium (Ir) above the background level at exactly the time of the Cretaceous-Tertiary (C-T) extinctions, 65 million years ago. As Ir is much depleted on the earth's surface with respect to the average solar system abundance, such an enrichment is not likely to have come from a terrestrial source. Extraterrestrial sources, however, contain three orders of magnitude more Ir than terrestrial sources and are much more likely to have produced the enrichment. Measurement of the isotopic ratio of the two iridium isotopes in the Italian C-T boundary layer showed that it was identical with that of a terrestrial source [1], and hence the iridium did not originate from a separate supernova. This suggests a solar system origin but does not rule out an interstellar origin for the asteroid as postulated by Napier and Clube for impacting planetesimals [29]. Many sources of extraterrestrial material from within the solar system were considered [1] and subjected to three tests. They had to be (1) able to deposit the Ir, (2) able to cause the C-T extinctions, and (3) likely to occur in a period of ∿65 million years. Only one source survived the tests.

Asteroid-Impact Hypothesis

We have developed an asteroid-impact theory that satisfied these conditions and many others [1]. The theory assumes the following events. An Apollo object (an earth-orbit-crossing asteroid) with about a 10 km diameter struck the earth 65 million years ago. It vaporized as did a much larger mass of terrestrial material. About 20 percent of this combined mass (modeled after the volcanic explosion of Krakatoa) ended up in the stratosphere, encircled the earth and blocked out nearly all of the direct sunlight. Photosynthesis then stopped. The larger animals which depended on the plant food chain died while some of the smaller ones (needing less food) survived. Some seeds could have survived for several years and then reestablished the plants after the dust cloud sank to the earth and the direct sunlight reappeared. Mass extinctions of this type have occurred about every 100 million years.

Killed the Dinosaurs?

Checking of Hypothesis

Four different ways of calculating the asteroid diameter all give a value of ~10 km and this consistency lends confidence to the asteroid-impact theory. The Ir anomaly was first observed by us in Italian rock. Our theory predicted that the unusually abundant Ir should appear all over the world where the C-T boundary is exposed (intact). Part of the hypothesis was confirmed when the anomalously high Ir abundance was found in the C-T boundary layers in Denmark, northern and south-east Spain, and half-way round the world in New Zealand. Another prediction of the theory is that a component of the clay layer at the C-T boundary would be different in composition from other clays in the same section because it contained a component from the impact site. This prediction was confirmed in measurements of the Italian and Danish sections [1].

Other Work

Dutch geologists, after learning of the Italian Ir data, found a similar Ir anomaly in a rock section from Caravaca (Barranco-del-Gredero) in south-east Spain [5]. On a sample kindly supplied by J. Smit from this section we were able to confirm the iridium abundances and determine the variation in abundance within a few centimeters of the C-T boundary [7]. Smit and Hertogen [5] also measured the abundances of another platinum group element, osmium, and it also exhibited an anomalously high abundance at the C-T boundary near Caravaca. These authors also stated the ratio of two osmium isotopes in the C-T boundary was indistinguishable within 0.1 percent from the terrestrial (i.e., solar system) ratio.

A concern has been expressed [8,9] about the correctness of the stratigraphy upon which the selection [1] of our Italian samples was based. Very recent work [10] on a deep-sea core from the southeastern Atlantic Ocean near South Africa, however, supported the Italian stratigraphy.

There have been a number of suggestions in the past that an extraterrestrial object impacting on the earth caused or could cause massive extinctions of life. E. J. Öpik [11], for example, discussed the lethal effects which could be caused by the heat generated from such objects striking the earth, and H. C. Urey [12] stated specifically that a comet was probably the cause of the Cretaceous-Tertiary extinctions. There have also been science fiction stories and a movie relating to the effects. The events likely to occur if the sunlight were temporarily "turned off" have also been discussed [13]. Our deduction in contrast to the others is based on physical science data (the iridium anomaly) and is the only explanation we found which explained the Ir anomaloy could cause the massive extinction of life and was likely to occur in a period of ~100 million years.

Smit and Hertogen [5] also concluded a large meteorite may have struck the earth and caused the extinctions. K. J. Hsü [14] suggested the extinctions were caused by cyanide poisoning resulting from a cometary impact, and Cesare Emiliani [15] had the suggestion that a sudden heat flash, possibly caused by an asteroid impact, could have caused the C-T extinctions. Ganapathy [19] measured the abundances of Pt group elements (Ru, Pd, Re, Os, Ir, Pt), Au, Co, and Ni in the Danish C-T boundary layer and confirmed that it contained a chondritic component.

Future Directions

The Ir anomaly has been observed in uplifted marine sediments in four locations in Italy, two in Spain, and one in New Zealand. It has not been found yet in continental sediments or deep-sea cores. If the asteroid-impact theory is valid, the Ir anomaly should be found wherever the C-T boundary is intact.

Additional measurements of the abundances of platinum group elements should be made in C-T boundary layer samples. Figure 1 shows the ratio of abundances of different elements to that of Ir in various mantle materials, divided by the same quantities for Cl chondrites [16]. The Pt and Au ratios approach the chondritic abundances as the degree of partial melting of the mantle gets higher. An ultra-basic nodule, [17] which may represent the mantle, has very close to the chondritic ratios for Ir and Pt. Some of our recent measurements of Pt, Au and non-volatile Os as well as Ir in the Danish C-T boundary layer are shown in figure 1 and Table 1. The Pt and Ir data are consistent with a chondritic origin as predicted by the asteroid-impact theory. Part of the data of Ganapathy is also included in figure 1 and Table 1. Besides additional measurements of Pt group elements on the Danish and other C-T boundary layers, such measurements are needed on nearby Cretaceous and Tertiary rock formations. $^{187}Os/^{186}Os$ isotopic ratios [20] might establish if the Os and Re have maintained chondritic relative abundances throughout the predominant part of their existence since coalescence from dispersed materials billions of years ago.

More studies of the clay mineralogy of the C-T boundary layer are needed. These may indicate if the boundary deposition was associated with an extraordinary event.

Rare earth abundance patterns, particularly of the clay fraction, may also help determine the origin of the terrestrial components. Rare earth patterns in clay fractions of sediments tend to inherit the patterns of the rocks from which they originated [24]. In figure 2 are shown several samples of the rare earth abundance patterns of nitric-acid-insoluble residues from the Danish boundary layer and the limestones above and below. Such patterns along with the other chemical data may indicate the

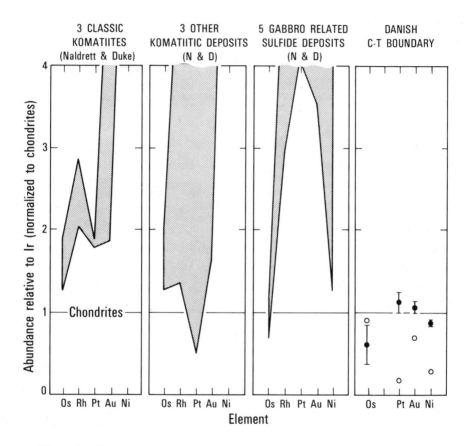

Figure 1. Relative abundances of some Pt group elements, Au, and Ni with respect to Ir.

Each abundance was divided by the abundance of that element (except for Rh) in Type I carbonaceous chondrites. Rh abundances were divided by Rh abundances in other types of chondrites as CI values were not available. Errors in the LBL measurements reflect 1 σ values of the counting errors, except for the Au error. The latter is the root-mean-square deviation of six measurements, because the six values were not consistent within counting errors. The Os measurement was on a HNO_3-insoluble residue that had been fired to 800°C. Key: ●, this work; and ○, previous work of Ganapathy.

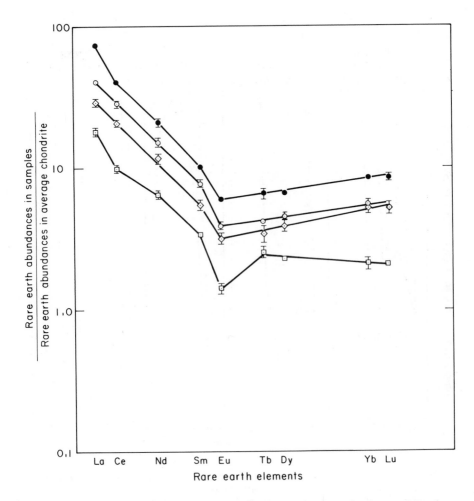

Figure 2.   Rare earth abundance patterns as normalized to the data of Masuda et al. (except for Tb) for the Leedy L6 meteorite (30).

The Tb abundance in meteorites is assumed to be 0.5 ppm. Errors for the Danish Cretaceous (3 samples) and Tertiary (3 samples) HNO₃-insoluble residues are root-mean-square deviations. Errors for the HNO₃-insoluble residues from the Gubbio and Danish boundary layers are 1 σ values of the counting errors. Key: ●, Gubbio boundary layer residue; ⊙, Danish Cretaceous residues; ◇, Danish Tertiary residues; and ⊡, Danish boundary layer residue.

Table 1.   On the Possible Chondritic Nature of the Danish
C-T Boundary[a].

	Ganapathy[b] (Average of 2)	Our Work	CI Chondrites
Pt/Ir	0.40	2.24 ± 0.23[c]	1.98[d]
Ni/1000 Ir	6.2	16.6 ± 0.5[c]	20[d]
Co/1000 Ir	0.82	2.07 ± 0.03[c]	1.01[e,d]
Au/Ir	0.21	0.319 ± 0.020[c]	0.296[d]

[a]If the C-T boundary includes an asteroid component with chondritic abundances, the ratios of the abundances of various elements to Ir should be larger than chondritic because of terrestrial contributions.

[b]Ratios were calculated from abundances given in Ganapathy [19].

[c]Pt and Au data are from unpublished work of F. Asaro, H. V. Michel, W. Alvarez, and L. W. Alvarez, 1980. The others were published in reference 1.

[d]Au and Ir data are from U. Krähenbühl. J. W. Morgan, R. Ganapathy, and E. Anders, Geochim. Cosmochim. Acta 37, 1353 (1973) [21]. Pt datum is from W. D. Ehmann and D. E. Gillum, Chem. Geol. 9, 1 (1972) [27].

[e]The Co abundance is a weighted average (507 ppm) of the data given by Carleton B. Moore in Handbook of Elemental Abundances in Meteorites edited by Brian Mason and published by Gordon and Breach Science Publishers (New York), 1971 [23].

type of rock from which the boundary layer originated. This may help indicate if the origin was continental, sea floor, rift or otherwise.

   The intensity of the C-T iridium anomaly as a function of geographical location along with the mineralogical and chemical studies of the boundary layers may suggest an impact location when considerably more data are obtained.

   If the asteroid-impact theory is correct, the extinctions should be repetitive and the Ir anomaly should be observed in other geological stratigraphic levels corresponding to known extinctions. About five other massive extinctions (besides the one at the end of the Cretaceous Period) have been noted [25]. These come at the end of the Cambrian (∼500 MY ago), the Ordovician (∼435 MY age), the Devonian (∼345 MY ago), the Permian (∼230 MY ago) and the Triassic (∼195 MY ago) Periods.

   In the next four years we hope to have tested the concept of worldwide distribution of the Ir anomaly, tested the premise of an extraordinary extraterrestrial origin of the anomaly, attempted to determine the location of the C-T impact site and tested the repetitive nature of the proposed impact. From the answers to these tests and the questions which will develop, the extent to which the theory is correct can be evaluated and the need and direction for future work can be assessed.

This work was done with support from the Department of Energy under Contract W-7405-ENG-48. Partial support was received from the California Space Group and the National Aeronautics and Space Administration Ames Research Center in the later stages of the work.

References

[1] Alvarez, L. W., Alvarez, W., Asaro, F., and Michel, H. V., Extraterrestrial Cause for the Cretaceous-Tertiary Extinction, Science 208, 1095-1108 (1980).

[2] Alvarez, W., Alvarez, L. W., Asaro, F. and Michel, H. V., Experimental Evidence in Support of an Extraterrestrial Trigger for the Cretaceous-Tertiary Extinctions, EOS Trans. Am. Geophys. Union 60, 734 (1979).

[3] Alvarez, W., Alvarez, L. W., Asaro, F. and Michel, H. V., Anomalous Iridium Levels at the Cretaceous-Tertiary Boundary at Gubbio, Italy: Negative Results of Tests for a Supernova Origin, Geol. Soc. America Abstract 11, 378 (1979).

[4] Russell, D., Alvarez, L. W., Alvarez, W., Michel, H. V. and Asaro, unpublished data (1980) cited in reference 1.

[5] Smit, J. and Hertogen, J., An extraterrestrial event at the Cretaceous Tertiary boundary, Nature 285, 198-200 (1980).

[6] Arthur, M., Alvarez, L. W., Alvarez, W., Michel, H. V. and Asaro, F., unpublished data, 1980.

[7] Alvarez, L. W., Alvarez, W., Michel, H. V. and Asaro, F., unpublished data, 1979. Ir values are plotted in reference 5.

[8] Wezel, F. C., The Scaglia Rossa Formation of Central Italy: results and problems emerging from a regional study, Ateneo Parmense, Acta Nat. 15, 243-259 (1979).

[9] Surlyk, Finn, The Cretaceous-Tertiary boundary event, Nature 285, 187-188 (1980).

[10] Private communication to Walter Alvarez from K. S. Hsü and News Release NSF PR 80-56 (1980).

[11] Öpik, E. S., On the catastrophic effects of collisions with celestial bodies, Irish Astronomical Journal 5, No. 1, 34, 1978.

[12] Urey, H. C., Cometary Collisions and Geological Periods, Nature 242, 32, 1973.

[13] K-TEC group (P. Beland et al.), Cretaceous-Tertiary extinctions and Possible Terrestrial and Extraterrestrial Causes, Proceedings of Workshop, National Museum of Natural Sciences, Ottawa, 16 and 17 November 1976, pp. 144-149.

[14] Hsü, K. J., Terrestrial catastrophe caused by cometary impact at the end of the Cretaceous, Nature 285, 201-203 (1980).

[15] Emiliani, Cesare, Death and Renovation at the End of the Mesozoic, EOS 61, No. 26, 505-506 (1980).
[16] Data for basic materials was taken from, Platinum Metals in Magmatic Sulfide Ores, by A. J. Naldrett and J. M. Duke, Science 208, 1417-1424 (1980).
[17] Greenland, L. P., Gottfried, D. and Tilling, R. I., Iridium in some calcic and calc-alkaline batholithic rocks of the western United States, Chem. Geol. 14, 117-122 (1974) reported in reference 18.
[18] Crocket, J. H., Platinum-Group Elements in Mafic and Ultramafic Rocks: A Survey, Canadian Mineralogist 17, 391-402 (1979).
[19] Ganapathy, R., A Major Meteorite Impact on the Earth 65 Million Years Ago, Evidence from the Cretaceous-Tertiary Boundary Clay, Science 209, 921-923 (1980).
[20] Allégre, C. J. and Luck, J. M., Osmium isotopes as petrogenetic and geological tracers, Earth Planet. Sci. Lett. 48, 148-154 (1980).
[21] Krähenbühl, U., Morgan, S. W., Ganapathy, R. and Anders, E., Abundance of 17 trace elements in carbonaceous chrondrites, Geochem. Cosmochim. Acta 37, 1353-1370 (1973).
[22] Ehmann, W. D. and Gillum, D. E., Platinum and gold in chrondritic meteorites, Chem. Geol. 9, 1-11 (1972).
[23] Handbook of Elemental Abundances in Meteorites, Brian Mason ed., Gordon and Breach Science Publishers, London, 1971.
[24] Cullers, R. L., Yeh, Long-Tsu, Chaudhuri, S. and Guidotti, C. V., Rare earth elements in silurian peltic schists from N. W. Maine, Geochim. Cosmochim. Acta 38, 389-400 (1974).
[25] Newell, N. D., Crisis in the History of Life, Scientific American 208, No. 2, 76-92 (1963).
[26] Perlman, I. and Asaro, F., Pottery Analysis by Neutron Activation, in Science and Archaeology, R. H. Brill, ed., MIT Press, Cambridge, Mass., Archaeometry 11, 21-52 (1969).
[27] Yellin, J., Perlman, I., Asaro, F., Michel, H. V. and Mosier, D. F., Comparison of Neutron Activation Analysis from the Lawrence Laboratory and the Hebrew University, Archaeometry 20, 95-100 (1978).
[28] Asaro, F., Applied Gamma-Ray Spectrometry and Neutron Activation Analysis, Proceedings of the XX. Colloquium Spectroscopicum Internationale and 7. International Conference on Atomic Spectroscopy Praha 1977 Invited Lectures II, 413-426.
[29] Napier, W. M. and Clube, S. V. M., A theory of terrestrial catastrophism, Nature 282, London, 455-459 (1979).
[30] Masuda, A., Nakamura, N. and Tanaka, T., Fine structures of mutually normalized rare-earth patterns of chondrites, Geochim. Cosmochim. Acta 37, 239-248 (1973).

RECEIVED April 27, 1981.

A Ceramic Compositional Interpretation of Incense-Burner Trade in the Palenque Area, Mexico

RONALD L. BISHOP[1]

Museum of Fine Arts, Research Laboratory, Boston, MA 02115

ROBERT L. RANDS

Southern Illinois University—Carbondale, Department of Anthropology, Carbondale, IL 62901

GARMAN HARBOTTLE

Brookhaven National Laboratory, Department of Chemistry, Upton, NY 11973

The Classic Maya culture of southern Mesoamerica had a strong theocratic orientation. Notable aspects of ceremonialism in the Palenque area include incense-burning, expressed archaeologically in ceramic supports and receptacles (incensarios). The supports are notable for their technological construction and ornate iconographic content. Incensarios form part of a much larger body of regional ceramics now being intensively studied. Objectives are to determine manufacturing centers and the directional flow of trading relationships; therefore paste composition is accorded special importance. Compositional data are derived through sampling that is successively less extensive but more intensive (binocular examination, petrography, and neutron activation). Focussing primarily on chemical composition, data reduction is achieved by a related set of vector manipulative techniques (iterative cluster analysis; discriminant functions; classification statistics). The resulting paste compositional reference units are evaluated by correlation with petrographic and archaeological information. Compositional data for incensarios are projected on the reference units for the Palenque region as a whole. Preliminary findings suggest that the ceremonial center of Palenque was the major focus of incensario manufacture. Apparently, exchange of incensarios to outlying communities, beyond the boundaries of intensive trace in other ceramics, helped to maintain Palenque's socio-religious primacy.

The study of ceramic trade is multidimensional, relating to cultural and physiochemical variables that can only be treated selectively in a progress report. The major lowland Maya site of

[1] Current address: Brookhaven National Laboratory, Department of Chemistry, Upton, NY 11973.

Palenque, Mexico, is the focal point of such an investigation
(figure 1). Although problem orientation is in part directed
toward chronological refinement [1][1], a primary goal has been
the delineation of a "sustaining area" or other spatial units
characterized by distinctive patterns of exchange [2,3,4,5].

In the present paper, attention is focussed on a particular
class of ceramic objects — incensarios or paraphernalia used in
the ritual burning of incense. In attempting to determine a
manufacturing source for this functionally specialized class of
pottery, it has been necessary to make compositional characteri-
zations within the broader context of the regional ceramics. In
the absence of such an approach, it would be far more difficult to
assess relative probabilities that the incense burners were
manufactured at a single center or in multiple centers and to
establish even the general location of the zone or zones of
production.

In an earlier petrographic study it was noted that one
regionally restricted class of incensarios — massive but elabor-
ately decorated flanged cylinders that apparently served as
supports for functional censers — were so similar mineralogically
as to suggest their manufacture at a single center, perhaps
Palenque [6]. It was recognized that patterned differences in
symbolic content might indicate the presence of different manufacturing centers or reflect temporal distinctions. As the
sensitivity of neutron activation analysis was brought to the
investigation of trade in the Palenque region, the incensario
supports and other classes of incense burners were analyzed as
one aspect of the research. In previous studies of Maya incensarios, it had been suggested, alternatively, that these objects
formed part of a folk cult [7] or, at least in the case of the
flanged cylinders, functioned on a hierarchal level of Maya
civilization [8]. Demonstration of patterns of exchange for
these various ceramic objects (incensario supports, functional
censers, and other classes of pottery) should contribute to an
understanding of this and wider problems.

In addition to the ornately decorated supports or stands,
incensario forms that were analyzed include pedestal censers,
ladles (often with effigy handles), bowls, and a conical to
trumpet-shaped container that appears to have been utilized in
conjunction with the supports, resting in the orifice of the
cylinder (figure 2). All form classes except the supports show
signs of interior burning; traces of resin are occasionally
present. Evidence linking flanged cylinders to incense burning
activities is discussed elsewhere [9,10,11]. Here we simply note
that the outcurved walls of the trumpet-shaped censer provide a
range in diameter so as to fit into one or another or the tubular

---

[1]Figures in brackets indicate the literature references at the end
of this paper.

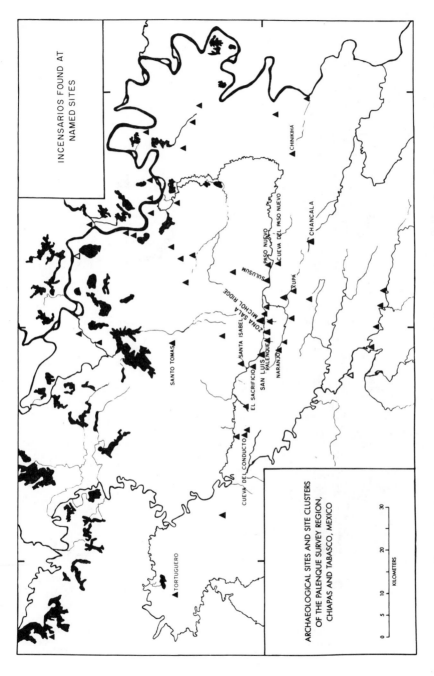

Figure 1. Sites of the Palenque region from which incensario fragments have been analyzed.

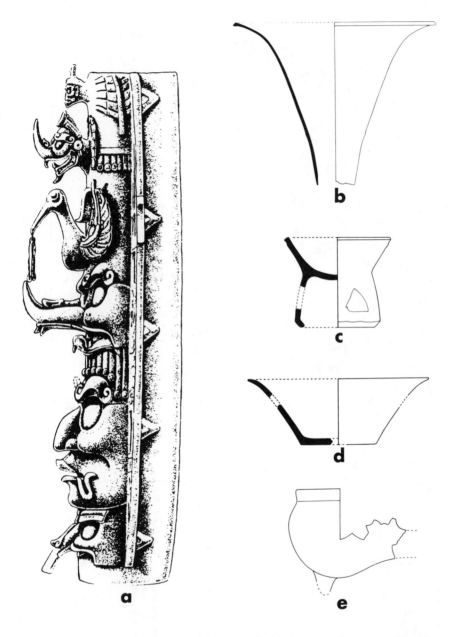

Figure 2.   Incensario form classes. Key: a, cylindrical support; b, trumpet-shaped receptacle; c, pedestal; d, bowl; and e, ladle. Specimens not to scale. Also, b, base of specimen missing; and a, after Rands 1969 (courtesy, Instituto Nacional de Antropologia e Historia), not analyzed.

supports, there being no necessity for a particular support and receptacle to be matched in diameter. These objects could have been used interchangably, being manufactured at different localities and still serving effectively in combination. The trumpet-shaped container is found consistently in cave and pyramid contexts that include the flanged cylindrical support, although the latter often occurs independently of the removable containers. Specimens are of Late Classic and perhaps Terminal Classic date (ca. A.D. 600-900).

Information about ceramic pastes has been derived through sampling that was successively less extensive but more intensive, ideally progressing from binocular examination to petrography to neutron activation as illustrated in figure 3 [12]. Chemical data were used for multivariate analysis, ordinal petrographic data being projected against the chemically characterized ceramic specimens. Steps in group formation are outlined below; we wish to make two points here. First, sharper inferences may be drawn about localities for clay procurement when chemical data are viewed from a geological-mineralogical perspective. This approach has been utilized in the ongoing research but enters only minimally into discussion in the present paper. Second, the analytical procedure enables mineralogical patterns which have strong chemical correlates to be seen in a body of ceramics which is much larger than the chemically sampled pottery, broadening the archaeological application of the compositional information.

The need for a broad compositional perspective was one factor in our sampling, which had the objective of distinguishing locally made pottery on an intraregional basis. Incensarios are uncommon relative to many other functional classes of ceramics and do not, therefore, comprise a reliable body of material on which to formulate groups that may be indigenous to a particular locality. Specialized functionally and in depositional context, incensarios were not represented in collections from a number of sites and so would contribute minimally to our initial understanding of basic compositional profiles within the Palenque region. Moreover, incensarios excavated at Palenque lack the time depth which is one indication of localized production according to the postulate of sustained, least-effort procurement of localized clay resources. In view of these factors, our initial chemical groups were largely based on jars, bowls, plates and other well represented form classes; only after groups had been formulated on this basis were most of the incensarios analyzed. It should also be noted that the capacity of our clustering program, CLUS, would not permit simultaneous analysis of all the incensario and non-incensario specimens that were chemically analyzed. In recognition of the formation of groups that permits the subsequent comparison of additional specimens, with the potential of increasing group membership, the term reference unit is employed.

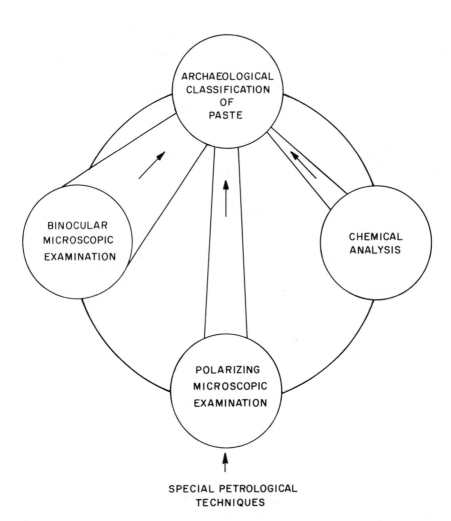

Figure 3. Starting with the binocular microscope, discriminatory power increases in a counterclockwise direction, as indicated by the circular background. Variations in flowline width reflect the differential sampling capabilities of the techniques. Special petrological techniques include X-ray diffraction, electron and proton microprobe, staining, and heavy mineral separation.

A further point requires explication. Just as archaeological problem orientations differ, leading to "lumping" in one case and "splitting" in another, so is it possible to use multivariately derived groups on a relatively broad or refined basis. The more refined groupings should tend to pinpoint microzonal differences but, because of sample size or other considerations, may be less robust statistically. Again, the investigator may find that the broader groups relate more readily to independent variables and thus have greater practical application to his problem, or the reverse may be true. The formation of paste compositional groups is a heuristic procedure.

Neutron Activation Analysis

Incensario and non-incensario ceramics were prepared for instrumental chemical analysis, following usual Brookhaven National Laboratory procedures [13]. Samples and standards were bombarded for 18 hours at a thermal neutron flux of 1×10^{14} n/cm^2 sec. After a 10-day cooling period, a one-minute bombardment at the same flux level was used to produce the short-lived isotopes. The activated samples were transferred to counting vials and placed in an automatic sample changer. The counting configuration consisted of a Princeton Gammatech Ge-Li detector (35 cm crystal with a better than 1.8 keV resolution on ^{60}Co) coupled to an ND-2400 4096 channel analyzer and magnetic tape readout. The gamma spectra were processed by the BRUTAL program on the Brookhaven CDC 7600 computer. Final elemental concentrations were obtained by use of the "in-house" programs ELCALC and SAMPCALC. The present investigation utilized the elements Na, K, Rb, Cs, Ba, Sc, Ce, Eu, Hf, Th, Cr, Mn, Fe, Co, and Ti, the last being determined by x-ray fluorescence. The elemental data are on file at the Department of Chemistry.

Statistical Procedures

Following neutron activation analysis, several steps were taken in the investigation of possible source areas for the Palenque-region incense burners. As has been seen, the investigation was part of a broader chemical characterization, and the effective compositional assignment of incensarios to Palenque or non-Palenque sources was necessarily made in relationship to other classes of ceramics. (1) The first step, therefore, was to formulate provisional chemical groups of non-censer ceramics for the region. Among these groups, some were comprised of pottery which was similar to the incensarios in temper and texture, being characterized by quartz sand in the medium to fine size range. Only those chemical groups in which the pottery is petrographically similar to the incensarios are relevant to the analytical steps discussed here. (2) The provisional groups were subjected to statistical evaluation and refinement. Divergent specimens were removed until internal taxonomic stability resulted. (3) Interpretation of the groups followed. Essentially, this was based on their correspondences with petrographic and distributional data. The latter took into consideration archaeological provenience, strong distributional patterns being noted for later reference. As actually implemented, considerable feedback between steps 2 and 3 took place. (4) Having statistically refined the groups (step 2) and noted that they possessed overall petrographic patterning and archaeological utility (step 3), we projected the compositional profiles of the incensarios against those of the reference units. The probability of incensario membership in the various reference units was determined by standard multivariate statistical techniques. These steps are now considered in greater detail.

Step 1. The initial partitioning of the non-incensario data set was achieved through the use of the program CLUS — an iterative clustering procedure [14]. CLUS seeks internal geometric evidence to arrive at the "best" number of groups and their membership. The clustering algorithm reflects the basic postulate that the total scatter or dispersion matrix T is comprised of the matrices of within-group dispersion W and between-group dispersion B; e.g., T = W + B [15,16]. For an "optimal" classification, since the total variation is held constant, one need only minimize W, thereby maximizing B. This identifies the clusters that are the most internally homogeneous and externally isolated. For a multivariate problem, the clustering criterion to be maximized becomes the ratio of the determinants T and W. This ratio will change as the number of groups is increased, and inspection of the rate of increase of the value log (max $\frac{T}{W}$) serves as an informal indicator of the "best" number of groups [17,18]. The data set was standardized and transformed to principal component space. The 10 largest components, retained to serve as the clustering variables, represented 93 percent of the total variation. It is, however, difficult for any grouping procedure to deal satisfactorily with the heterogeneity contained in the original data matrix, as produced by natural and cultural factors [19]. CLUS seeks a "global" solution; therefore, when the full data set was input, CLUS determined that three rather generalized groups were optimal, as evidenced by the criteria described above.

Geographic and petrographic correlations existed with the chemically derived groups but, as these were not strong, we attempted to see if refinement would generate statistically robust units having greater agreement with the mineralogical and cultural variables. Experience has shown us that to accept a finer partition of the whole data set after an "optimal" division has been obtained by the computer often results in a degenerative solution. Therefore, to ascertain if greater refinement was possible, each of the three groups was submitted separately to CLUS, allowing the full sensitivity of the clustering algorithm to be devoted to the restricted matrix. Seven partitions were selected, units 1 and 2 deriving from one of the original clusterings, units 3-6 from the second, and unit 7 from the third. The tentative groupings were usually those containing the largest number of specimens within an "optimal" partition and having the smallest internal elemental variation. Each unit was then inspected for its homogeneity as based on statistical, petrographic, and archaeological considerations. Better patterning with non-chemical data — hence a more viable and useful classification — was obtained.

Step 2. Operating in a space defined by linear discriminant functions, and under the assumption of a multivariate normal distribution, Bayesian classification techniques were used as a heuristic device to evaluate group memberships [20]. Different types of information were sought; one consideration was the likelihood that a given data point could exist as far as it did from the centroid of its group and yet be a member of that group. Specimens were removed that had group containment values so low as to suggest that the sherds might not belong to any of the groups under consideration. No absolute threshold was set; rather each group was individually inspected for the general level of compactness about the group centroid. It was found that removal of the specimens with "lower" conditional probabilities improved the internal petrographic homogeneity. A second consideration related to the probability of group membership. This involved the calculation of posterior probabilities, the samples necessarily being assigned to the group with the highest probability. The few specimens with posterior probabilities that were split between two or more groups were removed in order to provide more homogeneous and stable units. Group memberships were then reevaluated.

In an earlier draft [21] the assessment of group membership was based, in part, on the probabilities generated by SPSS, version 6, installed on an IBM 370 computer. Subsequently, we have learned that the IBM discriminant analysis routine was programmed in single precision. The resulting loss of precision at times created erroneous probabilities and incorrect group assignments. The probabilities and percentages reported here were obtained by ADCORR or SPSS (version 7) implemented on a CDC 7600. These data update the above mentioned paper and that of Rands, Bishop and Harbottle [11].

The classification statistics were based on a pooled covariance matrix; this has advantages but also weaknesses. Partitioning of a data set will frequently yield groups having only a few members. These small clusters — difficult to evaluate statistically — may archaeologically be as illuminating as the larger clusters; for example, long-distance trade may be indicated. Since derivation of discriminant functions requires the inversion of the covariance matrix, a group containing fewer samples than variates would necessarily have a singular matrix and cannot be inverted. Pooling allows the use of discriminant functions and related statistics to describe group separation and the linear combinations of variables (chemical elements) responsible for that separation. However, the unique configuration of a small group's scatter matrix could be overshadowed by the influence of larger groups. If a specified level of homogeneity is found over all groups, a pooled matrix may be acceptable. If significant heterogeneity is present, the actual probabilities of group membership will vary depending on whether the sample is evaluated relative to a pooled or a separate covariance matrix. If covariance matrices of unequal size are used, there is a greater chance of a specimen being assigned to the group having the largest overall dispersion [22]. Therefore, it is important to obtain as much information as possible about the group dispersions. The dispersion or "generalized variance" of a matrix was measured by the determinant of the group's covariance matrix, and a test for homogeneity of the separate matrices was performed using the method of Kendall and Stuart as programmed in SAS.76 [23,24]. Groups 2 and 7 were omitted from the homogeneity test because of insufficient members. The results for the remaining five units are given in Table 1, Column A, and indicate that the covariance matrices are not significantly heterogeneous (p<0.001). To meet most fully the archaeological objective of obtaining an adequate paste compositional coverage, it was necessary to retain all seven groups as reference units and as the basis of discriminant analysis.

Table 1. Test of Homogeneity of Within Group Covariance Matrices.

Log_e Determinant

Unit	N	(A)	(B) Units 1 and 2 Combined
1	68	92.30	94.33
2	11	singular	
3	21	88.81	88.81
4	20	85.51	85.51
5	22	92.52	92.52
6	39	92.05	92.05
7	13	singular	singular
Total N = 194			
Chi-square test		755.00	776.78
dfs		420	420
Prob. > Chi-square		0.0001	0.0001

Step 3. As indicated, evaluations based on correspondences in chemical, petrographic and archaeological patterning were made at various points in the investigation. As these do not require a formal description of statistical procedures, this step is not considered further at the present time.

Step 4. In order to obtain the probabilities of incensario membership in all the reference units, the program SPSS and a pooled covariance matrix were used. However, when the number of group members was sufficiently large, the probabilities of group containment were recalculated relative to the centroid and dispersion matrix of that single group. This was accomplished by using the program ADCORR [25]. While in part complementary, several major differences exist in the way SPSS and ADCORR determine group containment probabilities.

The basis of the SPSS classification statistics is the use of discriminant functions — weighted, linear combinations of the original standardized variables selected to differentiate maximally among the groups. The maximum number of functions is limited to the number of groups minus 1. ADCORR, however, assesses group containment quite differently. The original variables are first converted to log concentration space (for the rationale of this transformation, see Sayre [26].) The eigenvectors, equal in number to the rank of the original matrix, are obtained from the variance-covariance matrix and are used to determine the Mahalanobis distance of a sample from its centroid. The probability of a sample having a given Mahalanobis distance and yet belonging to the group is based upon the distribution of Hotelling's T^2. Due to the differences between SPSS and ADCORR, only rough correspondences in their respective calculations of group membership are to be expected. However, if the defined groups are fairly well isolated and homogeneous, the determination of peripheral or intermediate specimens should be possible with either program.

In effect, CLUS was used to form the seven provisional groups (step 1). The classification options of SPSS were primarily used to derive refined units from these initial groupings (step 2). At approximately a 10 percent probability of group containment, ADCORR served to support the SPSS classification. At this threshold, chemical, petrographic, and archaeological data appeared to converge in an archaeologically useful manner (step 3). Therefore, the 10 percent ADCORR level was retained as the somewhat arbitrary lower limit of the acceptable criterion of membership in the projection of incensario data against the reference units (step 4).

Data Interpretation for the Reference Units

Out of approximately 300 sandy-textured non-incensario ceramic samples, 194 were classifiable into seven stable groups; mean concentrations for the fifteen chemical elements are listed in Table 2. The other ceramics had sufficiently low group containment probabilities to suggest that they were not classifiable into any of these reference units, as currently defined. Each of the seven groups proved on inspection to have internal petrographic as well as chemical similarity. Based on sherd provenience (the archaeological criterion of abundance) two of the reference units have a pronounced locus at Palenque. This is seen in Palenque's sustained tradition of red-brown pottery, similar mineralogically to units 1 and 2, as determined by extensive microscopic sampling. The relatively long time span over which ceramics similar to those groups occurred is in contrast to low frequencies or lack of depth at Palenque of pottery resembling the other reference units. To minimize chance correspondences, all ceramics of non-Palenque provenience [7] were removed from the "indigenous" appearing units 1 and 2, reducing the total of the two groups to 79 specimens. As a result, still greater petrographic and chemical homogeneity was achieved within these units.

The goal of obtaining a chemical characterization sufficiently sensitive to differentiate pottery made in the immediate locality of Palenque from other regional ceramics guided the splitter's approach used in the preceeding analytical steps. It is recognized that chemically divergent clay resources might also occur in close geographical proximity to Palenque and so enter into "indigenous" ceramic production, constituting an uncontrolled aspect of the present research. We do not address this problem here aside from noting the ambiguous status of approximately 100 quartz-tempered specimens which are unclassified according to reference unit, some of which would have been included had boundaries been drawn less closely. We believe, however, that pottery has been isolated which can be ascribed with confidence to a microzone centered on Palenque. Although information regarding many specimens has been temporarily lost in the process, the multiple steps of data reduction have been useful in obtaining a chemical profile that is unlikely to be duplicated at manufacturing centers lying outside the immediate locality of Palenque. To attain this was our objective at this particular stage of the research program.

Total specimens for the seven reference units are given, by site provenience, in Table 3. The units are represented in five dimensional discriminant space, the coordinates for the first two dimensions being shown in figure 4. These two vectors account for 72 percent of the total discriminant power. For convenience, approximate territorial lines have been added to the plot. Apparently representing the compositional pattern of locally made Palenque pottery, units 1 and 2 are enclosed within a single territorial boundary. Less well represented but, as will be seen, of considerable interest for the incesario problem, is unit 7, which relates primarily to the site of Xupa. Observable separation of units 3, 4, and 5 is possible only when utilizing additional discriminant functions beyond the first and second. In

Table 2. Chemical Paste Compositional Reference Units: Mean Elemental Concentrations.

	Na₂O	K₂O	Rb₂O	Cs₂O	BaO	Sc₂O₃	CeO₂	Eu₂O₃	HfO₂	ThO₂	Cr₂O₃	MnO	Fe₂O₃	CoO	TiO₂
Unit 1	0.801	1.63	67.5	2.99	608	17.0	93.2	1.52	11.6	11.5	1190	325	4.64	26.6	0.968
Unit 2	0.552	1.66	68.5	3.52	647	17.8	138	2.51	8.77	13.8	986	360	5.52	41.0	0.899
Unit 3	0.213	0.785	47.0	3.12	448	19.3	93.2	1.41	12.0	12.0	1500	328	3.89	26.5	1.09
Unit 4	0.159	0.876	47.5	2.22	423	16.8	79.4	1.24	15.1	11.8	1600	251	3.68	21.5	1.21
Unit 5	0.230	1.00	52.6	2.30	436	18.6	108	1.61	10.8	12.6	1410	330	4.20	32.1	1.13
Unit 6	0.150	0.962	47.0	2.77	455	15.5	131	2.22	9.58	11.4	1090	231	2.53	19.0	1.01
Unit 7	0.991	1.48	62.2	2.44	516	22.0	106	2.14	5.26	9.22	1420	371	5.91	35.5	0.953

[a]Mean oxide concentration reported with three figure accuracy. Sodium, potassium, iron, and titanium are given in percent; others listed as parts per million. Standard deviations ranged from six percent for scandium in Unit 2 to 65 percent for sodium in Unit 3. Elements of lower mobility (e.g., Sc, Ce, Eu, Th) had standard deviations of 10-15 percent.

general, sherds low on the x-axis relate to the sierras (to the south in figure 1) and tend to be red-brown in paste color. In contrast, sherds especially high on the x-axis relate to the Chiapas plains and have lighter brown to buff pastes. Petrographically, a decrease in feldspar occurs in the ceramics as one moves from sierras to plains. A similar but less pronounced pattern is observed for the micaceous nature of the ceramic matrix. Conversely, opal phytoliths, present in the clay as the result of the decay of silica accumulator plants [27], have a plains rather than Palenque-sierras orientation. The observations are constant with earlier findings based on a Q-mode factor analysis [28].

Table 3. Provenience Distribution within the Seven Reference Units (non-incensario ceramics).

	1	2	3	4	5	6	7	Total
Belisario Dominguez Nansal			1	1	1			3
Chancala			1				1	2
Chinikiha					1	1		2
El Bari				1		2		3
El Retiro					8			8
Francisco Villa						1		1
Las Colmenas				3		1		4
Miraflores				1	4	1		6
Naranjo						2		2
Nututum			2			9		11
Palenque	68	11	14	6	6	9	3	117
Paso Nuevo						11	1	12
San Manuel				1				1
Santa Cruz			2		1			3
Santa Isabel			1	2				3
Santo Tomas				4				4
Xupa						2	8	10
Yoxiha					1			1
Zona Sala			1					1
Total	68	11	21	20	22	39	13	194

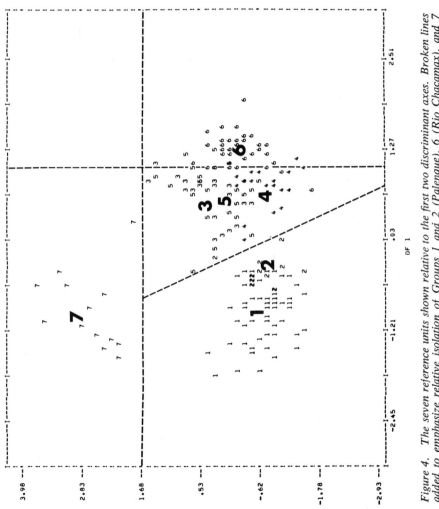

Figure 4. *The seven reference units shown relative to the first two discriminant axes. Broken lines added to emphasize relative isolation of Groups 1 and 2 (Palenque), 6 (Rio Chacamax), and 7 (Xupa orientation).*

Part of the evidence for the petrographic patterning summarized above is presented in figures 5 and 6. The coordinates
are held constant in the two-dimensional diagrams, permitting the
reference unit membership of a given ceramic specimen (figure 4)
to be compared with the petrographic variables of opal phytoliths
and feldspar. To facilitate recognition of major patterns,
abundance of the mineralogical variables, as recorded in the
original petrographic analysis, is presented in summary catego-
ries. The latter are expressed as rank ordered units for the opal
phytoliths (figure 5) and as approximate mean percentages for
feldspar (figure 6). The negative correlation observed between
opal phytoliths (attributable primarily to grassland or marsh
environments of the plains) and high feldspar (attributable to
steep gradients of the sierras, where erosion is constantly
exposing fresher materials) is entirely logical. General agreement
of the combined petrographic patterns with membership in the reference units, in combination with archaeological provenience of the
pottery, serves to set off "Sierras" and "Plains" groupings on the
first discriminant axis (see figure 4). Chemical data provide the
major basis for differentiating ceramic pastes within the Plains
and the Sierras. Nevertheless, exemplifying the contribution of
petrography on this more refined level, exceptionally abundant
phytoliths characterize Reference Unit 6, high on the x-axis, and,
within the Sierras, relatively abundant mica correlates with the
position of unit 7, high on the y-axis.

The contributions of the chemical variables to the SPSS-
derived discriminant functions are given in Table 4. Especially
notable is the very heavy loading of sodium on the first vector.
Although sodium is highly mobile in most environments [29], it
would nevertheless appear to be a good indicator of micro-
geographical differences in the Palenque region. If sodium is
omitted from the data analysis, potassium becomes the main contri-
butor to group separation. The sodium and potassium concentrations apparently reflect the abundance of alkali or plagioclase
feldspars in the pottery (figure 6), as well as qualitative
differences in clay mineralogy [30].

Suggested correspondences of the numbered reference units to
their approximate geographical locations are indicated in figure
7. It will be recalled that approximately one-third of the sandy-
textured pottery did not belong to any of the seven reference
units. These unplaced specimens may have loci in unworked or
poorly sampled zones on the map, perhaps in some cases reflecting
relatively long-distance trade. Some may yet be ascribed to new
or existing reference units through additional data accumulation
in the on-going research.

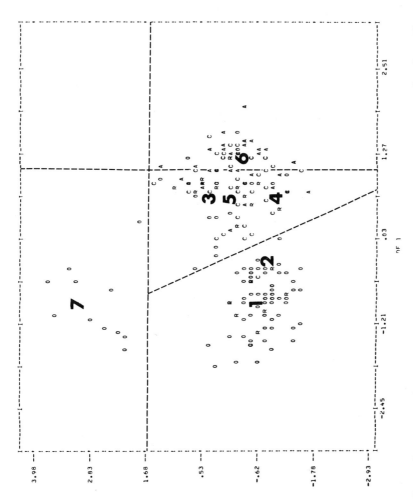

Figure 5. Abundance of opal phytoliths as summarized from original petrographic data. Key: 0, absent; R, rare; C, common; and A, abundant. Sample coordinates same as in Figure 3. Note essential absence of phytoliths in the Sierra units 1 and 2 (Palenque), and 7 (Xupa orientation).

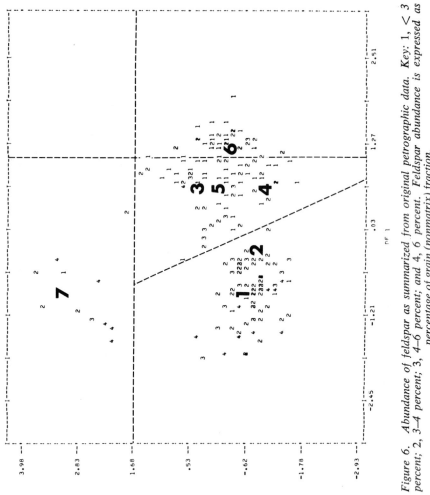

Figure 6. Abundance of feldspar as summarized from original petrographic data. Key: 1, < 3 percent; 2, 3–4 percent; 3, 4–6 percent; and 4, 6 percent. Feldspar abundance is expressed as percentage of grain (nonmatrix) fraction.

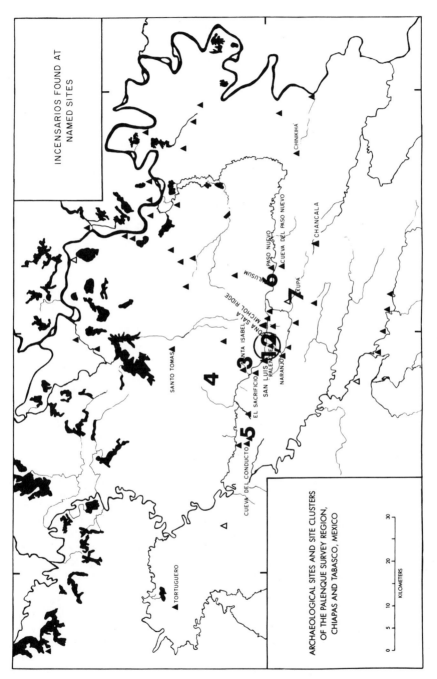

Figure 7. Map of suggested geographic loci for the seven reference units.

Table 4.   Standardized Linear Discriminant Function Coefficients
           for the Seven Reference Units.

	DF 1	DF 2	DF 3	DF 4	DF 5
Na	2.15	0.30	-0.50	0.55	0.08
K	0.42	-0.30	-0.14	-0.72	-0.73
Rb	-0.01	-0.01	0.36	-0.21	-0.58
Cs	0.27	-0.21	-0.08	0.49	0.86
Ba	0.05	-0.37	0.31	0.55	0.71
Sc	-0.20	0.67	0.62	0.57	0.06
Ce	-0.26	0.15	-0.74	-0.05	-0.10
Eu	-0.49	0.04	-0.51	0.26	-0.43
Hf	-0.07	-1.55	0.38	0.39	-0.47
Th	0.48	0.51	0.05	-0.93	0.54
Cr	0.18	1.13	0.76	-0.11	0.38
Mn	0.11	0.03	0.16	0.57	0.36
Fe	0.42	0.09	-0.16	-0.65	-0.01
Co	-0.08	-0.02	-0.17	-1.35	-0.24
Ti	-0.15	-0.17	0.11	0.01	-0.64
Eigenvalue	7.3	3.5	2.5	1.3	0.4
Percent of Trace	48.4	23.4	16.3	8.3	2.7

Projected Group Membership of the Incensarios

        Comparison of chemical profiles of the 82 incensarios with
those established for the reference units proved to be decisive.
Although proveniences of the flanged incensario supports lay in a
linear pattern along the front range of the Sierras (figure 1),
statistical evaluation shows concentration in only two of the
seven reference units. The incensario projection is shown for the
first two discriminant vector coordinates on a plot which also
contains the generalized territorial lines of the reference units
(figure 8, compare figure 4). The concentration of incensarios is
in units 1 and 2, which we believe to represent slightly divergent
patterns for pottery made in the vicinity of Palenque. Moreover,
the incensarios project close to the centroids of units 1 and 2 as
defined in the five-dimensional discriminant space. The strong
concentration of incensarios in these units is supported when
probabilities of group containment are based on separate

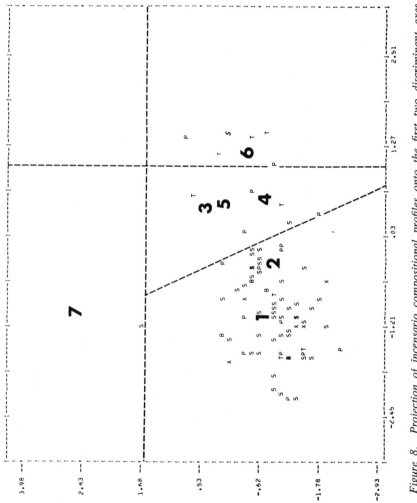

Figure 8. Projection of incensario compositional profiles onto the first two discriminant axes defined for the seven reference units. Key (symbols have been assigned to the incensario form classes): S, cylindrical support; P, pedestal; T, trumpet-shaped; B, bowl; and X, miscellaneous.

dispersion matrices. It will be recalled that unit 2 constituted
a small sample. In view of the indigenous appearance of pottery
represented in units 1 and 2 and their chemical similarity, it
appeared justified to combine them. A Palenque locus results that
has only slightly greater generalized variance than that present
in unit 1 (Table 1, Column B).
     Additional levels of interpretation are provided by relating
the incensario typology to the reference units (figure 8). As
indicated earlier, incensarios consist largely of cylindrical
flanged supports, pedestals, ladles, bowls, and conical (trumpet-
shaped) censers. Certain poorly-represented forms are given under
a single "miscellaneous" category.

Supports

     A strong tendency exists for chemical profiles of the ornate,
flanged cylindrical supports (figure 2a) to match the Palenque
compositional grouping, 77 percent having membership in reference
unit 1. This strengthens the probability, suggested elsewhere on
the basis of non-incensario materials, of a strong chemical
ceremonial orientation to unit 1 ceramics [31].
     Of the three cylindrical supports lying outside the "Palen-
que" compositional units 1 and 2 as shown in figure 8, one has
an archaeological provenience far to the west (Tortuguero) and is
stylistically atypical because of its lidded eye. The second is
intermediate chemically between the Palenque materials and unit
3. The third exception is atypical in its light buff color and
thumb-impressed treatment over the exterior surface of the entire
cylinder. On this level of interpretation, the data would seem
to support local Palenque manufacture of most of the sampled
cylindrical supports. When the statistical likelihood of group
membership is considered, 9 of the 47 supports have less than a
10 percent chance of any group containment as evaluated by ADCORR
(Table 5). Most of the nine are stylistically atypical of the
corpus of incensarios known from Palenque, although similarly
atypical features are occasionally present in those incensarios
which have acceptable group membership probabilities.
     In a few cases, the use of both ADCORR and SPSS pointed to
uncertainties of group membership for the incensario supports.
Four samples acceptable to SPSS were not acceptable to ADCORR;
conversely, three samples with greater than 10 percent probability
of group containment within the ADCORR-combined units 1 and 2
could not be considered members by SPSS. The source of these
specimens must for now be considered as indeterminant. As would
be anticipated, if the incensario supports with a projected group
membership probability of greater than 10 percent are added to
the reference units, the SPSS centroids are slightly shifted so
that those specimens previously peripheral tend to attain a higher
probability of membership in unit 1. This also holds true when
units 1 and 2 are combined with the incensario supports and
evaluated by ADCORR.

Table 5. Incensario Form Class and Provenience Distribution Within the Seven Reference Units[a].

	1	2	3	4	5	6	7	Split	None
SUPPORTS									
Chancala	1								
Cueva Conducto									1
Cueva Paso Nuevo	4								1
El Sacrificio	2								
Palenque	15							1	3
Paso Nuevo									1
San Luis	3								
Santo Tomas	1								
Sulusum	1								
Tortuguero									1
Xupa	4	1							
Zona Sala	3								
Unknown Provenience	2								1
PEDESTAL									
Chinikiha						1			2
Michol Ridge									1
Palenque	8	1							3
Xupa								1	
TRUMPET-SHAPED									
Cueva Conducto									1
Cueva Paso Nuevo			1						
Palenque	2								2
Xupa					1				
Zona Sala	1								
BOWLS									
Palenque	1								1
Tortuguero									1

Table 5 continued.

	1	2	3	4	5	6	7	Split	None
MISCELLANEOUS									
Palenque	5								2
Unknown									
Provenience	1								

[a]The projected unit memberships are based on the likelihood of
group containment as determined by SPSS version 7 using five
discriminant functions. In addition, the proposed membership in
Unit 1 was evaluated by ADCORR. In cases of ambiguity the
ADCORR probability of group containment was accepted, as the
probabilities were based upon a single variance-covariance
matrix.

Rare at Palenque but represented in three of the four incensario supports from the site of Xupa, a motif of some interest
is that of a standing anthropomorphic figure. All supports
having a Xupa provenience, with or without standing figures,
project into Palenque reference units 1 and 2, other forms of
Xupa censers having compositional affiliations with the plains-
oriented units (figure 9). In contrast, non-incensario sherds
from Xupa are especially associated with unit 7. Although
sampling is limited, this suggests either that this unit charac-
terizes the indigenous ceramics of the site or that close
trading relationships existed with a non-Palenque locality in the
sierras which produced the unit 7 pottery. Therefore, the directional flow of incensarios from Palenque to Xupa appears stronger
than that of other classes of pottery made at Palenque; perhaps
trade from Palenque in ceremonial items transcended exchange in
serving or domestic forms.

As noted, a thematic difference within incensario supports
sets off standing figures from the more typical tier of gro-
tesque heads that is seen in figure 2a. This does not seem to
be strongly reflected in paste composition, 82 percent of the
specimens showing standing figures having an acceptable projection into units 1 and 2 compared to 81 percent of the supports as
a whole. Ten of the 11 standing-figure supports which have been
sampled have a non-Palenque or unknown archaeological provenience,
eight of these lying east of Palenque at the sites of Xupa and
Cuevo de Paso Nuevo. On the basis of distributional considerations alone, a non-Palenque source for the standing-figure
incensarios would have been inferred. In most of the cases,
nevertheless, chemical data suggest a manufacturing source at or
near Palenque. Perhaps the differing style and motif may be
attributed to temporal changes in iconography, although this does

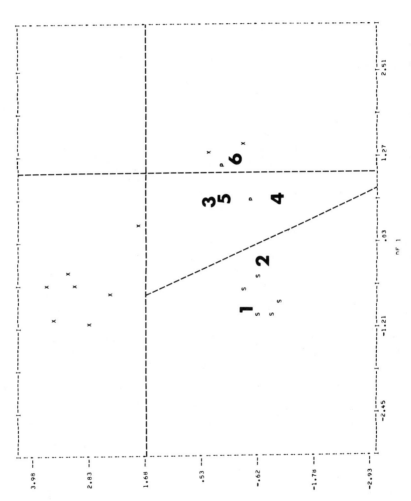

Figure 9. Incensarios and other ceramics of Xupa provenience shown relative to the first two discriminant axes defined for the seven reference units. Key: S, cylindrical support; P, pedestals; and X, nonincensario pottery.

not satisfactorily explain the rarity of the standing figure support among incensarios found at Palenque. Other explanations might focus on the specialized production of ceremonial goods by Palenque-based craftsmen to meet the requirements of a religious cult that varied regionally in its iconographic expression.

Functional Incensarios

Although all incensario form classes project into reference unit 1, less patterning is evident among the functional censers than with the incensario supports. This is especially evident for certain of the classes (figure 8, Table 5).

As evaluated by SPSS and ADCORR, nine of the 17 pedestal incensarios (figure 2c) project into Palenque reference units 1 or 1 and 2. All nine have a Palenque provenience, suggesting that pedestal censers were regularly produced at Palenque as an item for local consumption but not for export, differing markedly in the latter respect from the incensario supports. Of the four specimens which were unassigned by ADCORR, three are placed in units 1 or 2 by SPSS; the fourth, having split membership probabilities between units 4 and 6, indicates a general Plains orientation. Two additional specimens have Plains affiliations as indicated by both programs.

The small class of trumpet-shaped receptacles (figure 2b) is divided between Palenque and Plains orientations. Specimens relating to the latter have low probabilities of group containment (SPSS) and are unassigned to any reference unit by ADCORR.

Bowls (figure 2d) and members of the "Miscellaneous" category are sparsely represented, for the most part being strongly assignable to unit 1 by SPSS. However, using the ADCORR evaluation, only one of the three bowls is assignable (units 1-2). Better agreement is achieved for the "Miscellaneous" forms, six of the eight, mostly from the site of Palenque, being assigned to the Palenque compositional units by both evaluative procedures.

Relative to the number of incensarios sampled, it is more difficult to assign group membership to the functional censers than it is for the highly decorated cylindrical supports. For some of the functional censers there is virtually no probability of belonging to any of the defined reference units; for other, membership probabilities are split across two or more of the units 3, 4 and 6, forcing one to turn to a generalized "Plains" conceptual orientation. The ability to assign a large majority of the incensario supports unambiguously to Unit 1 (Palenque) strongly suggests the existence of a localized clay procurement and manufacturing center within or immediately adjacent to the ceremonial center.

Summary

Ceramic paste compositional units were first derived which, on the basis of chemical, petrographic and distributional criteria, appear to represent distinct resource procurement and manufacturing zones in the region of the Maya site of Palenque. The seven groups provide a basis of comparison with pottery used in burning incense. Assignment of incensarios to manufacturing locality could not effectively have been made in the absence of these reference units, as the latter relate to a much broader corpus of chemically unsampled but archaeologically and mineralogically analyzed ceramic data.

Representing the locally manufactured pottery at or near Palenque are units 1 and 2. Numerically largest of the reference groups, unit 1 provides the main chemical profile for Palenque. Determination of probable regions of manufacture for the non-Palenque groups has been most successfully achieved for units 6 and 7 (referable to the Chacamax Plains and locality of Xupa, respectively). If Xupa itself was not a production center for unit 7, the site was in close contact with such a center, at least during the beginning of the Late Classic period. Units 1, 2, and 7 have their loci in the sierras. In contrast, unit 3 seems in some way related to the Michol drainage, unit 4 having an apparent focus farther out in the plains, away from Palenque. Relative to Palenque, a western orientation is indicated for unit 5. Although some compositional agreement has been observed between chemically analyzed ceramics and specifically localized clay deposits [30], the geographic positioning of resource procurement and production localities suggested in figure 7 relies heavily on the assumption that, in general, pottery was used in greater relative abundance in its zone of manufacture than in any other zone to which it might have been exported.

The groups are not exhaustive in representing all possible clay procurement and manufacturing localities for pottery sampled in the archaeological survey. Approximately one-third of the initially considered sherds — those characterized by medium to fine quartz sand — were not amenable to independent grouping or to membership in the established reference units.

Projection of the incensario chemical profiles reveals strong resemblance of the flanged cylindrical supports to the Palenque reference units. The use of similar clays is indicated and, inferentially, the dispersal of these ornate objects from a production center located at Palenque. Occasional deviation from Palenque reference unit membership may, of course, be ascribed to the existence of additional places of clay procurement and manufacture but no strongly patterned compositional relationship is discernible among the divergent specimens. We infer that incensario supports were occasionally brought to Palenque from outside sources but that this was an uncommon practice. Evidence is lacking which might suggest that, in the region under consideration, a single souce other than Palenque regularly served as a center for production and exchange of the supports. Palenque, the dominant ceremonial and political center in the region [32], apparently held a monopoly on the manufacture for trade of these ritual items, which were distributed more widely than most of the site's locally made pottery. On the other hand, certain paraphernalia involved in incense burning, such as receptacles which were used in connection with the supports, show a less highly structured pattern of exchange. Culture-historical implications about the nature of the regional exchange system are examined in related studies [11,33].

We are grateful to the American Museum of Natural History and the Metropolitan Museum of Art for permission to sample specimens for compositional analysis. This material is based upon work supported by the National Science Foundation under Grant No. BNS76--3397. Aspects of the investigation were carried out under the auspices of the U. S. Department of Energy.

Literature Cited

[1] Rands, R. L., The Ceramic Sequence at Palenque, Chiapas, In: Mesoamerican Archaeology, New Approaches, University of Texas, Austin, pp. 51-76, 1974.

[2] Rands, R. L., Ceramic Technology and Trade in the Palenque Region, Mexico, In: American Historical Anthropology: Essays in Honor of Leslier Spier, Southern Illinois University, Carbondale, pp. 137-151, 1967.

[3] Rands, R. L., Comparative Data from the Palenque Zone on Maya Civilization, In: Actes du XLII Congres International des Americanistes, 8, Paris, pp. 135-145, 1979.

[4] Bishop, R. L., Western Lowland Maya Ceramic Trade: An Archaeological Application of Nuclear Chemistry and Geological Data Analysis, PhD Thesis, Southern Illinois University, Carbondale, 1975.

[5] Rands, R. L., Bishop, R. L., Resource Procurement Zones and Patterns of Ceramic Exchange in the Palenque Region, Mexico, In: Models and Methods in Regional Exchange, SAA Papers No. 1, Society for American Archaeology, Washington, pp. 19-46, 1980.

[6] Rands, R. L., Mayan Ecology and Trade: 1967-1968, Mesoamerican Studies, Research Records of the University Museum, University Museum, Southern Illinois University, Carbondale, p. 51, 1969.

[7] Borhegyi, S. F. de., The Development of Folk and Complex Cultures in the Southern Maya Area, American Antiquity, 21, 343-356 (1956).

[8] Rands, R. L., and Rands, B. C., The Incensario Complex of Palenque, Chiapas, American Antiquity, 25, 225-236 (1959).
[9] Borhegyi, S. F. de., The Composite or 'Assemble-it-yourself' Censer: A New Lowland Maya Variety of the Three-pronged Incense Burner, American Antiquity, 25, 51-58 (1959).
[10] Goldstein, M., The Ceremonial Role of the Maya Flanged Censer, Man, 12, 405-420 (1977).
[11] Rands, R. L., Bishop, R. L., Harbottle, G., Thematic and Compositional Variation in Palenque Region Incensarios, In: Tercera Mesa Redonda de Palenque, Vol. IV, Pre-Columbian Art Research Center, Herald Printer, Monterey, pp. 19-31, 1979.
[12] Rands, R. L., Benson, P. H., Bishop, R. L., Chen, P-y, Harbottle, G., Rands, B. C., Sayre, E. V., Western Maya Fine Paste Pottery: Chemical and Petrographic Correlations, In: Actas del XLI Congreso Internacional de Americanistas, 1, Mexico, pp. 534-541, 1975.
[13] Abascal-M., R., Harbottle, G., Sayre, E. V., Correlation Between Terra Cotta Figurines and Pottery from the Valley of Mexico and Source Clays by Activation Analysis, In: Archaeological Chemistry, Adv. Chem. Ser., 138, 81-99 (1974).
[14] Rubin, J., Friedman, H. P., A Cluster Analysis and Taxonomy System for Grouping and Classifying Data, IBM Corporation, Scientific Center, New York, 1967.
[15] Wilks, S. S., Mathematical Statistics, Wiley, New York, 1962.
[16] Cooley, W. W., Lohnes, P. R., Multivariate Data Analysis, Wiley, New York, p. 225, 1971.
[17] Rubin, J., Friedman, H. P., A Cluster Analysis and Taxonomy System for Grouping and Classifying Data, IBM Corporation, Scientific Center, New York, p. 12, 1967.
[18] Bishop, R. L., Textural and Chemical Variation in a Numerical Study of Western Lowland Maya Ceramic Trade, In: Archaeological Frontiers: Papers on New World High Cultures in Honor of J. Charles Kelley, University Museum Studies Research Records, Number 4, University Museum, Southern Illinois University, Carbondale, pp. 1-44, 1976.
[19] Bishop, R. L., Western Lowland Maya Ceramic Trade: An Archaeological Application of Nuclear Chemistry and Geological Data Analysis, Chapter III, PhD Thesis, Southern Illinois University, Carbondale, 1975.
[20] Nie, N. H., Hull, C. H., Jenkins, F. G., Steinbrenner, K., Bent, D. H., Statistical Package for the Social Sciences, McGraw Hill, New York, 1975.
[21] Bishop, R. L., Rands, R. L., Harbottle, G., Incense-burner Trade in the Palenque Area, Methodological Approaches, BNL-24645, Brookhaven National Laboratory, Upton, 1978.
[22] Van De Geer, J. P., Introduction to Multivariate Analysis for the Social Sciences, Freeman, San Francisco, p. 263, 1971.

[23] Kendall, M. G., Stuart, A., The Advanced Theory of Statistics, 2, Griffin, London, p. 282, 1966.

[24] Barr, A. J., Goodnight, J. H., Sall, J. P., Helwig, J. T., A User's Guide to SAS.76, Sparks, Raleigh, 1976.

[25] Sayre, E. V., Brookhaven Procedures for Statistical Analyses of Multivariate Archaeometric Data, BNL, Brookhaven National Laboratory, Upton, 1975.

[26] Sayre, E. V., Brookhaven Procedures for Statistical Analyses of Multivariate Archaeometric Data, BNL, Brookhaven National Laboratory, Upton, 1975.

[27] Rovner, I., Potential of Opal Phytoliths for Use in Paleocological Reconstruction, Quaternary Research, 1, 343-359 (1971).

[28] Rands, R. L., Bishop, R. L., Aspects of Ceramic Paste Composition and Trade in the Palenque Region, Chiapas and Tabasco, figures 4e, f, University Museum, Southern Illinois University, Carbondale, 1975.

[29] Levinson, A. A., Introduction to Exploration Geochemistry, Applied, Maywood, p. 143, 1974.

[30] Bishop, R. L., Aspects of Ceramic Compositional Modeling, In: Models and Methods in Regional Exchange, SAA Papers No. 1, pp. 47-66, Society for American Archaeology, Washington, 1980.

[31] Rands, R. L., Bishop, R. L., Aspects of Ceramic Paste Composition and Trade in the Palenque Region, Chiapas and Tabasco, figures 4c, 6f, g, University Museum, Southern Illinois University, Carbondale, 1975.

[32] Marcus, J., Emblem and State in the Classic Maya Lowlands, Dumbarton Oaks-Trustees for Harvard University, Washington, 1976.

[33] Rands, R. L., Bishop, R. L., Resource Procurement Zones and Patterns of Ceramic Exchange in the Palenque Region, Mexico, In: Models and Methods in Regional Exchange, SAA Papers No. 1, Society for American Archaeology, Washington, pp. 19-46, 1980.

RECEIVED April 27, 1981.

The Carbon-14 Dating of an Iron Bloom Associated with the Voyages of Sir Martin Frobisher

EDWARD V. SAYRE, GARMAN HARBOTTLE, and RAYMOND W. STOENNER—
Brookhaven National Laboratory, Department of Chemistry, Upton, NY 11973

WILCOMB WASHBURN—Smithsonian Institution, Office of American Studies,
Washington, D.C. 20560

JACQUELINE S. OLIN—Smithsonian Institution, Conservation-Analytical
Laboratory, Washington, D.C. 20560

WILLIAM FITZHUGH—Smithsonian Institution, Department of Anthropology,
Washington, D.C. 20560

There have recently been significant advances in the technique of carbon 14 measurements, which have permitted the determination of the concentration ratio C14/C12 in samples of small size. Two developments have occurred: the first is an entirely new mass-spectrometric separation of C14 and C12 ions and their subsequent estimation by counting [1-8][1], while the second is simply the extension of conventional proportional counter operation to very small size carbon samples [9]. The first method is very fast, precise, and capable of treating samples of even sub-milligram size, but requires an expensive installation. The second method is slow (counting times of two months or more are necessary), can probably be made sufficiently precise to handle most problems, works down to sample sizes of 10 mg carbon, and is relatively inexpensive, especially to install in already-existing radiocarbon laboratories.

Development of proportional counters to measure C14/C12 ratios in 10 mg carbon samples was undertaken in the Chemistry Department of Brookhaven National Laboratory in 1975 [10] for two reasons: (1) at the time, there was no other possibility in sight to accomplish the generally much-needed objective of small-sample C14 measurement, and (2) there was a particular carbon 14 dating problem at the Smithsonian Institution, which would only be solved if very small carbon samples could be handled. The development and testing of the counters has already been reported [9]; in the present paper we discuss the application of those counters to the actual dating problem which concerned the Smithsonian Institution, the dating of the "Frobisher iron bloom".

The "Frobisher bloom" has a curious history, which has been told in detail by Stefansson [11] but which bears recounting in brief here. The noted English Elizabethan explorer Frobisher made

---

[1]Figures in brackets indicate the literature references at the end of this paper.

0097-6156/82/0176-0441$05.00/0

three voyages, in 1576, 1577 and 1578, to the North American mainland in the vicinity of Baffin Island, Canada. The first voyage was intended to discover a short-cut to China, but the second and third voyages were after gold, based on a mistaken (positive) gold assay carried out on rock brought back from the first voyage [12]. The rocks found by Frobisher have been identified by Roy [13] as amphibolite and pyroxenite. Roy did not find "fool's gold" (pyrite) and surmises that the bronze-lustered mica might have been the basis for triggering the "gold rush" of 1577-1578.

The islands that dot Frobisher Bay, where the "gold ore" was mined on the second and third voyages, are in an out-of-the-way corner of an inhospitable land. It was only in 1861-1862 that the American explorer, Capt. C. F. Hall, heard from local Innuit (Eskimos) of earlier European sites and artifacts from this area. Following these leads Hall visited Kodlunarn Island and the surrounding area and discovered many traces and relics of Frobisher's ill-fated venture [11,13,14], including the deep mining trenches, bits of coal and flint, fragments of tile, glass, wood and pottery and ruins of three stone houses. On Countess of Warwick Island (Eskimo name "Kodlunarn" or "White men's island") Hall discovered "iron time-eaten, with ragged teeth, weighing from fifteen to twenty pounds, on the top of a granite rock, just within reach of high tide at full change of moon". Here he also discovered another object: "a piece of iron, semi-spherical in shape, weighing twenty pounds ... under the stone that had been excavated for the 'ship's way'". The "ship's way" was a long narrow excavation that had the appearance of a dry dock. This piece of iron Hall described as being of the same character as that found at Tikkoon, less than one mile from Kodlunarn in Countess of Warwick Sound and also as that obtained on "Look-out" Island, Field Bay. Hall sent the piece of iron found in the ship's trench to "the British government early in the year 1863, through the Royal Geographical Society of London" [15].

In 1927 Kodlunarn Island was again visited, in the course of the Rawson-MacMillan expedition sponsored by the Field Museum of Chicago [11,13,16]. Dr. Duncan Strong, archaeologist on the expedition, looked for Viking graves or other traces, but found none. Very evident, however, were the already-observed remains of the Frobisher voyages. Interestingly, the 1927 expedition did find "a fragment of porcelain ... with metallic slag adhering outside the rim as though the vessel had served as a crucible." [11]. This is of interest with regard to the record that Frobisher's team included miners and gold-refiners.

Hall was convinced that the origins of all of the artifacts he had found were the voyages made by Martin Frobisher. To safeguard them for posterity, and because he was an American and Frobisher an Englishman, he divided them between the Smithsonian Institution in Washington and the Royal Geographical Society in London. But by the time Stefansson was writing, in the mid 1930's, both organizations had totally lost or mislaid their

Frobisher collections: nothing could be found, despite his exhaustive inquiry. His frustration permeates his account of the search [11].

His first attempt to elicit information from the Smithsonian brought the response (August 12, 1935) that "... a very careful search has been made of our records but we find no evidence that any relics of this gallant explorer were ever deposited here." Since this response did not agree with the known evidence, Stefansson tried the Smithsonian again. This time (August 28, 1935) the reply came that

> ... a very thorough and careful search has been made of our records and of the specimens received by us from the Polaris [the name of Hall's ship] Expedition but unfortunately without any success so far as Frobisher relics are concerned. On one list there is mention of a single specimen as follows: "Iron bloom, obtained from Countess of Warwick Sound where it was made by Frobisher in 1578, searching for gold." A careful examination of the specimens, however, fails to reveal any object answering to this description.

In fact the accession record of the U. S. National Museum (#2157) contains a full listing of the Hall objects as they were returned from loan status at the International Exhibition held at Philadelphia to celebrate the centennial of American independence. Why they could not be located in 1935 is unknown, but in 1950, in gathering data for Remington Kellogg, Director of the U. S. National Museum, for a response to Mrs. H. S. Marlett who had raised the same questions Stefansson had earlier broached, the curator of the Division of Ethnology, H. W. Krieger, in a memo to Frank M. Setzler, head curator, Department of Anthropology, dated March 17, 1950, reported that the "Iron Bloom" (Cat.#10,291) "was recently located by Mr. R. Sirlouis in the Gallery storage of the Department of History" and "respectfully requested" that permission be given to transfer the catalog card describing it to the Department of History. As for the "Relics of Frobisher's Voyage" (Cat.#14,247) Krieger opined that it offered "a problem not readily solved." "If the pieces of coal, fragments of iron, glass and pottery were not numbered individually," he went on "it is unlikely that they will be found." Krieger concluded his memo by noting that "Practically all mistakes that a museum could make in accessioning, cataloguing, preservation, recording, classification, and correspondence are illustrated in our connection with the Charles F. Hall collections."

Unfortunately, Krieger, in a memo of January 26, 1953 to Dr. Kellogg through Mr. Setzler (correspondence File 197929) requested a condemnation committee to dispose of objects "fragmentary or broken beyond repair; in part, undocumented and unidentifiable." He noted that "a considerable number of these specimens were collected by exploring expeditions to the Arctic. Although the expeditions are historical, the relics submitted for disposal consist of driftwood fragments, stone pebbles, shells and animal bones. On inquiry, it was found that these items are not wanted either by History, Geology, Zoology or Anthropology." A condemnation committee chaired by Mr. Setzler, with Krieger and David H. Dunkle as members, recommended disposal of a variety of objects in a memo of March 27, 1953 to Dr. Kellogg, who approved the condemnation on March 30. Among the items discarded from the Hall collection were "minerals," "shot," "cross bar?," "quartzite," "iron pyrites," and "quartz rock," as well as "pebbles," "sand stone rock," "one lot sea shells," "drift wood," "plumbago used by natives," and "rib of spotted seal."

On his arrival at the Smithsonian in 1958 as curator of the Division of Political History, Wilcomb Washburn inherited some of the collections of the old Department of History, which in turn had broken off from the Division of Ethnology. Because of his interest in exploration and discovery, he was excited when in 1964 he discovered the iron bloom that is the subject of this paper in a remote part of the history storage area in the Arts and Industries Building (Figure 1). The object was not catalogued in his Division, but he filed a request through the Division of Naval History, to the new analytical and conservation laboratory to analyze the metal and determine its age. He hoped to be able to present technical evidence of the age of the speci- men in the course of a paper entitled "The Oriental Purpose of the Arctic Navigations" that he was scheduled to deliver in August 1970 in Moscow, at a meeting of the International Commission of Maritime History during the XIII International Congress of Historical Sciences. Because the information was not available by the time he left for Moscow, he was able to report in his paper, which was printed in Etudes D'Histoire Maritime, (Paris, 1970), pp. 132-146, merely that the Smithsonian Institution was working on the dating of the specimen.

The discovery of unmistakeable Viking remains at L'Anse aux Meadows by the Ingstads in 1960 had reoriented people's thinking concerning the reality of a Viking presence in the New World, and in 1967 Dr. Chauncey Loomis reported in an internal Smithsonian memorandum that "Dr. Melvin H. Jackson has suggested that the bloom might not be a Frobisher relic — since there was no mention of a smelter in any of the extensive Frobisher written records" [17]. We should note, however, that the Rawson-MacMillan expedi- tion in 1927 did report "the ruins of two furnaces, with evidence of old fires and clinkers in them." [13, p. 32] Perhaps this was where the "refiners" carried out their tests on the ore, mentioned earlier in connection with the crucible streaked with slag.

With the Frobisher bloom once more in hand, the problem of dating it became paramount. Owing to the circumstances of its

Figure 1.   An iron bloom in the collection of the Department of Naval History, National Museum of American History, Smithsonian Institution (Cat. No. 49459). The bloom weighs approximately 17 lb, after sampling, and is approximately 18 cm in diameter on the base as shown. (Smithsonian Institution negative No. 74836.)

discovery by C. F. Hall, there seemed to be little possibility
of carbon 14 dating by stratigraphic archaeological association
with, for example, charcoal, nor would the significance of such a
date have been very apparent. But in the 1960's Dr. N. Van der
Merwe had perfected a technique for the carbon 14 dating of iron
itself, through dating the carbon extracted from the metal [18].
It is important to restate the obvious; that the date obtained is
that of the carbon content of the iron and not necessarily that
of the smelting process. For example, if coke were used to smelt
the iron, the carbon would be quite inactive. If a very old
timber were converted to charcoal, the carbon taken up would
reflect that age. Likewise any admixture of limestone etc.,
containing dead inorganic carbon might make the age greater. Van
der Merwe had, however, obtained reliable dates on several pieces
of iron of known age, including bloomery iron from Roman times
[18]. Moreover, it is well known that in early metallurgy, bloom
iron makers ordinarily converted locally-cut wood to charcoal for
use in smelting, and this encouraged one to think that a carbon 14
measurement of the Frobisher bloom could yield useful information [19].

Bloomery iron is characterized, however, by very low carbon
content (Table 1) and if one wished to carry out a conventional
carbon 14 dating, consumption of the entire 20 lb. bloom would
have yielded only 4 or 5 grams of carbon, which is barely
enough. For this reason, the Conservation-Analytical Laboratory
of the Smithsonian Institution sponsored the development of the
micro-scale dating procedures at Brookhaven Laboratory already
referred to above [9,10].

The actual measurement proceeded as follows: Approximately
1 kilogram of the Frobisher bloom was milled into powder. Two
separate iron samples, differing only in their pretreatment,
were taken. The first sample of 40 grams was washed with three
100 mL portions of 0.1 $\underline{N}$ HCl by shaking vigorously in a polyethylene bottle, after which the sample was washed with distilled
water in the same manner and dried $\underline{in\ vacuo}$ at room temperature.
A second sample of 40 g was similarly treated except that the
washing with 0.1 $\underline{N}$ HCl was preceded by three washes with 0.1 $\underline{N}$
NaOH. As check samples, specimens of cast iron from Redding
Furnace (1761 A.D.) [17,21] and from Saugus (1648-1678 A.D.) [21]
were also measured. Both of these were bulk samples and encrusted
with corrosion. The corrosion was removed to reveal bright metal
with a small hand vibrating tool. Samples suitable for combustion
were taken by milling with a new tool which had been cleaned with
hexane.

A detailed procedure for combustion, purification, and filling of the small gas proportional counters may be obtained from
the Brookhaven authors. Since the apparatus for combustion had a
maximum capacity of 10 grams it was necessary to burn three
samples of Frobisher iron in order to get enough carbon dioxide to
fill the counter. The carbon dioxide from the three samples was

Table 1. Counting Data on Carbon 14 Extracted from Ancient Iron Specimens.

Sample	Percent Carbon[a]	^{13}C ref. to PDB per mil[b]	^{14}C Activity in cpd/mg-C[c]		Counter number[d]	Background in cpd	mg. Carbon	Age[e]		Days Counted[f]
			Sample	NBS Oxalic acid x 0.95				^{14}C years B.P.	Date, AD	
Frobisher	.051 .062 .127									
(combined)[g]		-23.6	14.44±0.19	15.68±0.15	C-5	51.1±1.1	10.34	679±133	1271±133	97/114
Frobisher	.049 .061 .048									
(combined)[h]		-24.2	14.69±0.15	16.16±0.13	C-12	73.6±1.6	16.16	792±107	1158±107	103/81
Saugus	3.73	-26.1	14.78±0.19	15.64±0.19	C-22	72.1±1.0	10.51	469±144	1481±144	79/69
Redding Furnace	3.98	-24.9	15.53±0.19	16.07±0.21	C-8	67.7±1.1	10.77	285±145	1665±145	83/57

[a] Because of the size limitations, three iron samples were separately combusted and the CO_2 combined. Carbon analyses for the three subsamples are separately listed.

[b] We thank Mr. A. P. Irsa for the mass spectrometric determination of ^{13}C.

[c] These count rates already include the background and isotopic corrections. Cpd = counts per day.

[d] C-5, C-22 and C-8 are 5 cc. counters, C-12 is a 7.5 cc. counter.

[e] Conventional ^{14}C date, $t_{1/2}$ = 5730 years, no calibration for Bristlecone pine.

[f] First number, sample, second number, oxalic acid standard.

[g] Pretreated with 0.1 N HCl and 0.1 N NaOH.

[h] Pretreated with 0.1 N HCl.

then combined before purification. Both Redding Furnace and Saugus contained such ample carbon that only one combustion sample was required. After the counting was completed, isotopic carbon abundances were measured so that appropriate corrections might be applied. The detailed analytical and counting data for the Frobisher iron samples are presented in Table 1, along with the summarized results for Saugus and Redding iron.

The approximate weight of the Frobisher bloom is nineteen pounds; this may be compared to the weights of English iron blooms reported by Tylecote, 1962 [22]. Up until the fourteenth century the weight of blooms did not exceed about thirty pounds; however, with the introduction of water-power in the fourteenth century, this increased to one to two hundred pounds. Nevertheless, if a bloom were produced under primitive conditions, it is likely that it could be smaller than those reported by Tylecote for the sixteenth century. This may have been the case if the Frobisher bloom was produced in Baffin Island. The phosphorus concentration in the metal of the bloom is also of interest. Tylecote reports on the changes in phosphorus concentrations that the highest concentrations in English blooms were during the period of about 800-1100 A.D. By the sixteenth century, the phosphorus concentration in the iron dropped to below 0.1 percent and this is attributed to a search in the Later Medieval period for ores containing lower phosphorus. We have analyzed a section of the Frobisher bloom using the electron microprobe and have found an average concentration of phosphorus in the iron phase of the bloom of approximately 0.6 percent. This information again suggests that the bloom is not characteristic of English blooms of the sixteenth century. It is interesting to compare the Frobisher bloom to an iron rivet, LaM 60, excavated at L'Anse aux Meadows, Newfoundland by Anne Stine Ingstad and discussed by Anna M. Rosenquist [23]. Although the rivet was examined with a microprobe, the concentration of phosphorus was not reported to have been measured; however, the author states from metallographic examination that the rivet "appears to have been forged from a piece of ferritic steel with a carbon content of less than 0.2 percent containing phosphorus." Further comparison of the Frobisher bloom and the rivet from L'Anse aux Meadows could be informative.

The average date of the carbon in the Frobisher bloom, 1214 A.D. ± 175, in the middle of the Norse occupation of Greenland, raises the question of its possible Norse relationship. The site at L'Anse aux Meadows has been dated to ca. 1000 A.D. [23], and Norse artifacts and artifacts made from Norse materials, such as iron, copper, bronze, European oak, and textiles have been found in several Dorset and Thule sites of the 12-13th Centuries [24-26]. One of these finds is a carved wooden figurine recovered from a Thule culture site on the south coast of Baffin Island near Lake Harbor [27]. The figurine, from a house whose artifact assemblage dates stylistically to the 12-13th Centuries, appears to be a Thule Eskimo portrayal of a Norseman encountered, perhaps, in the local area. Since Norse voyaging along this coast would require familiarity with Frobisher Bay, it is possible that they were in Countess of Warwick Sound around 1200 A.D. While this makes sense in terms of the bloom's size and composition, its discovery inside Frobisher's "ships trench", where repairs to his vessels would have required the use of iron, points strongly toward a Frobisher origin. It seems unlikely that this bloom, and possibly the others found by Hall nearby, were all recovered by Frobisher's men from Norse sites. Rather, it seems more likely that these blooms were produced locally by Frobisher's men under primitive field conditions that might have contributed to the differences noted from contemporary European blooms.

Such an interpretation, however, would require explanation of the early date. If the Bloom was produced at Frobisher's site, as suggested by the slag and crucible, early carbon could have been introduced if driftwood was the source of fuel. Today, Frobisher Bay and nearby regions of southeastern Baffin Island do not have trees or woody plants suitable for fueling a smelter, and there is no indication from palynology that living wood was more prevalent in the 13th Century [28]. However, driftwood, which is available today and certainly was also in the past, could have served this purpose admirably. Driftwood might reach Frobisher Bay from Hudson Bay or northern Quebec, but prevailing winds and currents carry most of this wood south along Labrador. Modern and fossil driftwood collected from the islands of the Eastern Arctic and Greenland comes primarily from Siberian sources via the slow Arctic Ocean circulation and subsequent transport south through the arctic islands or by the East Greenland, Irminger, and East Baffin Current. The time span of such a drift is unknown, but with interruptions and strandings it could take hundreds of years before a specimen came to rest in East Baffin. Modern driftwood samples collected from Jones Sound in southern Ellesmere Island have been dated between 30-250 years [29] while fossil driftwood from raised beaches here produced dates thousands of years old. Both modern and fossil driftwood and contemporary European wood fuels would have been available to Frobisher's men in Countess of Warwick Sound.

These and other factors of a metallurgical and archaeological nature must be considered in interpreting the date obtained from the Frobisher bloom. It would be especially useful to compare this specimen with samples of Norse iron and blooms from Greenland and Europe. Nevertheless, the fact that the date is at least two standard deviations removed from the date of the Frobisher voyages is of interest. Unless the carbon was not contemporary with the smelting of the iron, there is only a small chance that the bloom was "made by Frobisher in 1578, searching for gold".

The research performed at Brookhaven National Laboratory
was under contract with the U. S. Department of Energy and was
supported by its Division of Basic Energy Sciences.

References

[1] Muller, R. A., Science, 1977, 196, 489.
[2] Bennett, C. L., Beukens, R. P., Clover, M. R., Gove, H. E.,
Liebert, R. B., Litherland, A. E., Purser, K. H., Sondheim,
W. E., Science, 1977, 198, 508.
[3] Nelson, D. E., Korteling, R. G., Stott, W. R., Science, 1977,
198, 507.
[4] Muller, R. A., Stephenson, E. J., Mast, T. S., Science, 1978,
201, 347.
[5] Bennett, C. L., Beukens, R. P., Clover, M. R., Elmore, D.,
Gove, H. E., Kilius, L., Litherland, A. E., Purser, K. H.,
Science, 1978, 201, 345.
[6] Muller, R. A., Physics Today, Feb. 1979, 23.
[7] Gove, H. E., ed., Proceedings of the First Conference on
Radiocarbon Dating with Accelerators, University of Rochester,
1978.
[8] Stuiver, M., Science, 1978, 202, 881.
[9] Harbottle, G., Sayre, E. V., Stoenner, R. W., Science, 1979,
206, 683.
[10] Contract No. FC-5-53083 between the Smithsonian Institution
Conservation-Analytical Laboratory (Dr. R. M. Organ) and
Brookhaven National Laboratory, dated 6/24/75.
[11] Stefansson, Vilhjalmur, The Three Voyages of Martin
Frobisher, The Argonaut Press, London 1938, See Introduction
and Appendix 9.
[12] Morison, Samuel Eliot, The European Discovery of North
America: The Northern Voyages AD 500-1600, 1971 Oxford
Press, New York.
[13] Roy, Sharat K., Geological Series of (the) Field Museum of
Natural History 1937, VII (2) 21.
[14] Fox-Bourne, H. R., English Seamen under the Tudors, 2 Vols.
1868, London.
[15] Hall, Charles Francis, Life with the Esquimaux, Sampson
Low, Son, and Marston, London, 1865.
[16] (No author), Science, 1927, 66, 295-6.
[17] Conservation-Analytical Laboratory Requisition Folder
Number 0019.
[18] Van der Merwe, Nikolaas, J., The Carbon 14 Dating of Iron,
1969, University of Chicago Press.
[19] Shubert, H. R., History of the British Iron and Steel Indus-
try from ca. 450 BC to 1775 AD, 1957, London.

[20] The Redding Furnace specimen is a dated stoveplate which was provided through the courtesy of Mrs. Paul Sias, Director, Mercer Museum, Doylestown, Pa.
[21] The origin of the Saugus specimen is by courtesy of Mr. W. Glen Gray, Park Manager, Saugus Iron Works, National Historic Site.
[22] Tylecote, R. F., Metallurgy in Archaeology, London, pp. 296-298, 1962.
[23] Ingstad, Anne Stine, The Discovery of a Norse Settlement in America, Universitetsforlaget, Oslo, p. 381, 1977.
[24] Harp, Elmer, Jr., A late Dorset amulet from southeastern Hudson Bay, Folk, 16-17, 33-44 (1974/75).
[25] MacCartney, A. P. and Mack, D. J., Iron utilization by Thule Eskimos of Central Canada, American Antiquity, 38(3), 328-338 (1973).
[26] Schledermann, Peter, Notes on Norse finds from the east coast of Ellesmere Island, N.W.T. Arctic, 33(3), 454-463 (1980).
[27] Sabo, Deborah and Sabo III, George, A possible Thule carving of a Viking from Baffin Island, N.W.T., Canadian Journal of Archaeology, 2, 33-42 (1978).
[28] Short, Susan and Andrews, John T., Palynology of six Middle and Late Holocene peat sections, Baffin Island, Geographie Physique et Quaternaire, 34(1), 61-75 (1980).
[29] Blake, Weston, Jr., Radiocarbon age determinations and postglacial emergence at Cape Storm, southern Ellesmere Island, Arctic Canada, Geografiska Annaler, 57, Ser. A, 1-2, 1-71 (1975).

RECEIVED August 21, 1981.

23

s in the Radiocarbon Dating of Bone

R. E. TAYLOR

University of California, Department of Anthropology, Institute of Geophysics
and Planetary Physics, Riverside, CA 92521

Amino acid composition data and stable isotope ratios
are being evaluated as sources of information to indicate
the presence of non-indigenous organics in bone samples
intended for radiocarbon analyses. The study is being
conducted in the context of the planned ^{14}C measurement of
Pleistocene bone samples by high energy mass spectrometric
methods.

The assumptions of the radiocarbon method are widely known
and understood (Table 1). Each has been the subject of review and
comment over more than thirty years of ^{14}C studies [1-3][1].
Included in Table 1 under the heading contextual is an assumption
sometimes not emphasized in discussions of the accuracy of ^{14}C
determinations. It is essential that the field archaeologist,
geologist or geomorphologist document contextual relationships of
sample materials with a closely defined cultural or geological
event.
　　One way in which archaeologists have attempted to circumvent
the danger of possible misassociation of sample with cultural
event is to obtain a ^{14}C determination directly on the cultural
item itself. In certain contexts, however, an even more fundamental question is whether the phenomena in question can be related
to human behavior at all. This, for example, is one of the basic
questions that continues to arise in discussions concerning the
antiquity of Homo sapiens in the Western Hemisphere.
　　For more than a century, archaeologists and others have
debated the validity of proposed evidence for human occupation of
the New World during the Pleistocene. The issues generally
resolve themselves down to (i) can the alleged cultural materials
be conclusively demonstrated to be the product of human agency
and, (ii) what age assignment can be given to these items? Much

[1]Figures in brackets indicate the literature references at the end
of this paper.

0097-6156/82/0176-0453$05.25/0
© 1982 American Chemical Society

of the debate has been focused on the first question since, the proposed "cultural materials" have been, almost without exception, chipped stone [4]. The dating of human skeletal materials, would obviously render moot the question of human agency as well as the problem of association.

Table 1.   Assumptions of the Radiocarbon Method

Physical Assumptions

1.   The production of ^{14}C by cosmic rays has remained constant long enough to have established a steady-state or equilibrium in the $^{14}C/^{12}C$ ratio in the atmosphere;
2.   There has been a complete mixing of ^{14}C throughout the various carbon reservoirs on a relatively rapid time scale;
3.   The carbon isotope ratios in samples have not been altered except by the decay of ^{14}C;
4.   The total amount of carbon in the exchange system have remained constant;
5.   The decay constant (or half-life) of ^{14}C is known at an appropriate level of precision and accuracy;
6.   Natural levels of ^{14}C can be measured to appropriate levels of precision and accuracy.

Contextual Assumption

1.   There is a direct and specific association between a sample to be analyzed and event or phenomena to be dated.

Another area of current archaeological and paleoanthropological debate, where direct dating of skeletal materials is of crucial significance, involves the question of the origin of anatomically modern Homo sapiens. As a result of studies conducted over the last decade, previous views of the phylogenetic relationship of modern Homo sapiens to presumed ancestral forms, particularly the Neanderthal varients of Homo sapiens, have been undergoing major revision [5]. Crucial to these discussions are questions of the accuracy of the chronological framework for the relevant fossil and archaeological record. Unfortunately, the time period relevant to this issue is currently beyond the practical operational limit for standard decay counting ^{14}C systems. However, the development of direct ^{14}C counting methods, using high energy mass spectrometric analysis, is on the threshold of potentially providing the technology to permit the dating of organic materials for the last 10^5 years [6-8]. Many of the sites from which important fossil and archaeological evidence are derived contain significant quantities of faunal materials, including, of course, the hominid remains themselves. If the accuracy of ^{14}C determinations on relatively small quantities of Pleistocene age bone could be demonstrated, then this would permit direct age determinations on the critical fossil samples.

Until the development of the ^{14}C method, except for the use of fluorine and nitrogen values as relative temporal indicators, estimates of the age of human bone samples were based largely on indirect means. In New World studies, for example, general geological or paleontological considerations were employed, including the degree of morphological affinity of a given sample to modern Native American populations. It was initially assumed that any pre-Holocene age Homo sapiens in the New World would exhibit morphological characteristics similar to those observed on skeletal samples of Homo of similar age in Europe [9]. With the introduction of the ^{14}C method, one might have expected that the direct dating of bone samples would have provided a standard means of establishing age estimates. Unfortunately, this has not always proven to have been the case. The paper will review the problems of obtaining reliable ^{14}C determinations on bone particularly on bone samples of presumed Pleistocene age. Reliable age determinations on such samples would shed enormous light on the development of modern Homo sapiens in the Eastern Hemisphere and their subsequent migration to the New World.

Initial Radiocarbon Determinations on Bone

From the beginning of ^{14}C studies, bone was burdened with a marginal status as a sample type. It was missing from the list of sample materials which Libby initially recommended [10]. He and other researchers discouraged its use for the reason that the carbon content and specifically the organic carbon content, was low even in relatively recent bone and because it was a very porous structure potentially subject to chemical alteration and presumably to contamination. It was concluded that bone would systematically violate the third assumption of the ^{14}C method as listed in Table 1. (It should be noted that "burned bone" was highly recommended. However, the sample material was the carbonized hair, skin, and other tissue rather than the bone matrix itself.)

In 1963, Edwin Olsson surveyed the early ^{14}C literature as it related to problems of sample contamination [11]. The negative reputation of bone was, by then, firmly established as is evidenced in his graphical summary of opinions concerning the relative reliability of different sample types (figure 1). The consensus view was that ^{14}C dates based on bone were generally unreliable. At the same time, the differential reliability of the inorganic as opposed to the organic fraction of bone were recognized. On one hand, "Bone carbonate is worthless," [12], while Sinex and Faris [13] argued that the ^{14}C dating of "purified [bone] gelatin would be as reliable as charcoal dating or possibly more so, since analytical evidence may be obtained that the organic material is what it appears to be, namely, "bone protein". Berger et al. [14] published the first extended series of ^{14}C determinations on the collagen fraction of bone (Table 2). It

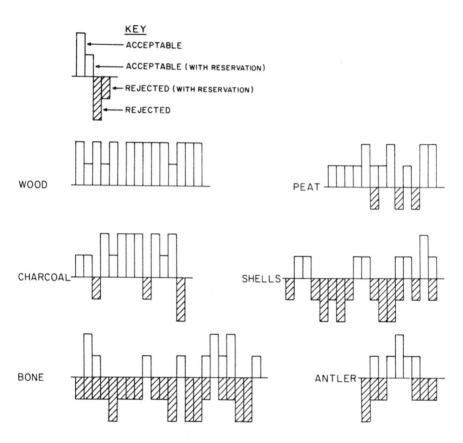

Figure 1.   Relative acceptability of six types of sample materials in early radio-carbon literature (adapted from Ref. 11).

was clear that a specific organic component of Holocene age bone could be prepared which yielded ^{14}C values in concordance with standard sample types. The discordant values were attributed to the lack of good association of bone with the control samples or, as was the case with the Chicago sample (C-302), the anomalous age was attributed to problems with the reagents used in the early phase of solid carbon preparations.

Table 2. Concordance of Suite of Radiocarbon Determinations on Bone Collagen and Associated Organics. Source: Berger et al. [14] and Personal Communication.

Sample No.	Provenience	Sample Type	Radiocarbon Age (^{14}C yrs. B.P.)
UCLA-684	Hastinapur, India	bone collagen	1800±100
TF-90		charcoal	2270±110
UCLA-689	Chimney Cave, Nevada	bone collagen	2500±80
UCLA-690		skin	2510±80
UCLA-692	Santa Rosa Island, Ca.	bone collagen	3970±100
UCLA-140(B)		charcoal	4260±85
UCLA-697A	Sage Creek, Wyoming	bone collagen	8750±120
UCLA-697B		charred tissue (burned bone)	8840±140
C-302		charred tissue (burned bone)	6876±250 (average)

Since that time, a sizable number of bone collagen ^{14}C determinations have been published. Several have figured predominantly in discussions concerning the antiquity of Homo sapiens in the New World. For example, ^{14}C determinations on two skeletal samples from southern California have yielded surprisingly early dates. ^{14}C determinations on the organic fraction of skeletal materials excavated in 1933 from Laguna Beach, a coastal suburb south of Los Angeles, yielded values of >14,800 (UCLA-1233B) and 17,150 ± 1470 (UCLA-1233A) ^{14}C years [15,16]. Likewise, ^{14}C values were obtained on skeletal materials excavated in 1936 by WPA crews in the Baldwin Hills section of west-central Los Angeles. The ^{14}C values on "Los Angeles Man" was >23,600 (UCLA-1430) ^{14}C years [16,17]. On the other hand, ^{14}C determinations on the collagen fraction of two skeletal samples from South America have resulted in a significant reduction in their previously

assumed age. On the basis of ^{14}C and thermoluminescence determinations on carbonates, an age of approximately 30,000 years had been
assigned to a skeleton excavated from Otavalo, Ecuador [18,19].
On the basis of ^{14}C determinations on the collagen fraction of two
bones, the age of the skeleton was determined to be no more than
about 3,000 ^{14}C years [20,21]. Likewise, a collagen-based [14]C
determination on a human cranium from Punin, Ecuador, thought to
be Pleistocene in age yielded on age of 6,900 ± 250 ^{14}C years
[22].

Although increasing confidence has been placed in the
validity of ^{14}C determinations on bone organic fractions, this
acceptance had tended to be less enthusiatic when Pleistocene age
bone samples are involved. When relatively large amounts of
collagen is available for processing, more confidence has been
placed in the resulting age. However, when the organic carbon
content of the bone slips below approximately the 0.5 percent
level, concern has been expressed as whether the recoverable
organic fraction is actually indigenous to the bone or might
be contaminated by migrating soil organics. For example, response
to the publication of the California paleoindian ^{14}C values
indicated a significant degree of reticence in accepting these
results. As one senior American archaeologist has commented, "the
collagen dating of human skeletons [in California] is suggestive —
but no more — of very early habitation in the area" [23].

Preparation of Bone Samples for [14]C Analysis

Table 3 lists the different inorganic and organic fractions
which have been utilized to obtain ^{14}C determinations on bone.
While there are variations between different species as well as
between different bones and bone structures within the same
animal, generally speaking, the inorganic fraction constitutes
about 70 to 80 percent of fresh, dry compact bone [24]. In the
majority of samples derived from most archaeological or geological
contexts, this fraction is composed of (i) the original apatite
component of the bone and (ii) diagenetic or secondary carbonate.
The ^{14}C activity of the secondary carbonate fraction generally
reflects the degree of isotopic exchange with groundwater. Thus
it is truly "worthless" in terms of dating the bone. By contrast,
several researchers have argued that the apatite fraction should
not have been subject to exchange. Experiments using the apatite
component in bone have unfortunately yielded inconsistent
results. Under certain conditions, ^{14}C analysis on this fraction
seem to yield accurate values [25]. However, geochemical and
mineralogical studies have revealed a number of mechanisms which
can significantly alter the carbon isotope values in apatite
structures [26,27]. Such obstacles may not completely exclude its
use for dating bone, as some other workers, more recently, have
been reporting more encouraging results [28].

Table 3. Components of Bone Used for Radiocarbon Determinations

Inorganic Fractions

Diagenetic or Secondary Carbonate

Apatite

Organic Fractions

Acid soluble, insoluble, undissolved fractions
(total and specific molecular weight ranges)

Collagen

Gelatin

Total Amino Acids

The questionable reliability of most ^{14}C determinations on most of the inorganic components of bone has led this, as well as many other researchers, to concentrate on one or more of the organic fractions [29-31]. In modern dry, fat-free bone, over 90 percent of the organics exist in the form of the protein collagen with the remaining non-collagenous proteins composed of a complex assortment of, as yet, incompletely characterized organic substances [32]. In living bone, collagen is deposited in a dense framework of laminated fibers with a highly specific physical structure [33]. In the case of most bone derived from the typical geological or archaeological site, however, it appears that one is not dealing with unaltered collagen [34]. Thus, it is probably more appropriate to use the term, "collagen-derived" when referring to this organic component in bone as it has been affected by diagenetic conditions. Emphasis should also be placed on the danger of using the terms, "collagen" or "collagen-derived" simply as synonyms for the acid soluble, insoluble or undissolved fractions. It should not be automatically assumed that any of these preparations, especially in Pleistocene age bone, will necessarily contain only, or even primarily, collagen-derived organics without independent confirmation.

Several procedures have been employed to prepare various organic fractions of bone for ^{14}C analyses. The most extensive published analysis of the relative merits of the different chemical pretreatment approaches is available as a result of the long term studies of Ingrid Olsson and her collaborators [35-40] and of Afifa Hassan [41-43]. All chemical pretreatments assume an initial physical examination of the external surface and fracture zones to insure the removal of preservatives, microorganisms, and humic materials, rootlets and other non-bone organic fragments [42]. Chemical processing involves initially the elimination of the inorganic carbonates. Both EDTA and HCl have been used for this purpose [14,36]. However, fear of contamination with "old" carbon (containing no ^{14}C) from the EDTA treatment has been expressed [36,41]. Such a problem can apparently be minimized or eliminated with sufficient washing [38]. Further preparations have included conversion to gelatin [13,44-46], treatment with NaOH to remove humates and other base soluble fractions [36,46, 47], and the separation of total amino acids or a single amino acid [30,48].

In every case, the goal of any pretreatment procedure is to isolate one or more "uncontaminated" organic fraction(s) which is (are) unambiguously indigenous to the original bone sample. The strategies employed to accomplish this, however, must conform to the reality of the degree of preservation of the organics in a particular sample. Where the organic carbon content is relatively high (e.g., >1 gram organic carbon depending on the counting system employed), one can use relatively rigorous sample pretreatment techniques. Where the organic carbon content is significantly less than this, care must be exercised so that the sample size following the completion of the pretreatment does not drop below acceptable limits in terms of counting requirements. Thus, an important requirement of any chemical or physical processing technique is that it can be flexibly applied to maximize the removal of non-indigenous organics while minimizing loss of autochthonous organics.

Various criteria have been employed in an attempt to identify those bone samples, of low organic carbon yield, where diagenetic effects may have contaminated the indigenous organic fraction(s). One quantitative approach might be called the "pseudomorph" test. Figure 2a illustrates where demineralization of a bone from a late Roman/Christian cemetary in England left behind an organic replica or pseudomorph of the original bone structure. By contrast, the same treatment of a bone sample from a Pleistocene site in North Africa yields a product in which the original physical structure of the bone is totally lost (figure 2b). Unfortunately, a high degree of physical degradation in the bone matrix may or may not correlate with the presence of significant amounts of non-in situ organics. Most bones, from the majority of environments with ages of several thousands or tens of thousands of years, would substantially "flunk" the pseudomorph test. Many of these samples, however, would yield apparently acceptable ^{14}C values. More quantitative methods need to be evaluated which can be employed to identify bones with potentially serious contamination problems.

Evaluating Contamination Effects

In reviewing various strategies that might be used to evaluate the differential effects of contamination, it might be helpful to briefly consider the relationship between the "actual ^{14}C age" and "apparent ^{14}C age" of samples to which have been added varying amounts of non-indigenous organic material. Two simple models can be employed to evaluate such conditions. The

Figure 2. Organic residue following demineralization of human bone sample from (a) late Roman/Christian cemetery near Poundbury, Dorchester, England (sample supplied by Theya Molleson, British Museum) and (b) Upper Paleolithic site in North Africa.

first involves plotting the effects of the addition of <u>modern</u> carbon to samples of varying "true or actual age". Figure 3 represents a summary of the relationship between actual and apparent ages in samples to which have been added varying percentages of <u>contemporary</u> carbon. Except at the level of five percent modern carbon, this plot shows that up to around 20,000 years, the effect on the dilution age is on the order of the typical statistical error for decay counting systems. Contamination with modern carbon on the order of 0.5 to 1 percent does, however, limit the confidence one can have with bone to about 35,000 ^{14}C years.

For those who are concerned with the preparation of samples for the direct ^{14}C counting systems, this summary vividly illustrates the effects on the final age value as a result of the addition of ppm levels of modern carbon for samples in different age ranges. For a sample with an actual age of 75,000 years, for example, a 100 ppm addition of modern carbon results in approximately a 5000-year error in the final measured age. Hopefully, contamination of a sample with modern carbon would be a relatively rare occurrence. A more probable situation would involve the addition of carbon differing in age from the original samples from several hundred up to several tens of thousands of years.

Olsson [37] previously provided a plot of the effects of introducing varying amounts of organic materials with age differences of up to 5,000 years between sample and contaminant. Figure 4 has extended her calculations to show the effect of the introduction of 0.5 percent to 30 percent contaminants with age differences of up to 50,000 years. These data show that for Holocene samples contaminated with up to five percent of material with a differential age of not more than 5,000 years, the effects generally are within the range of the typical one sigma counting errors for decay counting systems. Up to a 1.0 percent dilution with organics not more than about 20,000 years younger than the indigenous samples, the age reduction does not exceed about 1000 years. Figure 4 also shows the effects from 10 percent to 30 percent sample contamination with younger materials. Samples with this degree of contamination would almost certainly be identified during sample preparation steps and appropriate pretreatment strategies employed. Even with a 10 percent contamination factor, however, if the differential age of the contamination with respect to the autochthonous sample does not exceed about 10,000 years, the anomaly thus created is on the order of about 2,000 years.

In anticipation of the development to operational status of the ion or direct counting systems, it would be helpful if we could compare these values with projected counting errors for the two types of direct counting systems being developed. Table 4 lists projections for the Rochester Van de Graaff facility [49] and the University of California Lawrence Berkeley cyclotron system employing an external ion source [31,50]. Table 4 also lists the sample sizes and approximate measurement periods for both systems. This data illustrates the potential extention in dating

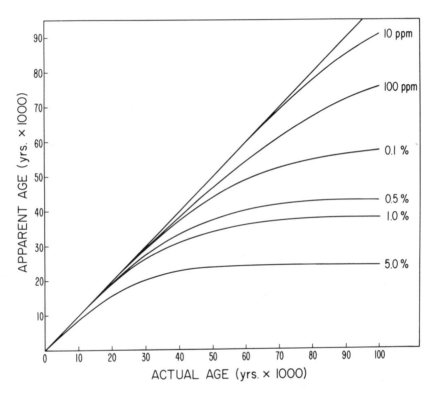

Figure 3. Relationship between actual age and apparent age for samples to which varying percentage of modern C has been added (31).

Table 4. Expected Counting Parameters for Direct Counting of Radiocarbon

	(VG)* Modern	(C)** 10,000	(C)** 25,000	(C)** 50,000	(VG)* 60,000	(C)** 75,000	(VG)* 100,000
Sample Age							
Sample Size	1 mg	50 mg	50 mg	50 mg	2-5 mg	50 mg	120 mg
Approximate Counting Error (1 sigma)	±25	±100	±240	±1100	±500	±5000	±2500
Time Needed/Measurement	15 min	100 min	100 min	100 min	2 hrs	100 min	1 day

* Van de Graaff System (Rochester): Source: reference 49.
** Cyclotron System (Berkeley): Source; T. S. Mast, personal communication.

range and major reduction in sample sizes that <u>theoretically</u>
should be possible with high energy spectrometric methods.
Obviously, both of these characteristics would offer significant
advantages in dating bone samples with low organic carbon yields
and documenting the time period on which our studies are particularly focussed. It should be noted that the counting error values
are calculated on the basis of statistics alone and do not take
into account any error introduced because of instrumental variability. Also, there is significant difference in error estimates
between the Van de Graaff and cyclotron facility because of
different assumptions concerning counting efficiencies and background levels. The background levels in the instruments are the
principal limiting factor in realizing the theoretical extention
in the dating range into the 10^6 year region [7]. Of equal
importance will be the ability to exclude contaminating organics,
as we have illustrated with the data in figures 3 and 4 [51].

Identification of Non-Autochthonous Organics in Bone Samples

Several methods are currently being evaluated to determine
whether one or more of them can be employed to identify bone
samples which contain unacceptable amounts on non-indigenous
organics. One approach is examining stable isotope data as a
means of monitoring the transport of organics from the surrounding
environment into the bone matrix. Although both hydrogen (D/H)
and carbon ($^{13}C/^{12}C$) isotope data on bone have been examined [31],
recent attention has been focussed on nitrogen ($^{15}N/^{14}N$) values.
Potential fractionation effects resulting from the acid hydrolysis
procedures used to isolate the total organic fraction have been
investigated (Table 5). Modern mammalian bone samples from
natural and captive (zoo) environments have been examined to
determine the effect of species, environmental, and dietary
variability, while measurements on Holocene and late Pleistocene
archaeological and paleontological samples have been obtained to
identify potential diagenetic processes affecting isotopic
composition (figure 5).

The basis of this study is the suggestion that organic fractions not indigenous to a bone will manifest distinctive stable
isotope patterns. It has been experimentally determined that ^{15}N
values in bone collagen is directly related to the ^{15}N value of
the diet with the collagen being enriched by some 2-3 per mil with
respect to the dietary source [54]. If it can be shown that the
isotope values attributable to initial environmental (including
dietary) and diagenetic processes can be distinguished from that
caused by as a result of the transport into the bone matrix of
organic material external to the bone itself, then such values can
be used as criteria to indicate the presence of contaminants.
Table 5 contrasts the ^{15}N values obtained from whole bone samples
(no pretreatment) as opposed to samples subjected to very rigorous
and extended acid hydrolysis [55]. Measurements conducted on

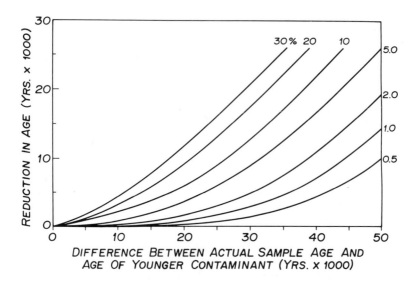

Figure 4. Reduction in age of samples resulting from addition of 0.5 to 30% of organics with 10,000 to 50,000 year younger age (after Ref. 37).

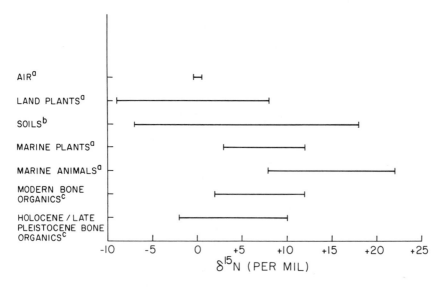

Figure 5. Nitrogen isotope measurements on various components of the natural environment. The ^{15}N values expressed with respect to AIR. (a) Data from Ref. 52; (b) data from Ref. 53; and (c) data from Ref. 31.

Table 5. Effect of Variation in Pretreatment Methods on ^{15}N Values of Contemporary and Late Pleistocene Bone Samples[a].

Sample	Pretreatment	Replication	^{15}N Values[b] Range	Average
Modern Bone (Bos taururus)	None (whole bone)	7	+5.56 - +6.21	+6.00
	Total Hydrolyses[c] (organic fraction)	3	+4.34 - +6.16	+5.67
Pleistocene Bone[d] (Mammuthus columbi)	None (whole bone)	2	+10.20 - +11.76	+10.98
	Total Hydrolyses[c] (organic fraction)	3	+6.67 - +8.38	+7.30

[a]G. Rau, personal communication.
[b]Per mil expressed with reference to AIR.
[c]Complete dissolution of bone matrix in 12N HCl for 12 hrs. See reference 31.
[d]Both organic and apatite fractions yielded age of approximately 18,000 ^{14}C years.

replicate samples, using techniques to be described elsewhere [56], indicate an actual precision for bone ^{15}N measurements of about ± 1.5 per mil. Measurements on modern bone show no significant effects as a result of the hydrolyses technique employed. These measurements confirm the expectation that essentially all of the nitrogen contained in contemporary bone is derived from the organic fraction which is predominantly collagen. Comparison of the ^{15}N values of the whole bone and organic fraction of a late Pleistocene bone (Mammuthus columbi) dated at about 18,000 ^{14}C years B.P. reveals differences on the order of 3 per mil. Whether this difference can be attributed to diagenetic effects (presence of non-organic nitrogen) or other factors is currently under study.

Figure 5 compares the range in ^{15}N values for four nitrogen containing segments of the natural environment compared with ^{15}N measurements on the total organic fraction of modern and archaeological/paleotological bone samples [31]. Measurements obtained to-date on modern bone suggest that the maximum variability in ^{15}N values attributable to species, environmental, or dietary causes is on the order of 10 per mil. Based on ^{15}N values obtained from land plants and soil samples, we would expect the range to be much greater than this especially if the diet of an animal reflected the exploitation of marine resources. Additional measurements on modern bone will be necessary to conclusively document non-diagenetic variability.

The use of amino acid composition and/or racemization values as a means of characterizing the organic constituents of fossil bone has been previously suggested by several researchers [31,32, 43]. Collagen in modern bone is characterized by a high glycine content, relatively high proline, and the presence of hydroxyproline and hydroxylyxine. Except for a limited distribution in some plant extracts, hydroxyproline has been found only in the hydrolyses products of collagen and its diagenetic derivatives. A characteristic of most amino acids is that they can exist in one or two possible optical forms designed as D- and L-. The amino acids in most living organisms consist only of L-amino acids. At the death, for example, of an animal, the amino acids contained in the proteins of the bone structure will begin to be subject to chemical changes which results in progressive alterations in the relative portions of the various amino acids as well, in the case of most amino acids, in the L-forms changing into their corresponding D-form. As is well-known, racemization (or epimerization as in the case of isoleucine) has been used as an independent means of dating fossil bone [57,58] with some however, expressing serious reservation as to the validity of many of the amino acid racemization dates [59,61].

The use of the relative proportions and racemization properties of amino acids as a means of identifying contaminated samples would have several advantages. First of all, modern amino acid analyzers are very sensitive analytical instruments. Systems are now available which can obtain quantitative amino acid composition analyses with simultaneous determination of D/L ratios on a wide assortment of amino acids employing extremely small amounts of sample [62]. This would permit measurements to be made on samples even when the preservation conditions are extremely poor. An additional advantage would be the ability to increase the number of bone samples for which there would be both direct ^{14}C and amino acid racemization determinations. Perhaps, this data set would contribute to resolving some of the issues surrounding the question of the validity of amino acid racemization deduced age values. Several studies are currently underway to examine the use of amino acid data as a means of identifying biochemically anomalous bone samples.

Probably the most direct approach to the problem of contamination in bone would be to obtain multiple ^{14}C determinations on different organic fractions of the same sample (cf. Table 3). The reliability of the ^{14}C determinations would be a function of the degree of concordance of the ages reflected in the different organic fractions. In cases where the degree of organic preservation permits, this approach can be and has been employed with success [38,39]. Until the advent of high energy mass spectrometric methods of measurement, however, it generally was not practicable to pursue this approach with most bone samples recovered from Pleistocene sites except where relatively large amounts of bone was available. When operational status for the direct ^{14}C dating systems are achieved, the extraction and dating of single amino acids such as hydroxyproline, known to have limited distribution except as a constituent amino acid of collagen, would provide the most secure approach in the dating of Pleistocene bone. Probably more common, however, will be the dating of acid soluble and insoluble fractions before and after base extractions and following conversion to gelatine [63]. This approach might be augmented by dating various fractions of different molecular weights. The basis of this is the suggestion that non-in situ organics in bone will generally be of lower molecular weight than the residual autochthonous organics. The ability to obtain ^{14}C determinations on 10-100 milligram size samples of geochemically characterized organic fractions of bone should allow this sample type to be routinely employed along with charcoal and other more standard sample types in future ^{14}C studies [64].

References

[1] Olsson, I. U., ed., Radiocarbon Variations and Absolute Chronology, Almqvist and Wiksell, Stockholm, 1970.
[2] Rafter, T. A., Grant-Taylor, T., eds., Proceedings of the 8th International Conference on Radiocarbon Dating, Royal Society, Wellington, 1973.
[3] Berger, R., Suess, H. E., eds., Radiocarbon Dating, University of California Press, Berkeley, 1979.

[4] Taylor, R. E., Payen, L. A., The Role of Archaeometry in American Archaeology: Approaches to the Evaluation of the Antiquity of Homo sapiens in California, In: Advances in Archaeological Method and Theory, Vol. 2, pp. 239-283, Academic Press, New York, 1979.

[5] Trinkhaus, E., Howells, W. W., The Neanderthals, Scientific American, 1979, 241, 94-105.

[6] Muller, R. A., Radioisotope Dating with a Cyclotron, Science 1977, 196, 489-494.

[7] Stuiver, M., Carbon-14 Dating; A Comparison of Beta and Ion Counting, Science, 1978, 202, 881-883.

[8] Berger, R., Radiocarbon Dating with Accelerators, Journal of Archaeological Science, 1979, 6, 101-104.

[9] Hrdlicka, A., Skeletal Remains Suggesting or Attributed to Early Man in North America, Smithsonian Institution, Bureau of American Ethnology. Bulletin 33. Washington, Government Printing Office, 1907.

[10] Libby, W. F., Radiocarbon Dating, p. 44, Chicago, University of Chicago Press, 1955.

[11] Olsson, E., The Problem of Sample Contamination in Radio-carbon Dating, unpublished doctoral dissertation, Columbia University, 1963.

[12] Olsson, E. A., Broecker, W. S., Lamont Natural Radiocarbon Measurements VII, Radiocarbon, 1961, 3, 141-175.

[13] Sinex, F. B., Faris, B., Isolation of Gelatin from Ancient Bones, Science, 129, 969.

[14] Berger, R., Horney, A. G., Libby, W. F., Radiocarbon Dating of Bone and Shell from Their Organic Components, Science, 1964, 144, 999-1001.

[15] Berger, R., Libby, W. F., UCLA Radiocarbon Dates IX, Radio-carbon, 1969, 11, 194-209.

[16] Berger, R., Protsch, R., Reynolds, R., Roxaire, C., Sackett, J. R., New Radiocarbon Dates Based on Bone Collagen in California Paleoindians, Contrib. Univ. Calif. Archaeol. Res. Fac., 1971, 12, 43.

[17] Berger, R., Advances and Results in Radiocarbon Dating: Early Man in America, World Archaeology, 1975, 7, 174-184.

[18] Davies, D. M., Fossil Man in Ecuador, Spectrum, 1973, 106, 4215.

[19] Davies, D. M., Some Observations on the Otavalo Skeleton and Other Ramains of Early Man in South America, Journal of Human Evolution, 1978, 7, 279-281.

[20] Brothwell, D., Burleigh, R., On Sinking Otavalo Man, Journal of Archaeological Science, 1977, 4, 291-294.

[21] Burleigh, R., On the Dating of a Human Skeleton from Otavalo, Ecuador, Journal of Human Evolution, 1980, 9, 153.

[22] Brothwell, D., Burleigh, R., The Human Cranium from Punin, Ecuador, with Particular Reference of Morphology and Dating, Journal of Archaeological Science, 1980, 7, 97-99.

[23] Willey, G. H., A Summary Scan, In: Chronologies in New World Archaeology, p. 528, New York, Academic Press, 1978.

[24] Herring, G. M., The Organic Matrix in Bone, In: Biochemistry and Physiology of Bone, 2nd Edition, pp. 128-189, New York, Academic Press, 1972.

[25] Haynes, C. V., Radiocarbon: Analysis of Inorganic Carbon of Fossil Bone and Enamel, Science, 1968, 161, 687-688.

[26] Hassan, A. A., Geochemical and Mineralogical Studies on Bone Material and their Implications for Radiocarbon Dating, unpublished doctoral dissertation, Southern Methodist University, 1976.

[27] Hassan, A. A., Termine, J. D., and Haynes, C. V., Mineralogical Studies on Bone Apatite and their Implications for Radiocarbon Dating, Radiocarbon, 1977, 19, 364-374.

[28] Hass, H., Banewics, J. J., Radiocarbon Dating of Bone Apatite Using Thermal Release of CO_2, Radiocarbon, 1980, in press.

[29] Taylor, R. E., Radiocarbon Dating: An Archaeological Perspective, In: Archaeological Chemistry II, pp. 33-69, Washington, D.C., American Chemical Society, 1978.

[30] Taylor, R. E., Slota, P., Fraction Studies on Marine Shell and Bone Samples for Radiocarbon Analyses, In: Radiocarbon Dating, pp. 422-432, Berkeley University of California Press, 1979.

[31] Taylor, R. E., Radiocarbon Dating of Pleistocene Bone: Toward Criteria for the Selection of Samples, Radiocarbon, 1980, in press.

[32] Hare, P. E., Organic Geochemistry of Bone and Its Relation to the Survival of Bone in the Natural Environment, In: Fossils in the Making: Vertebrate Taphonomy and Paleoecology, Chicago, University of Chicago Press, 1979.

[33] Miller, E. J., The Collagen of Bone and Cartilage, In: The Biochemistry and Physiology of Bone, 2nd Edition, pp. 1-20, New York, Academic Press, 1972.

[34] Tuross, N., Hare, P. E., Collagen in Fossil Bone, Carnegie Inst. of Washington Yearbook, 1978, 77, 891-895.

[35] Olsson, I. U., Modern Aspects of Radiocarbon Dating, Earth-Science Reviews, 1968, 4, 203-218.

[36] Olsson, I. U., El-Daoushy, M. F. A. F., Abd El-Mageed, A. I., Klasson, M., A Comparison of Different Methods for Pretreatment of Bones. I, Geol. Fören Stockholm Förh, 1974, 96, 171-181.

[37] Olsson, I. U., Some Problems in Connection with the Evaluation of ^{14}C Dates, Geol. Fören Stockholm Förh, 1974, 96, 311-320.

[38] Fakid, A. F., El-Daoushy, M. F. A. F., Olsson, I. U., and Oro, F. H., The EDTA and HCl Methods of Pre-treating Bones, Geol. Fören Stockholm Förh, 1978, 100, 213-219.

[39] Olsson, I. U., El-Daoushy, M. F. A. F., Uppsala Natural Radiocarbon Measurements XII, Radiocarbon, 1978, 20, 469-486.

[40] Follestad, B. A., Olsson, I. U., The ^{14}C Age of the 'Toten' Mammoth, Eastern Norway, Boreas, 1979, 8, 307-312.
[41] Hassan, A. A., Geochemical and Mineralogical Studies on Bone Material and Their Implications for Radiocarbon Dating, unpublished doctoral dissertation, Southern Methodist University, 1976.
[42] Hassan, A. A., Ortner, D. J., Inclusions in Bone Material as A Source of Error in Radiocarbon Dating, Archaeometry, 1977 19, 131-135.
[43] Hassan, A. A., Hare, P. E., Amino Acid Analysis in Radiocarbon Dating of Bone Collagen, In: Archaeological Chemistry II, p. 109-116, Washington, D.C., American Chemical Society, 1978.
[44] Longin, R., New Method of Collagen Extraction for Radiocarbon Dating, Nature, 1971, 230, 241-242.
[45] Protsch, R. R. R., The Dating of Upper Pleistocene Subsaharan Fossil Hominids and Their Place in Human Evolution, with Morphological and Archaeological Implications, unpublished doctural dissertation, University of California, Los Angeles, 1973.
[46] Berglund, B. E., Håkansson, S., Lagerlund, E., Radiocarbon-dated mammoth (Mammuthus primigenius Blumenbach) finds in South Sweden, Boreas, 1976, 5, 177-191.
[47] Haynes, C. V., Bone Organic Matter and Radiocarbon Dating, In: Radioactive Dating and Methods of Low-Level Counting, pp. 163-168, Vienna, International Atomic Energy Agency, 1967.
[48] Ho, T., Marcus, L. F., Berger, R., Radiocarbon Dating of Petroleum-impregnated Bone from Tar Pits at Rancho La Brea, California, Science, 1969, 164, 1051-1052.
[49] Purser, K. H., Accelerators-The Solution to Direct ^{14}C Detection, In: Proceedings of the First Conference on Radiocarbon Dating with Accelerators, pp. 1-32, Rochester, University of Rochester, 1978.
[50] Mast, Terry S., personal communication, 1980.
[51] Haynes, C. V., Applications of Radiocarbon Dating with Accelerators to Archaeology and Geology, In: Proceedings of the First Conference on Radiocarbon Dating with Accelerators, pp. 276-288, Rochester, University of Rochester, 1978.
[52] Sweeney, R. E., Liu, K. K., Kaplan, I. R., Oceanic Nitrogen Isotopes and Their Uses in Determining the Source of Sedimentary Nitrogen, In: Stable Isotopes in the Earth Sciences, pp. 9-26, Wellington, New Zealand, 1978.
[53] Rennie, D. A., Paul E. A. Johns, L. E., Natural Nitrogen-15 Abundance in a Wide Variety of Soils, Canadian Journal of Social Science, 1976, 56, 43-50.
[54] DeNiro, M. J., Epstein, S., Influence of Diet on the Distribution of Nitrogen Isotopes in Animals, Geochemica et Cosmochimica Acta, 1980, in press.

[55] In an earlier paper [31], the ^{15}N values for two fractions of this bone was reported as -7.84 and -7.66 per mil. Analytical problems, which were unrecognized during the earlier measurements, have now been resolved. The average values for whole bone and for the organic extract listed in Table 5 are now considered to better represent the correct values. G. Rao, personal communication.

[56] Rau, G., Kaplan, I. R., Taylor, R. E., manuscript in preparation.

[57] Bada, J. L., Schroeder, R. A., Amino Acid Racemization Reactions and Their Geochemical Implications, Die Naturwessenschaften, 1975, 62, 71-79.

[58] Bada, J. L., Master, P. M., Hoopes, E., Darling, The Dating of Fossil Bones Using Amino Acid Racemization, In: Radiocarbon Dating, pp. 740-756, Berkeley, University of California Press, 1979.

[59] Von Endt, D. W., Techniques of Amino Acid Dating, In: Pre-Llano Cultures of the Americas: Paradoxes and Possibilities, pp. 71-100, Washington, Anthropological Society of Washington, 1979.

[60] Hare, P. E., Organic Geochemistry of Bone and Its Relation to the Survival of Bone in the Natural Environment, In: Fossils in the Making: Vertebrate Taphonomy and Paleoecology, Chicago, University of Chicago Press, 1979.

[61] Bender, M. L., Reliability of Amino Acid Racemization Dating and Paleotemperature Analysis on Bones, Nature, 1974, 252, 378-379.

[62] Hare, P. E., Gil-Av., E., Separation of D and L Amino Acids by Liquid Chromatography: Use of Chiral Eluants, Science, 1979, 204, 1226-1228.

[63] Berglund, B. E., Håkansoon, S., Lagerlund, Radiocarbon-Dated Mammoth (Mammuthus primigenius Blumenback) finds in South Sweden, Boreas, 1976, 5, 177-191.

[64] The assistance of G. Rau and I. R. Kaplan (UCLA) in the stable isotope studies as well as the dedicated laboratory work of Peter Slota, L. A. Payen, and G. Prior (UCR) is gratefully acknowledged. This research was supported by a National Science Foundation grant (BNS-7815069).

RECEIVED August 11, 1981.

Absolute Dating of Travertines from Archaeological Sites

H. P. SCHWARCZ

McMaster University, Department of Geology, Hamilton, Ontario L8S 4MI Canada

Many archaeological sites of middle and upper Pleistocene age are found associated with calcium carbonate deposits formed at or near to the time of occupation of the site. In caves, stalagmitic layers are found interstratified between cultural deposits, sometimes, encrusting bones or tools. Fossil spring deposits were commonly camp sites for hominids, and the travertine layers are found to enclose tools and bones left by the ancient hunters. Such chemically precipitated carbonates can generally be dated by the measurement of radioactive disequilibrium between the daughters of ^{238}U and ^{235}U. The $^{230}Th/^{234}U$ method permits dating to about 400,000 years while the $^{231}Pa/^{235}U$ method allows dating to 300,000 years. In many prehistoric sites however the carbonate deposits are contaminated with detrital clay, limestone clasts, etc., which lead to errors in dating. Some methods for correcting for the effects of contaminants are presented. Travertine layers are also well suited to other physical dating methods including thermoluminescence, ESR and paleomagnetic correlation. Results are presented for the ages of sites in Israel, Greece, Hungary, Germany, and France, ranging in cultural level from the lower to upper Paleolithic.

Man has evolved over a period of more than three million years, from the earliest upright hominid, Australopithecus, to our own species, Homo sapiens sapiens, who first appeared about 40,000 years B.P. The dating of these evolutionary stages and the cultural changes that accompanied them has been largely accomplished through the use of various radiometric techniques originally devised for use in geophysics: $^{40}K/^{40}Ar$; ^{14}C; $^{230}Th/^{234}U$ and fission track analysis. Through the cooperation of archaeologists, geologists, and physicists, a wide variety of

0097-6156/82/0176-0475$05.00/0

methods has become available for the dating of prehistoric occupation sites. This article will review one of these methods which
has begun to provide us with a time scale for the interval from
about 400,000 y B.P. to a few thousand years ago (figure 1), thus
overlapping the range of ^{14}C dating at its lower limit and the
range of ^{40}K/^{40}Ar dating at its upper limit.

Chemically precipitated calcium carbonate is found associated
with hominid occupation sites in a wide variety of modes of occurrence, for example as stalagmitic floors and encrustations
interstratified with cave deposits (figure 2); as spring-deposited
mounds or terraces of travertine, enclosing traces of human
activity; as fossil lake deposits interstratified with shore
occupation sites; as calcreted soils under- or overlying archaeological deposits. In each of these modes the carbonate has been
precipitated from aqueous solutions containing traces of uranium
generally present as the $UO_2(CO_3)_2{}^{-2}$ complex [1][1]. Thorium, on
the other hand, is extremely insoluble in near-surface waters.
Therefore the freshly precipitated calcite contains a trace of
uranium, apparently in solid solution in the crystal lattice
(figure 3), but not accompanied by any of its short-lived daughter
isotopes, other than ^{234}U. The time elapsed since the deposition
of the calcite can be estimated from the extent to which the
various daughters have grown into equilibrium with the parent
uranium isotope. The relevant decay schemes are reviewed in
figure 4. The time needed for return to equilibrium is determined
by the decay constant of the daughter isotope; after about ten
half-lives, the activity of the daughter (in disintegrations per
unit time per gram of calcite) is approximately equal to that of
its corresponding parent.

The principal application of this method has been to the
dating of cave and spring deposits associated with lower and
middle Paleolithic occupations in Europe and Asia. Hominids have
apparently been living in limestone caves about 250,000 years
(although the site at Zhou Kuo Dien, China, may be more than twice
this old). Typically, such caves are partially filled with layers
of detrital sediment blown into the cave by wind, washed in by
streams, or fallen from the roof of the cave. During periods of
wet climate, water drips from the roof of the cave or seeps from
the walls and as it evaporates it leaves a chemical precipitate
of calcium carbonate. During such wet intervals, the cave may be
temporarily unoccupied; however the calcite deposits are found
either over-or underlying cultural layers or hominid skeletal
remains. Also, such deposits can infiltrate into underlying
porous deposits or form a cement between larger blocks of rock,
and thus provide a minimum age for these older strata. In some
caves only the remains of large vertebrates are found embedded in

---

[1]Figures in brackets indicate the literature references at the end
of this paper.

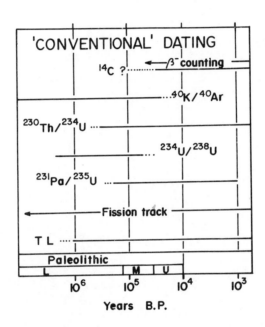

Figure 1. Absolute dating methods relevant to the period of man's evolution. TL (Thermoluminescence), $^{230}Th/^{234}U$ and $^{231}Pa/^{235}U$ dating are applicable to travertine deposits.

Figure 2. Stone artifact (hand-ax) embedded in travertine in the Zuttiyeh cave, Wadi Amud, Israel.

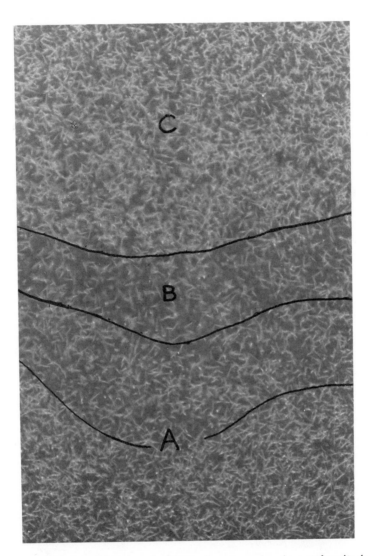

Figure 3.   Fission track map of the distribution of uranium in a stalagmite from Grotte Valerie, Nahanni National Park, N.W.T., Canada.

Uranium is distributed uniformly within individual growth layers, but varies greatly in concentration between successive layers (A, B, and C). The distribution of tracks suggests that U is distributed homogeneously in single calcite crystals, and is not concentrated at grain boundaries or in inclusions.

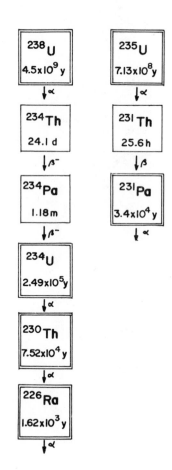

Figure 4. The decay schemes of ²³⁸U *and* ²³⁵U, *showing the longer-lived isotopes used in radiometric dating of Pleistocene and Holocene deposits.*

the travertine. However useful information can still be had if nearby archaeological sites are found to contain the same species.

Spring-deposited travertines are widespread in areas of Europe and Asia which are underlain by limestone and dolomite. Where ground waters emerge from carbonate-rock aquifers, they are usually supersaturated with $CaCO_3$ due to the high content of HCO_3^- ions introduced from soil at the site of groundwater recharge. Therefore, outgassing of CO_2 from these waters at the point of discharge results in the buildup of sheets and mounds of calcite that may be hundreds of meters in area and tens of meters thick. Such springs often served as watering holes for various animals and therefore also attracted the hominid game-hunters, who left traces in the form of stone tools; some hominid skeletons are also known from such deposits.

In addition to chemically precipitated carbonates, other types of deposits formed at the time of hominid occupation have been dated by uranium series methods, with varying degrees of success. The shells of various types of molluscs, found as refuse in middens, have been dated by the $^{230}Th/^{234}U$ method, but Kaufmann et al., [2] have shown that such dates commonly disagree with ^{14}C ages over time range in which these methods are both applicable. There has been post-mortem uptake of uranium into the shells that causes ages to be too young. Bones have been dated by both $^{230}Th/^{234}U$ and $^{231}Pa/^{235}U$ methods; Cherdyntsev [3] observed that there were gross discordancies between supposedly coeval bone dates. However, Szabo [4] has found that in many instances, there is good agreement between bone dates and archaeological estimates. He notes however that, like molluscs, the bone continues to adsorb uranium for about 3000 years after burial. This review will not consider these types of samples.

Analytical Methods

The analytical methods used have been described in detail by Gascoyne et al., [5] and basically follow methods devised for the dating of organically precipitated carbonates (corals, molluscs, etc.) as given, for example, by Kaufmann [6]. Essentially, the carbonate is taken into solution, along with a suitable isotopic tracer (generally ^{232}U in equilibrium with ^{228}Th) and Th and U are coprecipitated onto ferric hydroxide. Th and U are then separated from one another by ion exchange methods. The Th and U are finally plated onto metal discs and counted on an alpha spectrometer. The age of a carbonate is determined from the $^{230}Th/^{234}U$ and $^{234}U/^{238}U$ ratios. The latter ratio is commonly greater than unity because ^{234}U is preferentially leached from limestone due to various hot atom effects [7]. For samples less than 300,000 years old and containing more than 1 ppm of uranium, it is also possible to determine their age from the $^{231}Pa/^{230}Th$ or $^{231}Pa/^{235}U$ ratios. ^{231}Pa can be determined with adequate precision by measuring the activity of its daughter ^{227}Th; alternatively, Pa can be plated out and counted [8]. Rosholt and Szabo [9] have devised a method using neutron activation of ^{231}Pa to ^{232}U.

Problems of the Method

Dating of travertines is complicated by problems arising from the nature of the deposits and also from the chemical procedures used to analyze them. First, considering both spring and cave deposits, these materials tend to incorporate at the time of their formation minor amounts of detrital impurities which contribute various daughter isotopes, usually in excess of the amount of parent present. Therefore some correction must be devised to compensate for this contamination. Also, some travertine deposits are subject to post-depositional recrystallization, especially if the original deposit was relatively porous and permeable. This is particularly true of spring deposits, which commonly form around the stems of reeds, grasses, and mosses. Cave deposits may also be quite spongy if the calcite is forming due to rapid evaporation of water seeping into a cave of low humidity.

Corrections for detrital contamination have been largely made by taking advantage of the fact that ^{232}Th is absent in freshly precipitated calcite, because of its low solubility. However, detrital clays and sand contain significant amounts of this isotope accompanied by ^{230}Th present generally at an activity ratio of ^{230}Th/^{232}Th = 1.0 to 1.8. Thus, an amount of ^{230}Th suitably corrected for its decay since the time of deposition, can be subtracted from the total ionium activity, in proportion to the amount of ^{232}Th present. Such corrections are necessary if the observed ^{230}Th/^{232}Th of the sample is less than about 20. Other more elaborate schemes, have also been employed, making use of the correlation between ^{230}Th/^{232}Th and ^{234}U/^{232}Th that would be expected if the various samples from a site were admixtures of authigenic calcite plus varying amounts of the same detrital contaminant [10]. Schwarcz [11] has also suggested a method which corrects for ^{238}U and ^{234}U contributed by the detrital contaminant, as well. Ku et al., [12] in a study of calcrete ages, propose yet another method which depends on the analysis of many cogenetic samples.

The problem of porosity and migration of radioelements from the travertine after deposition is more difficult to contend with. We have found that spring deposits formed in desertic areas remain relatively unaltered by subsequent percolating waters, after cessation of activity of the spring. Thus, the active phase of many springs is very short compared with the total age of the deposit. Cherdyntsev et al., [13] believed that there had been extensive migration of uranium and/or thorium in the spring deposits at Ehringsdorf, DDR, found today in an area of moderate rainfall. However, our own studies of this site [14] indicate that the sequence displays systematically decreasing age from bottom to top, as expected.

Considering now the analytical procedures used in dating travertines, two general problems have been encountered. First, we observe that travertine associated with many archaeologically significant sites contains too little uranium to permit accurate measurement of the age. For example, sheets of speleothem precipitated on top of cave-wall engravings in the cave of Combarelles, France, contain less than 0.1 ppm U. Normally this is the lower limit below which a sample is considered to be undatable, due to the high probability of natural contamination. Furthermore, protactinium dating is only possible with samples containing 1 ppm or more U. A second analytical problem that is surprisingly common in archaeological deposits, is the low chemical yields of both thorium and uranium that are obtained during analysis. Typically, pure calcite (e.g., from speleothems deep inside caves) gives U and Th yields of 50 to 90 percent whereas the same procedures applied to some archaeological samples have on occasion given yields of five percent or less for both elements on replicate samples. Poor yields may in some instances be attributed to the presence of organic substances (humates, etc.) as evidenced by the fact that yields improve when the sample is ignited at 800 °C before dissolution. Phosphate admixed with the carbonate, from bone fragments or excrement, may also interfere with extraction of Th, unless special precautions are taken during coprecipitation.

Other Methods for Analysis of Travertines

Besides the methods of uranium series disequilibrium, mention should be made of other methods of dating of travertine that may be applicable to archaeological sites. Such travertine layers are chemically precipitated deposits which have remained as closed systems since the time of their formation. Therefore, many of the chemical or physical parameters that controlled their deposition, and many of their chemical and physical properties, should be considered as chronological and environmental markers for the period of deposition. For example, studies of the stable isotopic composition of speleothem from deep inside caves has been shown to be a useful paleoclimatic indicator for the site [15]. Similar studies of the oxygen isotopic composition of cold-spring deposits may also given clues as to changes in the ^{18}O content of spring waters, which are in turn largely controlled by climatic changes.

^{14}C-dating of travertines is possible in some instances; the initial $^{14}C/^{12}C$ ratio of speleothem or travertine will generally be less than that of the atmosphere at the time of deposition, due to admixture of dead carbon from the limestone from which the Ca and part of carbonate have been obtained. But, where active deposition is occurring, modern samples can be compared to give an estimate of this dilution effect. Hendy [16] has used this method for speleothems; no applications of this method to archaeologically significant travertines have yet been made.

Two other closely related methods of dating are thermoluminescence (TL) and electron spin resonance (ESR) both of which make use of the presence of trapped electrons and holes formed by ionizing radiation. TL has been applied to speleothem by Wintle [17] who finds that sample preparation has marked effects on the TL intensity. Ikeya [18] has applied ESR methods to the dating of stalagmites from an archaeological site at Petralona, Greece. In both methods, one must take account of the irradiation dose rate from the exterior and interior of the sample; the latter, arising from uranium in the calcite, will increase due to growth of the daughter isotopes into equilibrium with the uranium.

Fission track dating of calcite has been attempted by Macdougall and Price [19] who found that tracks in calcite anneal too quickly to preserve chronological information. Amino acid racemization dating is possible in principle because many chemically precipitated calcites contain minor amounts of organic residues. Other minor constituents of calcite (trace elements, pollen grains, fluid inclusions) may yield information about the environment at the time of deposition and thus give clues as to the time of formation by correlation with local glacial or pluvial sequences. Paleomagnetic records have been obtained from such deposits [20] and could likewise be used to determine times of deposition by correlation with local master curves from other dated deposits.

Application to Archaeological Sites

The first attempts to use uranium series dating on archaeological travertines were those of Cherdyntsev et al., [21] on various sites in eastern Europe. Most noteworthy, was his estimate of an age for the Vertesszollos site in Hungary, where spring-deposited travertines enclose primitive stone artifacts and where was found embedded the skull of a hominid related to Homo erectus. These authors obtained an age of 225 ± 35 ky. The nearby site of Tata, a Mousterian locality, gave an age of 116 ± 16 ky and 85 ± 10 ky by Io/U and Pa/^{235}U respectively. However, all the samples which he analyzed had very low ^{230}Th/^{232}Th ratios (1.0 - 4.2) making these ages rather suspect. Later, Fornaca-Rinaldi [22] succeeded in obtaining acceptable ages on upper and middle Paleolithic sites in Italy, measuring the ratio ^{230}Th/^{234}Th and assuming that ^{234}Th was in equilibrium with ^{238}U. Cherdyntsev et al., [13] attempted to determine the age of the travertine deposits at Ehringsdorf, DDR, in which were found embedded a Neanderthaloid cranial fragment and many artifacts. They found that radioisotopes had been partially mobilized in these deposits, resulting in discordant activities of ^{230}Th, ^{234}U,

^{231}Pa, and ^{226}Ra. The latter isotope, with a 1600 y half life, is a monitor of relatively recent disturbances of the deposits. Nevertheless, Cherdyntsev et al., believed that the data were consistent with an age of about 115 ky, which agreed with the estimates based on the faunal remains in the travertine.

The site of Caune de l'Arago in the French Pyrenees has yielded a skull of Homo erectus which is overlain by travertine deposits. Turekian and Nelson [23] attempted to date this deposit but found that samples of travertine were highly contaminated with ^{232}Th (^{230}Th/^{232}Th = 4 - 15). They were able to estimate the age from a correlation of the contaminant ^{232}Th activity with the observed ^{230}Th/^{234}U ratio. Schwarcz [11] has attempted a modified analysis of their data suggesting that the samples were formed about 114 ky B.P. This is too young to be consistent with the typology of the artifacts in the site or the grade of evolution of the hominid found there.

In Israel we have found a number of prehistoric sites associated with travertine deposits, both in caves and in fossil springs. In the northern Negev desert region, in the valley of Nahal Zin, a series of travertines deposited about 46.5 ± 2.9 ky B.P. contained artifacts associated with the transition from middle to upper Paleolithic culture. A few km away, at Nahal Aqev, an extinct spring-deposited mound of travertine enclosed artifacts of Levallois-Mousterian affinity, representing the middle Paleolithic of Israel. The travertine just underlying these artifacts yields an age of 80 ± 10 ky [24].

In Greece, the cave of Petralona has yielded a skull of a variety of Homo erectus. A number of workers including the present writer have attempted to determine the age of this site [25-27]. Although the samples of travertine cannot be very accurately located with respect to the former location of the skull, the various analysts concur that these travertine layers must be greater than 350 ky in age. Liritzis has attempted to estimate the age on the basis of ^{234}U/^{238}U ratios: 500 ± 50 ky; the assumptions in such a date are however very shaky.

In the Charente district of France, the cave of La Chaise has yielded lower and middle Paleolithic stone artifacts, and cranial fragments from a primitive Neanderthaloid hominid. Travertine deposits between these artifact-bearing layers have been dated by the ^{230}Th/^{234}U method [28]. The Neanderthaloids were apparently occupying this site from 185 to 145 ky B.P., while Mousterian artifacts are found dating from before 145 ky to a little after 106 ± 10 ky B.P., the age of an upper travertine layer that seals the deposit (figure 5).

We have also re-investigated the travertine deposits at Ehringsdorf. We find that apparent ^{230}Th/^{234}U ages decrease regularly from the bottom to the top of this stratigraphic sequence [14]. The hominid remains occur at a level corresponding to an age of about 210 ky, roughly coincident with the next-to-last interglacial. Earlier faunal studies and Cherdyntsev et

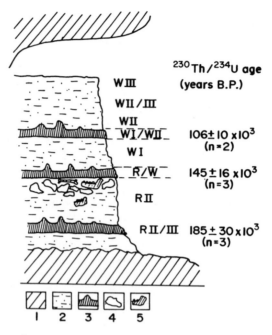

Figure 5. Schematic cross section of the cave of Lachaise-de-Vouthon, Charente, France, showing the relation between dated travertine layers and the artifact- and fossil-bearing layers.

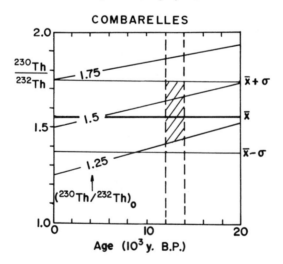

Figure 6.   Variation of the $^{230}Th/^{232}Th$ ratio for samples of travertine that contained contaminant Th at the time of deposition with varying initial $^{230}Th/^{232}Th$ ratios as shown.

The observed mean and standard deviation of nine analyses of travertine from the wall of the cave of Combarelles is shown on the right ($\bar{X} \pm \sigma$). The age of the wall inscriptions in this cave as deduced from archaeological data and ^{14}C ages elsewhere, is shown by the vertical dashed lines. The U-series data are consistent with this age estimate, for $(^{230}Th/^{232}Th)_0$ values in the range commonly observed in cave deposits (1.25 to 1.75).

al.'s, [13] work suggested that the deposits correspond to the last interglacial, but evidently the faunal evidence is equivocal [29]. Not far from Ehringsdorf, Harmon et al., [30] have dated a travertine layer containing a well preserved Homo erectus cranium, at the site of Bilzingsleben; they obtained an age of 228 ± 15 ky, which they also attribute to the penultimate interglacial; however the hominid which is associated with their site is apparently considerably more primitive in evolutionary grade than that at Ehringsdorf. This raises the interesting possibility that even as recently as 200,000 years ago there were significantly different evolutionary grades of the genus Homo co-existing in Europe.

Concluding Remarks

I have attempted to summarize some of the more interesting results which are arising from the radiometric dating of travertines at archaeological sites. These studies, together with other work in Israel [31] and France [32] to appear presently, will help define the time scale of the evolution of man from the level of Homo erectus to Homo sapiens sapiens. This interval of hominid evolution and cultural change is largely outside the range of dating of ^{14}C and archaeological sites are rarely associated with volcanic deposits datable by $^{40}Ar/^{40}K$. Therefore we should be grateful that so many of our ancestors chose to live in caves or to stop for warm baths at the ancient hot springs of Europe and Asia. One other line of inquiry which may prove fruitful is the dating of cave art. Many caves in southern France and northern Spain possess paintings or inscriptions on their walls. These have been dated by analogy with art mobilier (small, portable art works) which are found interstratified between cultural layers in these caves, and are stylistically similar to the parietal art. We are now proposing to date some of these art works directly, through analysis of the stalagmite layers that under-and overlie them. However, initial attempts in the cave of Combarelles were frustrated by the very low concentrations of uranium in the overlying deposits. However, the isotopic data at this site are not inconsistent with the archaeological estimate that engravings on the wall of this cave were made during the middle or late Magdalenian, approximately 14 to 14 ky B.P. (figure 6). These studies will soon be extended to caves in Northern Spain.

This research has benefited immeasurably from the advice and assistance of many archaeological colleagues including Paul Goldberg, Na'ama Goren, André Debenath, Francois Bordes, and Aris Poulianos. The laboratory work was assisted by Ada Dixon and Marija Russell. Research was supported by grants from the National Geographic Society, Social Sciences Research Council,

and the National Science and Engineering Research Council.
McMaster Isotopic and Nuclear Geochemistry Research contribution
no. 121.

References Cited

[1] Langmuir, D., Uranium solution-mineral equilibria at low
    temperatures with applications to sedimentary ore deposits,
    Geochim Cosmochim. Acta, 42, 547-569 (1978).
[2] Kaufmann, A., Broecker, W. S., Ku, T. -L., Thurber, D. L.,
    The status of U-series methods of mollusk dating, Geochim.
    Cosmochim. Acta, 35, 1155-1183 (1971).
[3] Cherdyntsev, V. V., Uranium-234. Isreal Program for
    Scientific Translations, Jerusalem, 234 pp.
[4] Szabo, B., Results and assessment of uranium-series dating
    of vertebrate fossils from Quaternary alluviums in Colorado,
    Arctic and Alpine Res., 12, 95-100 (1980).
[5] Gascoyne, M., Schwarcz, H. P., Ford, D. C., Uranium series
    dating and stable isotope studies of speleothems: Part I:
    Theory and techniques, Proc. British Cave Res. Assoc., 5,
    91-111 (1978).
[6] Kaufmann, A., Th^{230}/U^{234} dating of carbonates from Lake
    Lahontan and Bonneville, unpublished Ph.D. Thesis, Columbia
    University, New York, 1964.
[7] Osmond, J. K., Cowart, J. B., The theory and uses of natural
    uranium isotopic variations in hydrology, Atomic Energy
    Review, 14, 621-679 (1976).
[8] Ku, T. -L., Protactinium method of dating coral from
    Barbados Island, J. Geophys. Res., 73, 2271-2276 (1968).
[9] Rosholt, J., Szabo, B., Determination of protactinium by
    neutron activation and alpha spectrometry, in DeVoe, J. R.,
    ed., Modern Trends in Activation Analysis, U. S. Nat. Bur.
    Stds. Special Publication 312, 1, 327-333 (1968).
[10] Kaufmann, A., U-series dating of Dead Sea basin carbonates,
    Geochim. Cosmochim. Acta, 35, 1269-1281 (1971).
[11] Schwarcz, H. P., Absolute age determination of archaeological
    sites by uranium series dating of travertines, Archaeometry,
    22, 3-24 (1980).
[12] Ku, T. -L., Bull, W. G., Freeman, S. Thomas, Knauss, K. G.,
    Th^{230}/U^{234} dating of pedogenic carbonates in gravelly desert
    soils of Vidal Valley, Southeastern California, Bull. Geol.
    Soc. America, 90, 1063-1073 (1979).
[13] Cherdyntsev, V. V., Senina, N., Kuzmina, E., Die Alterbes-
    timmung der Travertine von Weimar-Ehringsdorf (Uber das
    Alter des Riss-Wurms-Interglazials). Abh. Zent. Geol. Inst.
    Palaontol, Abh., 23, 7-14 (1975).
[14] Schwarcz and Blackwell (in preparation.)

[15] Harmon, R. S., Thompson, P., Schwarcz, H. P., Ford, D. C., Late Pleistocene paleoclimates of North America as inferred from stable isotope studies of speleothem, Quaternary Res., 9, 54-70 (1978).

[16] Hendy, C., The use of C-14 in the study of cave processes In: I. Olsen, ed., Radiocarbon variations and absolute chronology, Proceedings of the 12th Nobel Symposium Uppsala, 1969, Wiley, New York, p. 419-443, 1970.

[17] Wintle, A., A thermoluminescence dating study of some Quaternary calcite: Potential and problems, Can. Jour. Earth Sci., 15, 1977-1986 (1978).

[18] Ikeya, M., Electron spin resonance as a method of dating, Archaeometry, 20, 147-158 (1978).

[19] Macdougall, D., Price, P. B., Attempt to date early South African hominids by using fission tracks in calcite, Science, 185, 943-944 (1974).

[20] Latham, A., Schwarcz, H. P., Ford, D. C., Pearce, G. W., Paleomagnetism of stalagmite deposits, Nature, 280, 383-385 (1979).

[21] Cherdyntsev, V., Kazachevskii, I., Kuzmina, E. A., Dating of Pleistocene carbonate formations by the thorium and uranium isotopes, Geochem. Internat., 2, 794-801 (1965).

[22] Fornaca-Ronaldi, G., ^{230}Th-^{234}Th Dating of Cave Concretions, Earth and Planet. Sci. Lett., 5, 120 (1968).

[23] Turkian, K. K., Nelson, E., Uranium series dating of the travertines of Caune de l'Arago (France) In: Colloque I, Datations Absolute et Analyse Isotopiques en Prehistoire, Methodes et Limites, Union des Sci. Prehist. Protohist. IX Congres, Nice, 172-179, 1976.

[24] Schwarcz, H. P., Blackwell, B., Goldberg, P., Marks, A., Uranium series dating of travertine from archaeological sites, Nahal Zin, Israel, Nature, 277, 558-560 (1979).

[25] Hennig, G., Bangert, U., Herr, W., Poulianos, A. N., Uranium series dating and thermoluminescence ages of speleothem from Petralona cave, Anthropos (Athens), 7, 174-214 (1980).

[26] Schwarcz, H. P., Liritzis, Y., Dixon, A., Absolute dating of travertines from Petralona cave, Khalkidiki Peninsula, Greece, Anthropos (Athens), 7, 152-173 (1980).

[27] Liritzis, Y., Th-230/U-234 dating of speleothems in Petralona, Anthropos (Athens), 7, 215-242 (1980).

[28] Schwarcz, H. P., Debenath, A., Datation absolute des restes humains de la Chaise-de-Vouthon (Charente) au moyen du desequilibre des séries d'uranium, C. R. Acad. Sci. Paris, 288, 1155-1157 (1979).

[29] Currant, A., personal communication, 1980.

[30] Harmon, R. S., Glazek, J., Nowak, K., ^{230}Th/^{234}U dating of travertine from the Bilzingsleben archaeological site, German Democratic Republic, Nature, 284, 132-135 (1980).

[31] Schwarcz, H. P., Liritzis, Y., Dixon, A., Absolute dating
     of travertines from Petralona cave, Khalkidiki Peninsula,
     Greece, <u>Anthropos</u> (Athens), <u>7</u>, 152-173 (1980).
[32] Blackwell, B., Uranium series dating of Prehistoric sites
     in France, unpublished M.Sc. Thesis, McMaster University,
     1980.

RECEIVED May 5, 1981.

APPENDIX

Guide to Chapter Contents — Techniques and Applications

Notes and References

1. The figures listed in the table indicate Chapter numbers. The letters refer to Symposium papers discussed in the Preface or in this Appendix and abstracted in the <u>Book</u> <u>of</u> <u>Abstracts</u> published by the American Chemical Society for the 179th National Meeting (March 1980); ACS abstract numbers are indicated in brackets.

    a.   Isotope Abundances and Anomalies and the Measurement of Geologic and Solar System Time [NUCL-004].
          G. J. Wasserburg

    b.   Trace Geochronology [NUCL-012].
          R. M. Walker

    c.   Stable Isotopes: Tools for Reconstructing the Past [NUCL-033].
          J. C. Lerman

    d.   Advances in Amino Acid Dating Techniques [NUCL-014].
          P. E. Hare

    e.   Present Status and Future Prospects for Thermo-luminescence Dating [NUCL-015].
          R. M. Walker

    f.   State-of-the-art of Obsidian Hydration Dating [NUCL-013].
          I. Friedman and F. Trembour.

2. In addition to Chapters 3-5 in this book, two other symposium papers were presented on the characteristics and applications of <u>Accelerator</u> <u>Mass</u> <u>Spectrometry</u> (AMS):

        Radioisotope Dating with Accelerators: Present Status and Future Prospects...R. A. Muller,

        and

        Cosmogenic Radionuclides in the 10^5-10^7 Year Region: Results and Prospects for High-Energy Ion Counting...J. R. Arnold.

Table 1. Content Guide: Techniques and Applications.

	Methodology	Cosmochemistry	Geochemistry	Archaeology
Nuclear Techniques				
Accelerator mass spectrometry	3-5	2-4	2-4, 11	3, 23
Isotope abundance mass spectrometry	6-8, a	6-8, a	7, 8, a	---
Uranium disequilibrium	24	---	11	24
Low level counting; resonance ionization spectroscopy	9, 10, 16, 23	2, 18	9-11, 13, 16-18	22-24
Nuclear track measurement	b	---	---	24
Physicochemical Techniques				
Light element (isotope ratio) mass spectrometry	14, c	---	14, 19	24
Amino acid racemization	d	19	11, 19	23, 24
Thermoluminescence	e	---	---	23, 24
Diffusion	11	---	11	23
Hydration	f	---	---	---
Chemical patterns	2, 10, 15	19	2, 10, 15, 19, 20	21, 23

The first of these papers has subsequently been published by
T. Mast, R. Muller, and P. Tans in Proc. Conf. on the Ancient Sun,
R. O. Pepin, J. A. Eddy, R. B. Merrill, eds., Geochim. et
Cosmochim. Acta, 1980, Suppl. 13, 191. In his paper, Prof. Arnold
noted that AMS may be better adapted to the measurement of long-
lived cosmic ray and solar particle produced nuclides (^{59}Ni,
^{36}Cl, ^{10}Be, ^{26}Al, ^{129}I) than to ^{14}C, but that a major effort is
called for to build up a fundamental information base comparable
to that which exists for ^{14}C dating.

3. Chemical Dating is a central issue in the papers of Walker
[b,e], Hare [d], and Friedman and Trembour [f]. In all of these
papers chemical transformations (reaction, excitation) must be
considered directly in order to interpret a "date". Brief
extracts from the papers follow.

The first technique discussed by Walker [b] may be considered
nuclear dating, in that the rate at which nuclear fission tracks
are formed depends upon the nuclear decay (spontaneous fission)
of ^{238}U. As noted by Walker, however, the interpretation of a
fission track age (derived from the accumulated track density)
involves understanding of the problems (and opportunities) posed
by samples having complex thermal histories. Regarding extraterrestrial samples, Walker pointed out that those "whose formation
dates back to the beginning of the solar system, contain an
important contribution of fission tracks from the decay of the
now-extinct isotope ^{244}Pu. These can give information about the
thermal histories of various bodies as well as setting certain
constraints on the nucleo-synthetic history of the solar system
material".
Thermal history is central also to the interpretation of
Thermoluminescence (TL) Dates [e]. In this method, age is deduced
from the accumulation of "chemical" excitation energy due to
external radiation. Essentially, this "dating method consists of
measuring the light emitted by the recombination of charge
carriers liberated during rapid heating of a sample. The samples
are subsequently given standard doses of radiation to calibrate
their response. The radiation dose in nature is estimated by
measuring the content of radioactive elements in the sample. The
technique has been applied to a wide variety of materials ranging
from human ceramic artifacts to soils heated by contact with lava
flows". Walker goes on to suggest that "in concert with ^{14}C
determinations,...TL may yield information on variation of ^{14}C
production prior to 8,000 B.P.".
Amino Acid Dating Techniques depend on the "rates of
hydrolysis reactions of proteins and racemization, epimerization,
and decomposition reactions of amino acids; [they have] been
applied to the age-dating of fossil bone, teeth, and shell.
Activation energies range from near 20 kcal per mole for
hydrolysis reactions to around 30 kcal per mole for racemization

reactions" [d].   Unlike nuclear techniques which include fixed
time constants, absolute amino acid dates require calibration by
other dating techniques (such as radiocarbon), and interpretation
once again depends upon knowledge of other environmental factors
such as ground water leaching and thermal history.   A major
impetus for the method has come from advances in liquid chromatographic separation of D and L enantiomers using optically active
eluants.

    The rate of hydration of obsidian, which is diffusion limited,
forms the basis for Obsidian Hydration Dating [f].   A date refers
to "the time a fresh surface of obsidian was created, either
naturally or by man.   ...Laboratory and field studies have confirmed that the time indicated by a hydrated layer is proportional
to the thickness squared of the layer.   The hydration rate is
independent of the relative humidity of the environment, but the
chemical composition of the obsidians affect the rate by orders of
magnitude. SiO_2 increases the rate, whereas CaO, MgO, and
H_2O decrease it.   A 6 - 8 °C temperature increase causes doubling
of the rate."   The method is quite inexpensive, and it is applicable to ages between a few hundred and several million years.

INDEX

Platinum group elements with
respect to Ir abundances,
404, 405$\underline{f}$
Pleistocene
human occupation of New
World during, 453
bone samples, effect of
pretreatment on N-15
values, 467$\underline{t}$
deposits, isotopes used in
radiometric dating, 476,
480$\underline{f}$
waters, 198
Polar ice, dating by C-14 and
Rn-222, 319-328
Pottery, 412
abundance of alkali or
plagioclase feldspars,
426, 428$\underline{f}$
Preboreal, 33
Precambrian era, biological
activity, 397
Precipitation, worldwide,
deuterium vs. oxygen
isotope ratio, 248$\underline{f}$
Protactinium dating, 483
Protein, bone, 455
Protein collagen in bone, 459
Polar ice caps, measurements
of radioisotopes of
solids, 37-39
Polar ice cores, chemical
composition, 304
Polar ice sheets, 17
Polar snow
chemical composition, 313
nitrate, 307
Pollutant sources, chemical
and radiochemical
fingerprinting techniques
used to identify, 170
Pollutants, 308
Polycyclic aromatic
hydrocarbons (PAH) in
sediment, 172
Postglacial, 33

Q

Quercus patraea, German oaks,
isotope ratios, 259

R

Ra-226, 323
decay chain, 332
Rn-222
counting, 324
dating polar ice, 319-328
RIS schemes for periodic
table, 152$\underline{f}$
RISA cycle, ion-molecule,
using NO, 157$\underline{f}$
RISA process, 156$\underline{f}$
Racemization amino acids, 210
dating, 484
properties, 468
Radiation, cosmic, constancy,
12
Radiation/modulations and
interactions, cosmic, 10$\underline{f}$
Radioactive atoms, direct
detection, 46
Radioactive chronometers, 332
Radioactive coral debris
transport, 341
Radioactive counting
techniques, 164-168
Radioactive dating,
principles, 7
Radioactive decay,
accumulation of products,
201-207
Radioactive decay rate, 46
Radioactive disposal sites,
deposition of Pb-210 and
Pu-239,240 in sediment,
353-359
Radioactive gases
nuclear weapons, 210
power reactors, 210
Radioactive isotope, number of
nuclei, 12
Radioactive nobel gas
isotopes, 13
Radioactive waste disposal
sites, sedimentation
rates of Pb-210
measurements, 349$\underline{t}$
Radioactivity, natural, rare
earth elements, 1